analysis of biological systems

analysis of biological systems

corrado priami

melissa j. morine

The Microsoft Research – University of Trento Centre for Computational and Systems Biology, Italy
and
University of Trento, Italy

ICP

Imperial College Press

Published by

Imperial College Press
57 Shelton Street
Covent Garden
London WC2H 9HE

Distributed by

World Scientific Publishing Co. Pte. Ltd.

5 Toh Tuck Link, Singapore 596224

USA office: 27 Warren Street, Suite 401-402, Hackensack, NJ 07601

UK office: 57 Shelton Street, Covent Garden, London WC2H 9HE

Library of Congress Cataloging-in-Publication Data
Priami, Corrado.
Analysis of biological systems / Corrado Priami, Melissa J. Morine, The Microsoft Research University of Trento Centre for Computational and Systems Biology, Italy, and University of Trento, Italy.
pages cm
Includes bibliographical references and index.
ISBN 978-1-78326-687-6 (hardcover : alk. paper)
1. Biological systems--Mathematical models. 2. Biological systems--Computer simulation. 3. Biological systems--Analysis. I. Morine, Melissa J. II. Title.
QH323.5.P73 2015
571.7--dc23
2014047042

British Library Cataloguing-in-Publication Data
A catalogue record for this book is available from the British Library.

Typeset by Stallion Press
Email: enquiries@stallionpress.com

Printed in Singapore

To Leonardo and Silvia

c. priami

To Alex, my mother, my father, John, David and Leonardo

m.j. morine

Preface

Formal modeling of biological systems is becoming a fundamental step in designing experiments and interpreting their outcome. Omics technologies produce data that challenge standard analytical approaches, thus calling for new methods to analyze and interpret the overwhelming mass of data produced. Simultaneously, the outcome of static omic analysis often provides knowledge to build dynamic models that can uncover the mechanistic details governing the biological system being studied. Many techniques have been developed in the last decade for dynamic simulation of biological systems and the emerging field of language-based modeling is becoming well established in the community. We felt the lack of a comprehensive textbook unifying the static and dynamic approaches that define systems biology, as well as the foundational knowledge in molecular biology that is essential to adequately analyze biological systems.

This book is intended for practitioners of the systems biology field, both with a biological and mathematical/computing background, who want to understand algorithmic modeling and algorithmic systems biology. Some knowledge of basic molecular biology and basic computer science can help, but the aim of the book is to be a self-contained approach to the field. The second chapter introduces some basic molecular biology and experimental techniques to understand how data are produced. The Appendices report basic mathematics and computing needed to understand fully the material in the book. All chapters propose further reading about the topics introduced, to drive the reader to deeper treatments of the topics in the book. All of these references are collected in the bibliography reported at the end of the book.

The book can also be used in advanced undergraduate courses on modeling and analysis of biological systems. It contains many examples that can help students gain a practical grasp of the main concepts throughout the book. These examples are also templates for approaching exercises and problems that may be encountered outside of the book. The organization of the material is the result of ten years teaching a modeling course at the University of Trento.

The book collects classical material on analysis, modeling and simulation, acting as a unique source of reference. Additionally, it expands on language-based modeling with recent research advances, and therefore can also be considered a research text introducing both the state of the art and recent developments. Therefore, it can also be useful in graduate programs and for researchers that want to approach systems biology from a well-rounded standpoint.

As a final remark, we stress that we were forced to choose among many different formalisms and methods to constrain the book to a reasonable size. The choice was driven by our experience both as researchers and teachers working in the field. We are aware that there are many other excellent solutions to the problems addressed in the book that we were not able to include. The references are intended to manage this issue at least partially.

Acknowledgments

We thank all the people that helped directly and indirectly with this project.

All the co-authors of our papers that contributed to develop ideas, methods and examples discussed in the book. The whole COSBI team for many fruitful discussions on most of the central topics of the book.

Gianfranco Balbo, Luca Cardelli, Pierpaolo Degano, Ozan Kahramanogullari, Paola Lecca, Rosario Lombardo, Luca Marchetti, Isar Nassiri, Federico Reali, Simone Rizzetto, Adelinde Uhrmacher and Hong Thanh Vo for a careful reading of a draft of the book and their useful comments.

Bianca Baldacci who produced most of the figures in the book.

The students of the courses Algorithmic Modeling and Modeling and Simulation of Biological Systems at the University of Trento in the academic years 2012–2013 and 2013–2014, respectively, who read the draft of this book and helped to remove inconsistencies and improve its presentation and examples.

Contents

Chapter 1

Algorithmic systems biology

The convergence between computer science and biology has occurred in successive waves, involving increasingly deeper concepts of computing. Since its early days, computer science has taken inspiration from nature, with the works of Turing, Von Neumann and Minsky. These milestones led to extraordinary results, some of which recall biology even in their names: cellular automata, neural networks, genetic algorithms. The current situation makes computer science a suitable candidate for becoming a foundation for systems biology with the same importance as mathematics, chemistry and physics. The same applies to systems biology domains related to the healthcare sector such as systems nutrition (which studies the impact of nutrients on cellular machinery) or pharmacology.

Biology is experiencing tremendous growth in the capacity to measure biological molecules and processes with unprecedented depth and precision. The resulting inundation of data presents as many challenges as it does opportunities, such that organization and analysis of these data requires specialized training and dedicated research. Statistics and computer science are now fundamental to biological data analysis, as it has become commonplace to analyze terabytes of data that comprehensively characterize complex biological systems. Another defining feature of modern biological research is a growing interest in interpreting living systems as dynamic information manipulators, and as such it is moving toward systems biology.

1.1 Converging sciences

Computing and biology have been converging ever more closely for the past two decades, but with a vision of computing as a resource for biology that has propelled bioinformatics. Bioinformatics addresses structural and static aspects of biology and has produced databases, pattern manipulation and comparison, search tools and data mining techniques. Computational approaches in biology are now moving toward systems biology (see *Systems biology* box, below). This poses both challenges and opportunities to describing the step-by-step mechanistic behavior that underlies complex phenotypes.

Systems biology

There is no universal agreement on a definition of systems biology. We consider systems biology a transition from:

- qualitative biology to a quantitative science;
- reductionism to system-level understanding of biological phenomena;
- structural, static descriptions to functional, dynamic properties; and
- descriptive biology to mechanistic/causal biology.

Some of the main aspects of interest in the transition are causality between events, temporal ordering of interactions and spatial distribution of components within the reference volume of reactions.

The current approach in biological data collection is to take snapshots of biological systems and subsequently try to model the variation of measures in the snapshots through equations. Such snapshots may represent a wide variety of systems and states, from a simple *in vitro* cellular model at multiple time points, to a series of human tissue samples across a range of disease states. These snapshots contain a great deal of information about the state of a given system, but can be increasingly complicated to analyze and interpret, and thus require dedicated statistical and bioinformatic strategies. Naturally, system snapshots do not directly capture the dynamics that carry the system from one state to the next. Algorithms describe the steps between system states in a causal continuum of actions that make the measures change, thus providing a dynamic view of the system under question. Through algorithms and the (programming) languages used to specify them, we can recover temporal, spatial and causal information on the modeled systems by using well-established computing techniques that deal with program analysis, composition and verification, integrated software development environments and debugging tools, as well as algorithm animation.

Hereafter we call algorithmic systems biology the specification of biological models through algorithms to describe their step-by-step dynamics.

Algorithm

An algorithm is a finite list of well-defined instructions for allowing an executor to perform a task without ambiguity. Programming languages are used to express algorithms when the executor is a digital computer. Algorithms need a syntax to be described and a semantics to associate them with their intended meaning so that an executor can precisely perform the steps needed to implement the algorithms with no ambiguity. Algorithms are quantitative when the selection mechanism of the next step is determined according to probabilistic/temporal distributions associated with either the rules or the components of the system modeled.

The main difference between algorithmic systems biology and other techniques used to model biological systems stems from the intrinsic difference between algorithms (operational descriptions) and equations (denotational descriptions). An equation might be an elegant way of describing the result of the execution of an algorithm. Furthermore, equations specify dynamic processes by abstracting the steps performed by the executor, thus hiding from the user the causal, spatial and temporal relationships between the elementary steps. Equations describe the variation of variables (usually concentrations of species) from one state to another of a system, while algorithms highlight why and how a system moves from one state to another one. Algorithms force modelers/biologists to think about the mechanisms governing the behavior of the system under question. Algorithms can serve to coherently extract general biological principles that drive the data produced in systems biology, and are a practical tool for expressing and favoring computational thinking. Therefore, they are also a conceptual tool that helps to understand fundamental biological principles. Statistical modeling and analysis are essential in this task for extracting the relevant biological signal from high-throughput data that may initially be too complex to be modeled by algorithms.

The notion of simulation needs some consideration. Algorithmic simulations are executable on computers and rely on deep computing theories, while mathematical simulations are solved with the support of computer programs (hence, computing here is just a service). Execution of algorithms exhibits the emergent behavior produced at system level through the set of local interactions between components without the need for specifying it from the beginning. The complex interaction of the concentrations of the species, the sensitivity of their interactions expressed through stochastic parameters, the localization of the components in a three-dimensional hierarchical space, and hence the dynamic evolution of a system can all be modeled through computational simulation.

Simulation

A simulation is the process of model solution/execution to reproduce an approximation of the dynamics of a system through a solver/executor (usually a digital computer).
The outcome of a simulation is an approximation of the time behavior of a system.

Overall, the integration of many different disciplines like mathematics and statistics to extract knowledge from data, network theory to identify functional biological subnetworks from data, computing and mathematics to describe the dynamics of systems and simulate their temporal behavior, and biochemistry, biology and medicine to provide the background knowledge for interpretation of modeling results are all mandatory to succeed in systems biology.

1.2 The approach

In this section we introduce the steps of the algorithmic systems biology approach, which we are presenting in the rest of the book in detail.

The first step in approaching a biological problem is to collect data relating to the phenotypic phenomenon (i.e., biological system) we are interested in. This is usually accomplished by sampling clinical markers on a population of subjects and producing high-throughput data from tissue or blood samples (the *Biological system of interest* part in Fig. 1.1).

Measurements in the lab result in a vast collection of numbers representing multiscale phenomena. Knowledge extraction from these datasets requires dedicated statistical and bioinformatic approaches, with the goal of identifying the key patterns of variance and covariance among the elements of the system that define the phenotype of interest (the *what* part in Fig. 1.1).

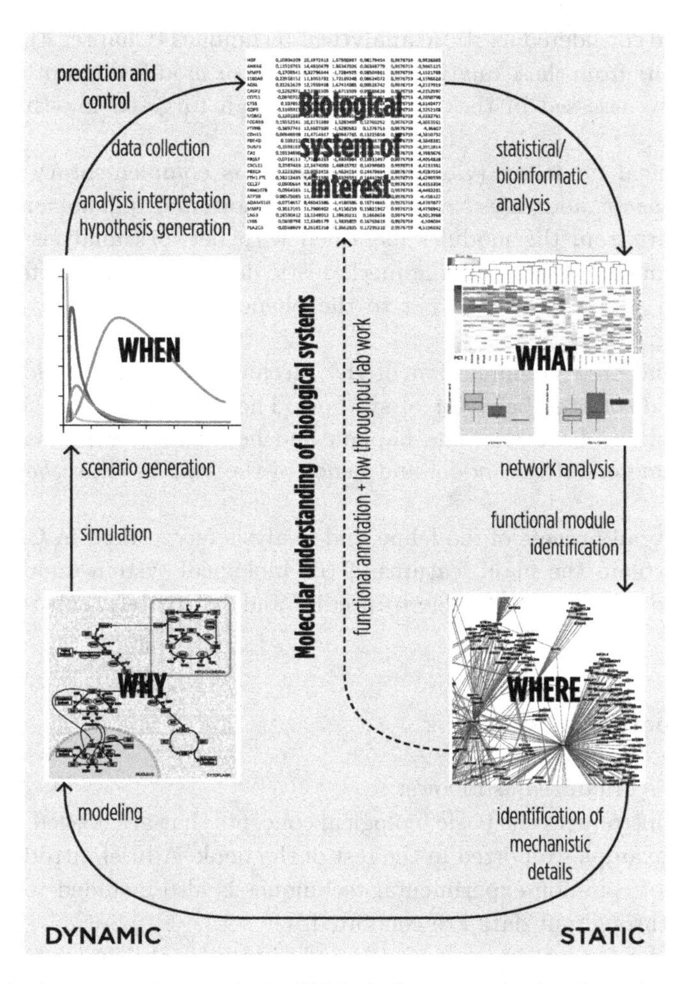

Figure 1.1 An approach to analysis of biological systems, leading from data collection to statistical and network analysis, then mechanistic modeling and dynamic simulation.

The variables (i.e., molecules) in the datasets may then be analyzed in the context of a molecular interaction network, in order to identify sub-networks/modules that characterize the molecular basis of the phenotype being studied (the *where* part in Fig. 1.1).

The steps described thus far serve to characterize a complex system based on statistical and bioinformatic analysis of a high-throughput dataset. The goal of these approaches is not to model dynamics of systems, and thus can be considered as static analytical techniques (Chapter 4). In many cases, results from these analyses generate new or modified hypotheses that are directly assessed in the laboratory through targeted low-throughput approaches.

Dynamical modeling (Chapters 5 and 6) is complementary to static modeling as it addresses *why* and *when* molecular phenomena occur. It may start from the modules identified with network analysis and produce dynamic models by adding mechanistic details and parameters to the interaction subnetworks relevant to the phenotype under study (the *why* part in Fig. 1.1).

The final step is simulation of the mechanistic model, which reveals how time affects the behavior of systems. The outcome of the simulation is a prediction of when events happen and how they are affected by the main parameters of the model and hence of the system (the *when* part in Fig. 1.1).

The overall process of modeling and analysis (structured in Chapter 7) seeks to explain the main features of the biological system under investigation, and use this knowledge to predict and, ultimately, control system behavior.

1.3 Structure of the book

The book is organized as follows:
Chapter 2 introduces the basic biological concepts that are needed to understand the examples reported in the rest of the book. A brief introduction to a number of common experimental techniques is also included to describe how high-throughput data are generated.
Chapter 3 defines systems and models and introduces their main properties.
Chapter 4 provides an overview of fundamental statistical and bioinformatic techniques used to analyze biological data, including regression, dimensionality reduction, clustering and network analysis.
Chapter 5 introduces the main dynamic modeling techniques starting from equation-based to computing approaches such as network-based and automata-based formalisms. Rule or reaction-based formalisms are also introduced in this chapter.
Chapter 6 introduces the core of algorithmic systems biology by covering modeling languages.
Chapter 7 describes a structured process to approaching real biological problems. It provides a guideline for the reader on how to step from

the identification of a problem to an actual model to be used as a predictive tool.
Chapter 8 defines the main aspects of simulation techniques and introduces random number generators and stochastic simulation algorithms.
Chapter 9 briefly discusses the prevailing challenges and perspectives of the field of systems biology.
Appendix A recalls the main mathematical concepts that are needed for the treatment in the book including functions and relations, logics and algebra.
Appendix B briefly recalls the basic concepts of probability and statistics needed to understand data analysis and simulation.
Appendix C defines the basic notions of languages and semantics, and reports the formal definitions of the modeling languages dealt with in Chapter 6.

1.4 Summary

In this chapter we introduced the concept of algorithmic systems biology and discussed how a range of disciplines must be integrated to succeed in systems biology. We then introduced the steps of the algorithmic systems biology approach. Each of the steps denotes different techniques that are discussed in detail in the next chapters of the book.

1.5 Further reading

Algorithmic systems biology is introduced in [Priami (2009a)], and similar concepts are also addressed in [Fisher *et al.* (2011)]. Primary bioinformatics references are [Spengler (2000); Roos (2001)]. Systems biology has been introduced in [Kitano (2002a)] following the interpretation of biological systems as information processing systems [Hood and Galas (2003)]. For the main differences between bioinformatics and systems biology we refer to [Cassman *et al.* (2005)]. Computational thinking is discussed in [Wing (2006); Cohen (2008)].

Examples of the static analytical approach described in Fig. 1.1 are in [Morine *et al.* (2010, 2011); Nguyen *et al.* (2011); Morine *et al.* (2013)]. The dynamic modeling approach is described in [Dematté *et al.* (2008); Kahramanoğulları *et al.* (2012)].

Chapter 2

Setting the context

This chapter introduces the molecular components and processes of the cell, which form the *systems* that are modeled in systems biology. We also present a number of laboratory techniques that are commonly used to produce data in systems biology. The material presented here focuses on *eukaryotic* cells, and specifically on human systems wherever species-specific detail is provided. As this is intended to be an introduction to cell and molecular biology, this chapter does not delve deeply into these topics, and also does not strictly require background knowledge in biology. Additional references are provided at the end of the chapter to provide the reader with resources for further information on these topics.

2.1 The structure of the cell

Eukaryotic cells are distinguished by the presence of a *nucleus*, which encloses genetic material called *deoxyribonucleic acid* (DNA). The nucleus is delimited by a double membrane that is composed of the *nuclear envelope* surrounding the *nuclear lamina* (see Fig. 2.1). The nuclear membrane has openings called *nuclear pores* that serve to allow cellular components

Prokaryotic, eukaryotic

Eukaryotic cells contain a cellular structure called a *nucleus*, which encloses the cellular DNA. *Prokaryotic* cells lack a nucleus and any other membrane-bound organelle, such that all cellular components are contained in a single cellular compartment. Bacteria and archaea are prokaryotes, whereas animals, plants and fungi are eukaryotes.

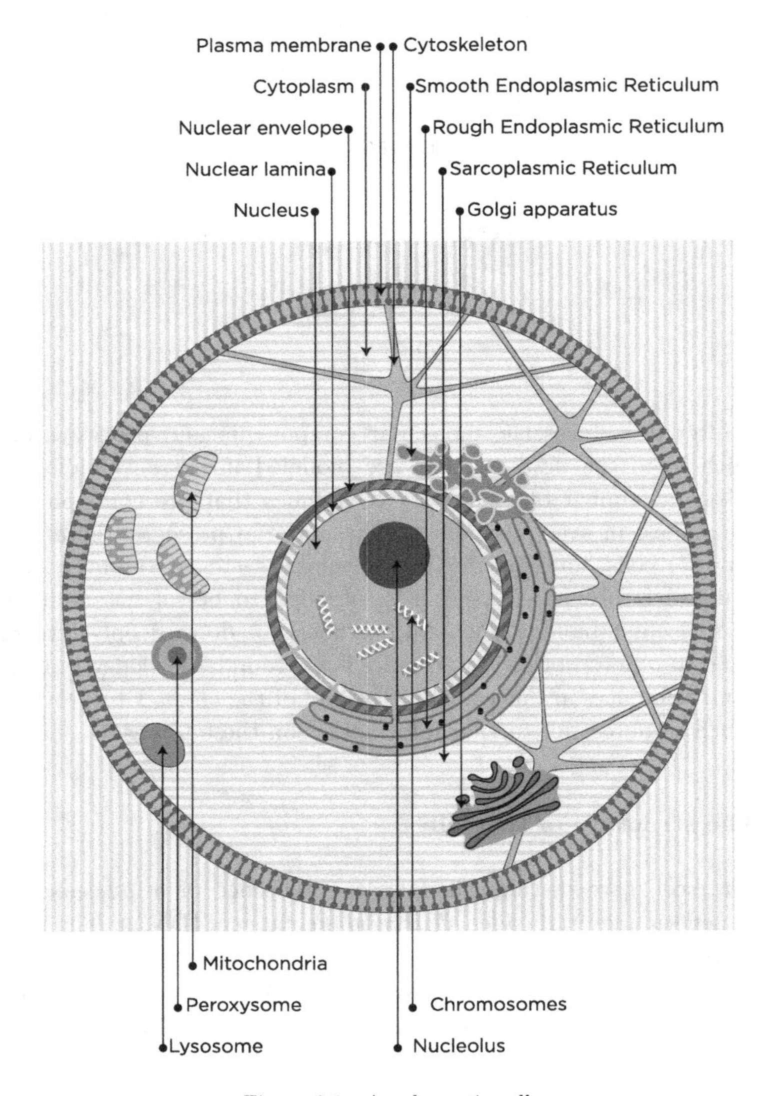

Figure 2.1 A eukaryotic cell.

in and out of the nucleus selectively. Apart from DNA, the nucleus contains another type of structural element called *nucleoli*. Nucleoli are not membrane-enclosed organelles, but rather are composed of *proteins* and *ribonucleic acids* (RNA; i.e., single-stranded copies of DNA regions, as described below), and are responsible for the synthesis of *ribosomal RNA* (rRNA). Although there is considerable range, human cells have an average volume of approximately $4 \times 10^{-9}\,\text{cm}^3$, 10% of which is the nucleus (diameter approximately $6\,\mu$m).

The space surrounding the nucleus, and delimited by the *plasma membrane* (also called the cellular or cytoplasmic membrane), contains a range of cellular organelles contained in a gel-like fluid called *cytosol*; these components are collectively referred to as *cytoplasm.* The plasma membrane is a phospholipid bilayer with embedded proteins that controls the flow of ions and organic molecules in and out of the cell. The plasma membrane is not simply a barrier, but also plays critical roles in cell adhesion and signaling.

The *endoplasmic reticulum* (ER) is an organelle with an extensive phospholipid membrane network of *cisternae* (sac-like structures). The membrane is connected to the nuclear membrane and encloses the cisternal space (or lumen), which is continuous with the perinuclear space but separate from the cytosol. The three types of ER are called *rough endoplasmic reticulum* (RER), *smooth endoplasmic reticulum* (SER) and *sarcoplasmic reticulum* (SR). The membrane of the RER contains many ribosomes (described below) that are continuously being bound and released. RER synthesizes proteins, while SER synthesizes lipids and steroids, metabolizes carbohydrates and steroids, and regulates calcium concentration, drug detoxification and attachment of receptors on cell membrane proteins. SR solely regulates calcium levels.

Ribosomes are fundamental components of cells as they implement the translation of RNA to protein in the cytoplasm. Specifically, ribosomes translate the genetic information encoded in *messenger RNA* (mRNA) into *polypeptide* chains composed of amino acids, which are delivered by *transfer RNA* (tRNA). Ribosomes are divided into two subunits that split when the translation of an mRNA is completed. The smaller subunit binds to the mRNA, while the larger subunit binds to the tRNA and the amino acids. Ribosomes can be either free or membrane-bound. Free ribosomes float in the cytosol, but are not contained in the cell nucleus and other organelles. If a ribosome needs to synthesize a protein that will be targeted to a specific organelle, it binds to a membrane in the RER and then the protein produced is delivered by a specific secretory pathway to the final destination.

The *Golgi apparatus* is fundamental for managing the vesicular transport in the cell. This organelle is composed of stacks of membrane-bound structures known as cisternae. A mammalian cell typically contains 40 to 100 stacks, each of them containing from four to eight cisternae. Each stack has four functional regions: the cis-Golgi network, medial-Golgi, endo-Golgi

Amino acids, Genetic code

Amino acids are organic molecules composed of an amine, a carboxylic acid, and a variable side chain that is unique to each type of amino acid. There are 20 standard amino acids that are utilized in vastly varying combinations to produce cellular proteins.
The *genetic code* is the code by which genetic sequences are translated to amino acids, in order to form proteins. Each triplet of nucleotides, called a *codon*, is associated with (i.e., codes for) a specific amino acid. The genetic code is widely conserved across species and is degenerate/redundant, such that multiple codons are associated with a single amino acid.

and trans-Golgi network. Each cisterna comprises a flat, membrane enclosed disc that includes special Golgi enzymes which modify or help to modify cargo proteins that travel through it.

The shape of the cell and its structure is supported by a matrix of filaments globally known as the *cytoskeleton*. The most abundant filaments are *microtubules* and *actin* microfilaments, both of which maintain a polarity. Microtubules bind together in groups of nine triplets to form cylinder-shaped *centrioles*, of which two join in a perpendicular fashion to form the *centrosome*. The centrosome is a central cytoskeletal organelle that determines the position of the nucleus and overall spatial organization of the cell, and is thought to play a central role in cell division. Actin microfilaments undergo continuous rearrangements to allow cell motility. Actin also participates in cell division and cytokinesis, vesicle and organelle movement, cell signaling, and the establishment and maintenance of cell junctions and cell shape.

The cytoplasm contains double membrane-enclosed organelles called *mitochondria* that act as energy plants for the cell, producing energy in the form of *adenosine triphosphate* (ATP). Mitochondria possess their own genome (arranged in a circular structure) which makes this organelle unique in the cell. The external mitochondrial membrane is smooth, while the inner membrane forms folds called *cristae* where sugar is oxidized. Mitochondria have five compartments: the outer membrane, the intermembrane space, the inner membrane, the cristae and the space within the inner membrane called the matrix. The matrix contains a highly concentrated number of enzymes, special mitochondrial ribosomes, tRNA, and several copies of the mitochondrial genome. The human mitochondrial genome codes for

13 peptides, 2 ribosomal RNA and 22 tRNA. The peptides merge with proteins encoded by the host cell to form mitochondrial proteins. Similar to cellular division, mitochondria replicate by fission, coordinated with DNA replication. It is hypothesized in [Gray *et al.* (2001); Wallen (1927)] that mitochondria have their origin in primordial endosymbiotic prokaryotes that invaded larger cells as parasites, and gradually evolved a symbiotic relationship.

Peroxisomes are another basic cellular organelle, and contain a range of enzymes that are involved in the catabolism of fatty acids, D-amino acids, polyamines and biosynthesis of ether lipids and cholesterol. Peroxisomes also play a role in the production of bile acid, which is essential for absorption of fats and fat-soluble vitamins from the diet. Although peroxisomes do not contain an internal genome (peroxisomal proteins are imported from the cytoplasm) they replicate by fission, like mitochondria.

Also playing an important metabolic role in the cell are *lysosomes*. These organelles contain over 50 different enzymes, and are capable of degrading lipids, nucleic acids, proteins and carbohydrates. As such, they are known as a sort of intracellular digestive system, breaking down both obsolete cellular components and material taken up from outside the cell (such as bacteria or viruses). All of the enzymatic activity in lysosomes is carried out by a family of proteins called acid hydrolyses. As these enzymes require an acidic environment, lysosomes maintain an interior pH of 4.8 by acquiring protons from the surrounding cytosol. This pH difference between the lysosomal interior and cytoplasm helps to avoid the unconstrained digestion of cellular components.

2.2 DNA, RNA and genes

All cells store their genetic information in a double-stranded filament called *DNA* (Fig. 2.2). Each strand is made up of *deoxyribose* sugars that form strong covalent bonds through -OH groups. DNA strands possess a specific orientation, such that one end (called the $5'$ end) contains a phosphate group bound to the fifth carbon of the deoxyribose sugar, while the opposite end (called $3'$ end) contains an -OH group attached to the third carbon of the deoxyribose sugar. The two strands join together in anti-parallel orientation, such that a $3'$ end is always paired with a $5'$ end, to form the familiar double-stranded helix configuration. Each deoxyribose sugar carries one of four *nucleic acids* (also referred to as *bases*): adenine (A),

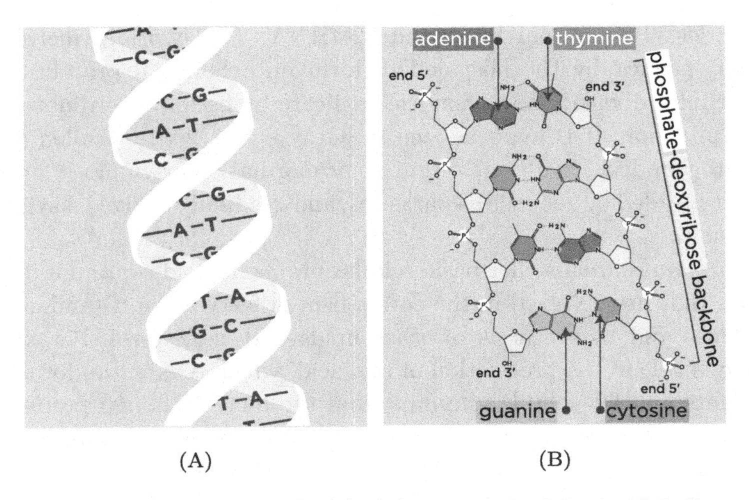

(A) (B)

Figure 2.2 Deoxyribonucleic acid (DNA). (A) Illustration of the double-helix structure of DNA; (B) Detailed view of the molecular constituents of DNA.

thymine (T), cytosine (C) and guanine (G), which form complementary pairwise hydrogen-bonds (A with T and C with G). All four bases are aromatic heterocyclic organic compounds; A and G are part of the subclass *purines*, while C and T are *pyrimidines*. The dimer formed by a deoxyribose sugar and base is called a *nucleotide* and is the basic unit of hereditary information. Due to complementary base pairing, the two strands in a DNA

Homologous chromosomes

Homologous chromosomes are a pair of 'matching' chromosomes, wherein one was derived from the maternal lineage and the other from the paternal lineage.

molecule are complementary: each nucleotide of one strand corresponds to a nucleotide with a complementary base on the other strand. Each human cell contains about two meters of DNA (one meter for sex cells), totaling about 100 trillion meters of DNA per human. Fitting all of this material into subcellular structures therefore requires highly efficient packaging. When DNA is not being copied, the long strands fold into the familiar double helix configuration, which is further coiled around specialized proteins called *histones* in the nucleus. These DNA-histone complexes are referred to as nucleosomes, which are further packed to form chromatin threads. Finally, the chromatin thread is coiled once again to form chromosomes, of which each human cell contains 23 homologous pairs (or 23 single chromosomes for sex cells).

Mitosis, Meiosis

Mitosis describes the division of genetic content and cellular components that produces two genetically identical daughter cells from a single progenitor cell; *meiosis* refers to cellular division that results in sex cells (i.e., gametes), which contain half of the genetic content of progenitor cells. Mitosis involves a single cell division of doubled DNA, whereas meiosis involves two successive divisions.

During cell division DNA is copied and passed from one cell generation to the other in a process called *DNA replication.* It is possible that bases may be incorrectly interred during this process (i.e., *mutations*), producing genetic variation in the form of *single nucleotide polymorphisms* (SNPs). In eukaryotes, the introduction of such mutations in sex cells and subsequent passage to the next generation (via sexual reproduction) produces genetic variation, which is naturally observed in populations. While many mutations will have no effect on overall cellular/organism function, some may have a negative or positive effect (many may even be lethal). The process of *natural selection* will ultimately favor the cells/organisms that better function in a given environment (i.e., those with higher *genetic fitness*). Mutation and selection on the basis of fitness are thus the processes on which *evolution* is based.

Genes are the core of the *central dogma* of biology, which describes the flow of information between DNA and proteins (which perform the work of the cell). The first step of this process is the *transcription* of genes into single-stranded filaments called *messenger RNA* (mRNA). The second step is the *translation* of the mRNA into a polypeptide chain. Finally the polypeptide folds into a protein: a three-dimensional structure that is functionally active. Both transcription and translation can be activated by the availability of free energy (usually in the form of ATP molecules) and of the raw material needed to create the products (nucleotides for mRNA and amino acids for proteins).

RNA is a single-stranded flexible filament that can create weak self-bonds, allowing it to fold into a variety of forms both as single and paired

Gene, SNP, Allele

A fragment of DNA (i.e., a sequence of nucleotides) that encodes the information for producing an organic, functional component of the cell is called a *gene*. A gene is thus a region of the DNA that encodes a discrete hereditary unit. Bacterial genes are continuous sequences of coding DNA, while eukaryotic genes are small sequences of coding DNA (*exons*) interleaved with non-coding sequences (*introns*).
An *SNP* is a point of single base-pair variation in DNA occurring in a population, whereas an *allele* is a term to describe an alternate variant of a gene — such as the alternative variants resulting from introduction of an SNP in a population.

molecules. It has the same general structure as DNA, but the sugar forming the RNA backbone is *ribose* and the base *uracil* (U) takes the place of *thymine* (T). The cell may contain many copies of RNA corresponding to a given gene, which allows for genes to be dynamically up-regulated and down-regulated based on the requirements of the cell.

Although most genes code for proteins (and hence their transcripts are mRNA) there are also genes that code for tRNA and rRNA. tRNAs contain a region complementary to a specific codon (called an anticodon) to bind mRNA, as well as a binding site for the amino acid corresponding to the tRNA codon. tRNA is usually about 80 nucleotides (nt) long, and there is a unique tRNA for each possible codon. The role of tRNA is to transfer amino acids to a growing polypeptide chain for protein translation. Ribosomal RNA is the catalytic component of a protein complex called the ribosome, and is used to synthesize proteins. The total RNA in a cell is about 10 mg/ml, 80% of which is rRNA and 3–5% is mRNA.

MicroRNA (miRNA) derives its name from its small size, typically about 20–22 nt. A complex of miRNAs and enzymes can break down mRNA in parts complementary to miRNA, thus blocking further translation into proteins. This process is also known as *RNA interference* (RNAi). Two other types of small RNA filaments involved in nucleotide modification of other RNA strands are *small nuclear RNA* (snRNA, about 150 nt) and small nucleolar RNA (snoRNA, about 60–300 nt). Finally, there may also be RNA material of viral origin in eukaryotic cells, present as double stranded filaments similar to DNA (*double-strand RNA*, dsRNA) or in small filaments (*small interfering RNA*, siRNA, about 20–25 nt). Viral RNA can be responsible for triggering RNAi.

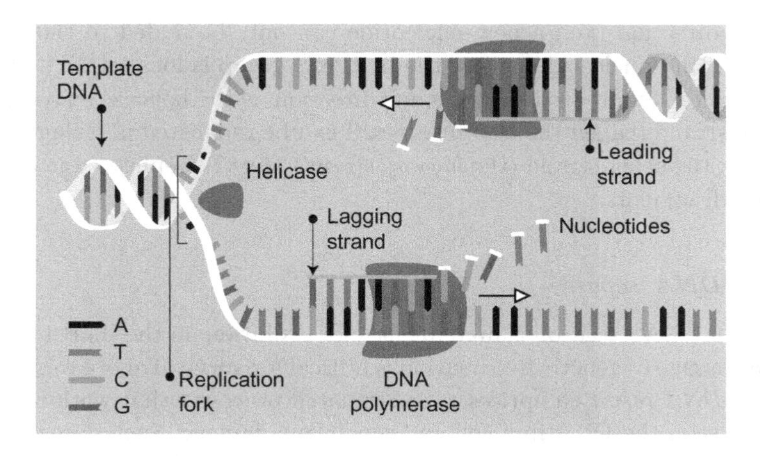

Figure 2.3 The primary components of DNA replication.

2.2.1 *DNA replication*

DNA replication (see Fig. 2.3) is a fundamental cellular process that duplicates DNA in order to copy the genetic information of a cell and pass it to the genetically identical daughter cells. This process ensures the passage of genetic information, both within the context of cell division as well as in the production of gametes for sexual reproduction. Understanding DNA replication is therefore critical for appreciation of cell division, as well as the mechanistic underpinnings of microevolution. There are a number of players and subprocesses involved in DNA replication, each of which must function with high accuracy and precision to ensure proper passage of genetic information from progenitor to daughter cell.

Under cellular growth conditions (i.e., when the cell is not dividing) DNA remains tightly packed in chromatin strands. Upon initiation of cell division, a type of protein called *helicase* travels along the double-stranded DNA and breaks the hydrogen bonds between nucleotide bases, thus unwinding the double helix into single strands that are each stabilized by *single-stranded binding proteins* (Fig. 2.3). As it travels along the DNA, helicase thus creates a forked shape, leaving behind one single strand running in $5' \rightarrow 3'$ orientation and the other in the $3' \rightarrow 5'$ orientation. *DNA polymerase* is the critical enzyme responsible for traveling along each single strand and adding complementary bases from the nuclear pool of free nucleotides, thus forming two double stranded molecules. However, there is an obstacle in this process, in that nucleotides can only be added in the

$5' \rightarrow 3'$ direction (i.e., a new nucleotide can only be added to the $3'$ end of an existing nucleotide, where the hydroxyl group is located). Therefore, since replication proceeds in the same direction as the helicase enzyme, the $3' \rightarrow 5'$ strand (called the *leading strand*) can be replicated in a single pass; however, the other strand (the *lagging strand*) must be replicated in a series of short fragments.

2.2.2 *DNA repair*

Although DNA variation is fundamental for evolution, in the short term the stable passage of genetic information is critical for survival of organisms and species. *DNA repair* comprises a set of related processes that work continuously to scan the DNA in a cell and repair any damage. Such damage may occur through various sources such as errors in nucleotide addition by DNA polymerase, damage by reactive oxygen species, or environmental damage (e.g., UV radiation). Such damage may manifest as sequence errors, modified bases, sugar-phosphate backbone breaks or covalent cross-linking of DNA strands. If DNA damage produces uncorrected changes in nucleotide sequence, this may result in dysfunctional activity in critical proteins, and these deleterious genetic changes can then be passed to daughter cells. Cancer represents a well-studied consequence of accumulation of DNA mutations in critical cell cycle regulation genes.

The specific machinery for repairing DNA depends on the type of damage. During DNA replication, proofreading is carried out by DNA polymerase, which excises and replaces incorrectly incorporated bases. In cases where these incorporation errors evade proofreading, the process of *DNA mismatch repair* works to correct the DNA sequence. Although the precise process differs between prokaryotic and eukaryotic cells, the general procedure is the same, such that misincorporated bases are detected by a family of *Mut* proteins, which then bind the daughter strand of DNA and recruit *DNA helicase*. The helicase then separates the two strands, *DNA exonuclease* removes the incorrect base (plus a variable segment of additional bases) and finally, *DNA polymerase* and *DNA ligase* reinsert correct bases and repair the sugar-phosphate backbone, respectively.

Base excision repair involves a group of enzymes called *DNA glycosylases*, which scan the DNA for damaged bases (e.g., methylated guanine, deaminated cytosine) and cleave these bases from the sugar-phosphase DNA backbone. Base excision by DNA glycosylases is then followed by removal of the segment of sugar-phosphate backbone by *AP endonuclease* and *phosphodiesterase*, then base reinsertion by DNA polymerase and backbone repair by DNA ligase. Larger anomalies in the DNA molecule are instead managed by the process of *nucleotide excision repair*. Such anomalies may include pyrimidine dimers and carcinogenic molecules covalently bonded to DNA bases. Upon recognition of such an error, the sugar-phosphate backbone is cleaved on either side of the damaged region, then the entire segment is removed by DNA helicase. The resulting gap is repaired, as before, by DNA polymerase and ligase.

2.2.3 *DNA recombination*

In addition to SNPs, the process of *DNA recombination* is responsible for broader changes in genetic sequence that may alter, for instance, the order of genes in the genome, or the specific set of genes/alleles allocated to a given set of sex cells (i.e., eggs, sperm) during meiosis. One form of genetic recombination, *homologous recombination*, occurs during cell division. During meiosis, when homologous chromosome pairs are aligned in the middle of the cell before the first division, *chromosomal crossover* may occur (see Fig. 2.4) between two homologous chromosome segments, resulting in an exchange of genetic material between the chromosomes.

Chromosomal crossover involves the breakage of the DNA helix, followed by rejoining with the partner chromosomal homolog. When chromosomal crossover occurs, the daughter cells will therefore each carry a unique complement of progenitor alleles. And if this process occurs during meiosis, the resulting sex cells will carry the capacity to produce offspring that is a mosaic of each parent.

A second common form of genetic recombination is *site-specific recombination*, which involves movement of genetic material to non-homologous sites (i.e., to target sites either on the same or a different chromosome).

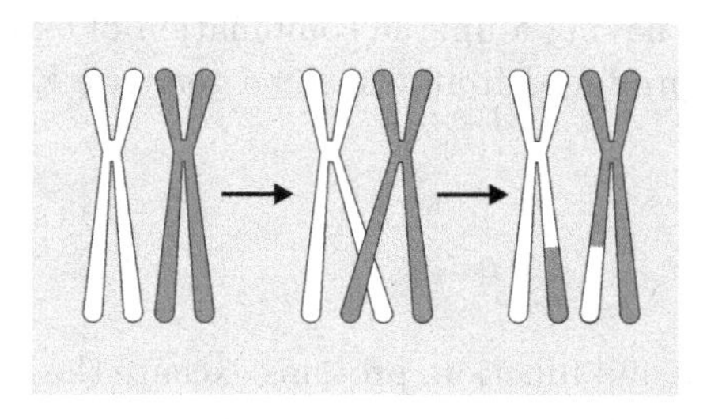

Figure 2.4 Illustration of the process of chromosomal crossover during meiosis.

These mobile segments of DNA, called *mobile genetic elements*, may range from hundreds to tens of thousands of base pairs in length. Movement of these elements requires a class of enzymes called *recombinases*, which are typically coded within the sequence of the mobile genetic element.

2.2.4 *Transcription*

Transcription is the fundamental biological process that creates an RNA strand complementary to a piece of DNA. Many copies of RNA can be produced from the same gene, so that many copies of the same protein can be produced in parallel. The number of RNA molecules produced partially regulates the rate at which the concentration of a given protein grows within a cell. The entire set of transcripts in a cell under a given condition is called the *transcriptome.*

RNA polymerases perform the transcription of DNA into RNA with the help of transcriptional factors, activators and mediators. The transcription process can occur in parallel, such that many RNA fragments can be simultaneously transcribed from different portions of the same DNA molecule. In eukaryotic cells, RNA polymerase is of three types: I, II and III. The majority of genes, including those coding for proteins, are transcribed by RNA polymerase II (hereafter we consider this polymerase, unless stated otherwise). The speed of RNA polymerase is about 20 nt s^{-1} in eukaryotes. mRNA is distinguished from other types of RNA by the presence of non-coding RNA at the $5'$ end and a polyadenylation signal (i.e., multiple adenine bases) at the $3'$ end. During eukaryotic transcription, both exons and introns are transcribed to pre-mRNA. *Alternative splicing* is performed by the *spliceosome* (a complex of about 50 proteins and snRNAs) and removes intronic and often exonic sequences from pre-mRNA to produce mature mRNA. Alternative splicing provides the ability to produce a range of different mature mRNAs (and hence different proteins) from the same gene, each having a unique combination of exons. These variants of mature mRNAs produced from the same gene are known as *alternative transcripts.*

2.3 Proteins

Diverse in both form and function, proteins execute the molecular functions of the cell. In mammalian cells, proteins typically account for 18% of total cell weight, making them the second most predominant cellular component after water. We have already discussed a range of proteins that perform basic functions in the cell, such as *DNA polymerase*, *helicase* and *ligase.* The entire set of proteins in a cell under a given condition is called the *proteome.* The proteome of a cell thus varies widely, depending on the type of cell and its environmental conditions. For instance, a muscle cell in the condition of intense physical activity would be expected to have a very different proteome than a cancerous liver cell.

Proteins consist of chains of amino acids, called *polypeptides*, each bound together by *peptide bonds.* The polypeptide is then intricately folded, sometimes in combination with other polypeptides, to form a mature protein. As previously indicated, amino acids are composed of an amine, a carboxylic acid and a variable functional side chain (see Fig. 2.5). Each amino acid (and thus, each protein) possesses unique chemical characteristics (e.g., polarity, hydrophobicity) depending on its variable side chain.

Broadly, proteins can be divided into four functional classes: structural proteins, enzymes, signaling proteins and transport proteins. Structural proteins provide the framework for every cell, giving shape and stability to the fluid environment of the cytoplasm. As introduced in Sect. 2.1, this framework is called the *cytoskeleton*, and consists of: microfilaments (the smallest components, composed of double helices of *actin* proteins), intermediate filaments (composed of various proteins, depending on cell type, including *vimentin*, *keratin* and *laminin*) and microtubules (the largest component of the cytoskeleton, composed of α- and β-*tubulin*). In addition to offering structural stability, the cytoskeleton also serves as a scaffold for the transport of proteins and small molecules throughout the cell.

The metabolic work of the cell is performed by a class of proteins called *enzymes.* From the set of compounds taken in from the environment (e.g., through diet or pharmaceutical intervention) enzymes catalyze the chemical

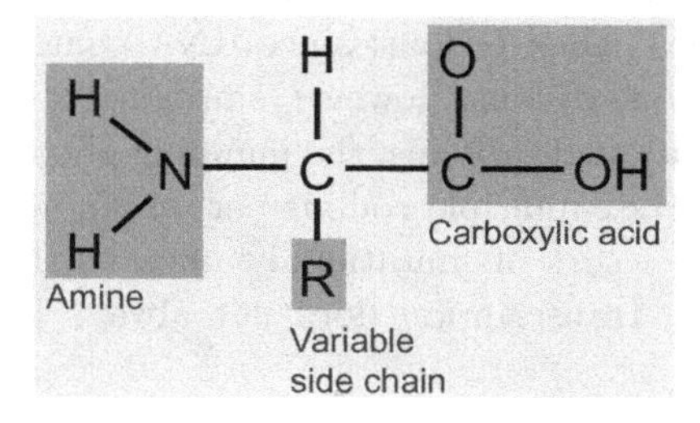

Figure 2.5 Generic illustration of an amino acid.

reactions that produce the vast range of compounds required for cellular function. Organized into serialized groups of reactions, enzymes and their associated metabolites form *metabolic pathways*, such as the pathway producing energy, CO_2 and H_2O from simple sugar. Enzymes are highly specific to their associated substrate, and often require additional metabolites (called *cofactors*) to function efficiently.

Signaling proteins act as the messengers of the cell, forming an information system that is critical for appropriate cellular response to changing circumstances. For instance, consumption of a carbohydrate-rich meal produces a rapid rise in blood sugar levels, requiring adipose and muscle cells to shift their activities toward uptake and metabolism of glucose. This shift is controlled by the insulin signaling pathway. Dysfunction of the signaling proteins in this pathway is the direct precursor to the rapidly growing health crisis of diabetes mellitus. There are a variety of specific protein types that can be considered signaling proteins, including hormones, neurotransmitters and cytokines, although it is important to note that other chemical classes (primarily lipids) may also act as signaling molecules.

The fourth class of proteins is transport proteins. Biomolecules do not simply move in and out of cells passively, due to the hydrophobic interior of lipid bilayer membranes that block free passage of most molecules. The transport of molecules is thus tightly controlled by transport proteins embedded in cellular membranes. There are two main types of membrane transport proteins: *carrier proteins* and *channel proteins*, with the difference that carrier proteins dynamically change shape in facilitating movement of transported molecules, whereas channel proteins remain static.

2.3.1 *Translation*

Translation describes the biological process that creates a polypeptide chain of amino acids from an mRNA strand. mRNA is translated to protein in triplets of nucleotides called codons; since RNA contains 4 unique bases, there are $4^3 = 64$ unique codons. However, since only 20 unique amino acids are used in the formation of proteins, the mapping of codons to amino acids must be degenerate (i.e., multiple codons per amino acid). This feature of translation provides a sort of mutational buffer to the cell, such that a mutation or error in transcription does not always lead to a change in amino acid.

The *genetic code* is the schema for translating DNA sequence information into amino acids, and is highly conserved across species. As illustrated

Silent, Missense, Nonsense mutations

Silent mutations describe those DNA/RNA changes that do not result in a change in amino acid. *Missense mutations* are those that result in an amino acid change, while *nonsense mutations* involve the substitution of an amino acid-coding codon with a stop codon.

TTT	Phe	TCT	Ser	TAT	Tyr	TGT	Cys
TTC	Phe	TCC	Ser	TAC	Tyr	TGC	Cys
TTA	Leu	TCA	Ser	TAA	Stp	TGA	Stp
TTG	Leu	TCG	Ser	TAG	Stp	TGG	Trp
CTT	Leu	CCT	Pro	CAT	His	CGT	Arg
CTC	Leu	CCC	Pro	CAC	His	CGC	Arg
CTA	Leu	CCA	Pro	CAA	Gln	CGA	Arg
CTG	Leu	CCG	Pro	CAG	Gln	CGG	Arg
ATT	Ile	ACT	Thr	AAT	Asn	AGT	Ser
ATC	Ile	ACC	Thr	AAC	Asn	AGC	Ser
ATA	Ile	ACA	Thr	AAA	Lys	AGA	Arg
ATG	Met	ACG	Thr	AAG	Lys	AGG	Arg
GTT	Val	GCT	Ala	GAT	Asp	GGT	Gly
GTC	Val	GCC	Ala	GAC	Asp	GGC	Gly
GTA	Val	GCA	Ala	GAA	Glu	GGA	Gly
GTG	Val	GCG	Ala	GAG	Glu	GGG	Gly

(A)

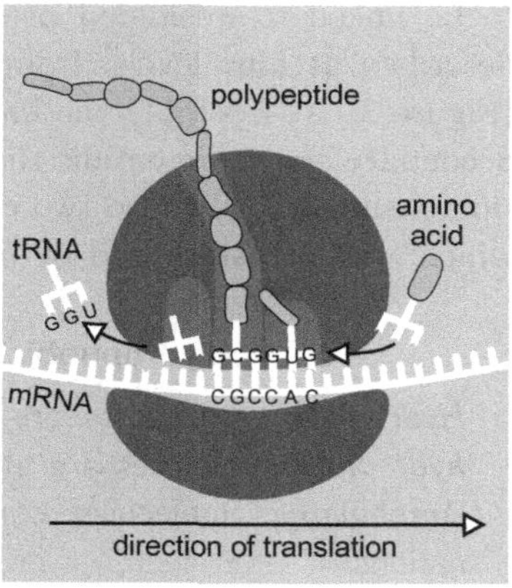

polypeptide
amino acid
tRNA
GGU
GCGGUG
mRNA
CGCCAC
direction of translation

(B)

Figure 2.6 Illustration of (A) the mapping of codons to amino acids, and (B) protein translation by the ribosome.

in Fig. 2.6A, each codon non-uniquely codes for a specific amino acid, with the exception of TAA, TAG and TGA, which signal the termination of the polypeptide strand (also referred to as *stop codons*). A mutation from an amino acid-coding codon to a stop codon is likely to be highly deleterious, as it would result in a truncated (and most likely non-functional) protein product.

Ribosomes play a critical role in the process of translation as they travel along the mRNA strand, reading the codons and recruiting the appropriate amino acids in production of the nascent polypeptide chain. Specifically, upon initiation of translation of a given mRNA molecule, the ribosome binds to the 5′ end of the strand. This activates the recruitment of tRNA carrying the amino acid corresponding to the codon at the site of the ribosome. As the ribosome travels along the mRNA strand, each recruited amino acid is attached to the previous amino acid with a peptide bond, ultimately forming a polypeptide chain (Fig. 2.6B).

2.3.2 *Protein folding*

Proteins do not function in the cell as simple linear polypeptides but as complex three-dimensional structures, and the specific function of a protein is directly linked to its shape. In fact, a number of diseases are known to be linked to misfolded proteins. The overall structure of a protein is described at four levels, from primary to quaternary. Primary structure (Figure 2.7A) is simply the linear chain of amino acids (i.e., polypeptide). Secondary structure entails the immediate shape adopted by the polypeptide chain, and falls into two classes: α-helices and β-sheets (Figure 2.7B), which are both stabilized by hydrogen bonds. α-helices and β-sheets then

Hydrophobic, Hydrophilic, Amphipathic

Hydrophobic molecules are those that are repelled by water, while *hydrophilic* molecules are attracted to and generally soluble in water. *Amphipathic* molecules contain both hydrophobic and hydrophilic regions.

combine to form tertiary structure (Fig. 2.7C), and finally, entire tertiary subunits are combined to form quaternary structure as multisubunit complexes (Fig. 2.7D). The final shape of a protein is defined by the biochemical features of its constituent amino acids, as hydrophobic amino acids will tend to be protected in the interior of the structure, and amino acids with complementary charge will tend to interact. The process of protein folding is aided by a specialized class of proteins called *chaperones*, and often occurs during protein translation.

2.4 Metabolites

Whereas proteins constitute the molecular machines of the cell, *metabolites* represent the molecular substrates and products of cellular metabolism. The entire set of metabolites present in a cell (i.e., the *metabolome*) thus provides a sort of biochemical signature of cellular state. As of the publication date of this text, the human metabolome database describes 41,815 unique

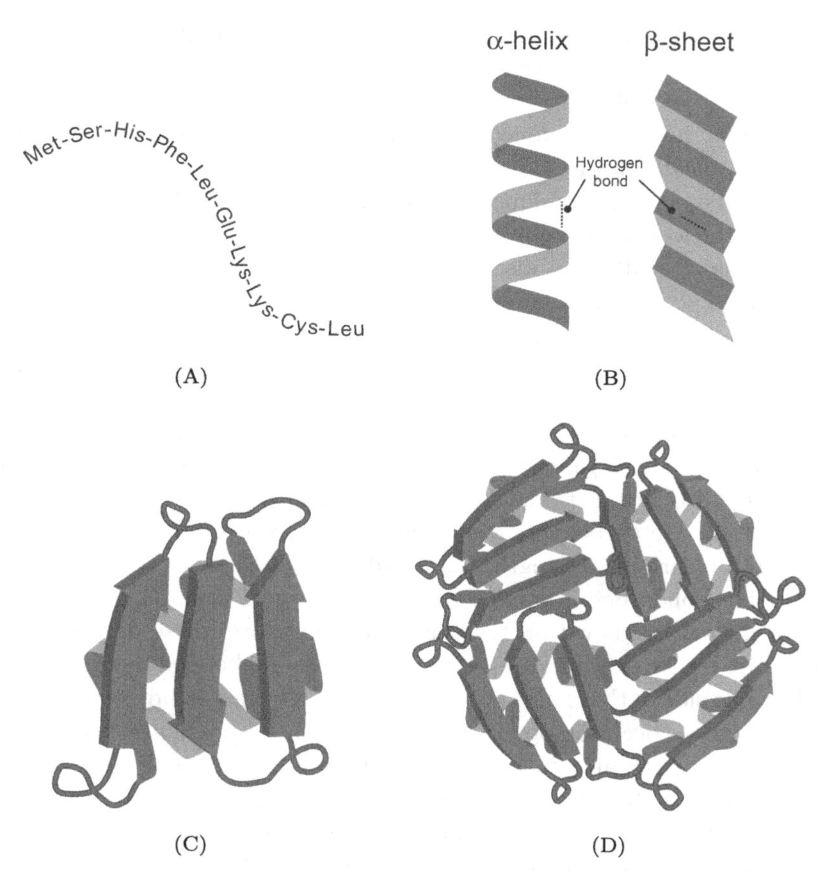

Figure 2.7 Illustration of (A) primary, (B) secondary, (C) tertiary and (D) quaternary protein structure.

metabolites found in the human body, spanning a wide range of biological classes (see www.hmdb.ca/classes for an overview). Two major classes of metabolites, nucleotides and amino acids, have already been discussed in this text; however there are others that also play critical structural and functional roles in the cell.

Beyond acting as simple energy-storage molecules, *lipids* form the basis of cellular membranes and act as potent signaling molecules. Lipids also carry important roles in clinical diagnostics, such as the routine quantification of cholesterol and triglycerides in assessment of cardiovascular health. There is a vast range of known lipid species, spanning the broad classes of fatty acids, glycerolipids, glycerophospholipids, sphingolipids, sterol lipids, prenol lipids, saccharolipids and polyketides. The primary unifying chemical characteristic of lipids is that they are either amphipathic or completely hydrophobic; however, beyond this common feature lipids are structurally and biochemically diverse. Attempts to define the human plasma lipidome have identified over 500 unique lipids in blood alone. Lipidomic profiling is proving to be a powerful tool in the comprehensive characterization of lipid homeostasis in health and disease. However, there remain obstacles in the functional analysis of lipidomic profiles, including a lack of knowledge about the unique functional properties of each lipid metabolite, and incomplete characterization of the metabolic and signaling pathways in which they are involved.

Fats and oils constitute the primary long-term energy stores in mammals, and are mainly sequestered in *adipocytes* (specialized cells containing a large lipid droplet). Fatty acids, the principal class of lipids that form these fats and oils, are carboxylic acids containing hydrocarbon chains ranging from 4 to 36 carbons in length. Each hydrocarbon chain may be *saturated* (no double bonds between carbons) or *unsaturated* (one or more double carbon-carbon bonds within the chain). Furthermore, these double bonds may be in *cis* (hydrogen atoms on adjacent double-bonded carbons occur on the same side of the chain) or *trans* (hydrogen atoms on adjacent double-bonded carbons occur on opposite sides of the chain) configuration. Within the class of unsaturated fatty acids, *monounsaturated* fatty acids carry a single double bond, and *polyunsaturated* fatty acids contain multiple double bonds. Polyunsaturated fatty acids (including the well-known 'omega 3' and 'omega 6' fatty acids) carry particular importance in human nutrition, and have been strongly linked to cognitive function and cardiovascular health.

In addition to common names such as palmitic acid and oleic acid, a number of more descriptive naming conventions exist for fatty acids. One of these naming schemes is the simple indication of the number of carbons and number of unsaturated carbon-carbon bonds, separated by a colon (e.g., 18:3, referring to a fatty acid with 3 unsaturated bonds along a chain of 18 carbons, or 16:0, referring to a 16 carbon chain with no desaturations). Since the position of the unsaturated carbons is of biochemical relevance, this may also be indicated in the abbreviation — such as $18{:}3(\Delta^{9,12,15})$, referring to an 18:3 fatty acid with unsaturations at the 9th, 12th and 15th carbons, counting from the carboxylic acid end. For storage in the cell as long-term energy, fatty acids are typically linked in groups of three via an ester linkage between the carboxyl end of each fatty acid and a glycerol molecule, forming *triacylglycerol* (also known as *triglyceride*).

Lipids also play important structural properties in the cell. All cellular membranes are composed of a double lipid layer (with embedded proteins), which is selectively permeable to ions and organic molecules moving in and out of the cell. The primary types of lipids that comprise membrane lipids are *phospholipids*, *glycolipids* and *sterols*, each sharing the important property of being amphipathic, such that the hydrophilic region of the lipid faces the cytoplasm or extracellular space and the hydrophobic region is protected in the interior of the bilayer. Broadly, phospholipids can be categorized into *glycerophospholipids* and *phosphosphingolipids*, wherein glycerophospholipids consist of a glycerol backbone, two fatty acid chains attached to the first two carbons of the glycerol molecule, and a polar group attached via phosphodiester linkage to the third. Phosphosphingolipids consist of a backbone of sphingosine (or a sphingosine derivative) attached to a single fatty acid and a polar head group, the latter being attached by phosphodiester linkage. Similarly, glycolipids can be divided into glyceroglycolipids and glycerosphingolipids, which differ in their glycerol or sphingosine backbone. The distinction from phospholipids is that glycolipids possess a carbohydrate polar head group in place of the phosphate-linked polar head group. Sterols are structurally distinct from both phospholipids and glycolipids, consisting of four fused carbon rings (three with six carbons and one with five carbons) and a polar head group attached to the third carbon of the first six-carbon ring. This structure makes sterols relatively rigid, compared to phospho- and glycolipids, providing stability to the lipid bilayer.

A third major function for lipids in the cell is their role as potent signaling molecules. *Phosphatidylinositol* is a family of glycerophospholipids that act as signaling molecules via a fundamental second messenger signaling system wherein phosphatidylinositol 4,5-bisphosphate, located at the cell membrane, is hydrolyzed by the enzyme phospholipase C in response to extracellular signals to form inositol 1,4,5-trisphosphate (IP_3) and diacylglycerol. The presence of IP_3 triggers the release of Ca^{2+} from the endoplasmic reticulum, which in turn activates a variety of calcium-dependent signaling cascades. Diacylglycerol activates protein kinase C, a family of protein kinases that mediate a variety of tissue-dependent functions including muscle contraction, glycogenolysis, gluconeogenesis, and gastric acid secretion. Sphingolipids also carry important signaling roles in the cell. In particular, sphingomyelin and ceramide drive cellular activities such as cell division, migration and apoptosis, as well as potential roles in insulin signaling. Vitamins A, D, E and K are all fat-soluble isoprenoid compounds with well-studied roles in cellular signaling. In particular, vitamins A and D serve as hormone precursors, in that vitamin D is metabolized to the bioactive 1,25-dihydroxycholecalciferol hormone, and vitamin A to retinoic acid. 1,25-dihydroxycholecalciferol plays a role in calcium uptake, while retinoid acid is a strong regulator of gene expression in a variety of cell types.

In addition to nucleic acids, amino acids and lipids, *carbohydrates* represent the fourth major macromolecule found in the cell. Carbohydrates (also referred to as *sugars*, *glycans* or *saccharides*) consist of carbon, oxygen and hydrogen, and like lipids and proteins, are diverse in both form and function. The collection of all carbohydrates in the cell is referred to as the *glycome.* Structurally, carbohydrates are divided into monosaccharides, disaccharides, oligosaccharides and polysaccharides. *Monosaccharides* are the simplest building blocks of carbohydrates, and are typically composed of C, H and O in the proportions $C_x(H_2O)_y$. Carbohydrates are referred to as *complex* if they consist of more than a single type of monosaccharide building block. Common examples of monosaccharides include glucose and fructose (both $C_6H_{12}O_6$). Monosaccharides combine to form more complex structures such as the disaccharide sucrose ($C_{12}H_{22}O_{11}$) which is formed by one glucose and one fructose bound by a *glycosidic linkage*, and the polysaccharide *amylopectin*, which is a variable length branching chain of glucose, and is one of the two components of *starch* (with the non-branching glucose chain *amylose* representing the other component).

Nutritionally, carbohydrates are not essential for survival (and thus are not considered to be *essential nutrients*), but nonetheless play an important physiological role as an energy source. Carbohydrates posses similar energy density as proteins, but less than fats; whereas fats contain on average 8.8 kilocalories (kcal) per gram, carbohydrates and proteins contain about 4.1 kcal per gram. Although the primary energy storage in humans is in the form of fats stored in adipose tissue, *glycogen* is a glucose polysaccharide that is stored in muscle and liver tissue, and acts as a short-term energy source that can quickly be converted into glucose for energy production.

In mammalian systems, carbohydrates are thought to primarily function in metabolism, cell signaling and housekeeping functions; however, *glycobiology* is an active area of research, with regular discoveries on the diverse functions of carbohydrates in the cell. As signaling and housekeeping molecules, carbohydrates are typically linked to other biomolecules to form *glycolipids*, *glycoproteins* and *proteoglycans* (collectively referred to as *glycoconjugates*). Glycoconjugates are most commonly formed by n-linked or o-linked glycosylation, where n-linked glycosylation involves attachment of a carbohydrate to the nitrogen atom of asparagine and arginine amino acid side chains, and o-linked glycosylation involves attachment of a carbohydrate to an oxygen atom on certain classes of lipid (such as ceramide) or on serine, threonine, tyrosine, hydroxylysine or hydroxyproline amino acid side chains.

2.5 Cellular processes

This section describes three major cellular processes: metabolism, signaling and trafficking.

2.5.1 *Metabolism*

Broadly, cellular metabolism is the process of energy conversion in the cell through enzymatic reactions. The two essential components of metabolic reactions are *substrates* (i.e., small molecules) and *metabolic enzymes*, which are proteins that catalyze the conversion of the substrates to form metabolic products, which in turn may act as substrates for subsequent reactions. Metabolic reactions can be generally classified into *anabolic* (entailing synthesis of complex structures from simple building blocks, typically requiring energy) and *catabolic* (entailing the breakdown of molecules, typically releasing energy in the process). The common unit of energetic currency in the cell is ATP, which is consumed by a catabolic reaction to produce $ADP + P + energy$ in the form of heat.

The modern convention in molecular biology is to represent metabolism with discrete *pathway* models, describing a series of reactions that occur in the cell. In nature, however, metabolic pathways overlap and intersect to form complex *metabolic networks*. Metabolic pathways possess a simplicity in representation, and thus may be favorable from an interpretive and analytical standpoint. It is nonetheless important to appreciate that metabolic networks, while often highly complex, more accurately capture the system in its entirety. Figure 2.8 describes an example of a metabolic pathway, illustrating a series of enzymatic reactions converting D-glucose 1-phosphate to pyruvate. This pathway contains both anabolic and catabolic reactions. For instance, the conversion of β-D-fructose 6-phosphate to β-D-fructose

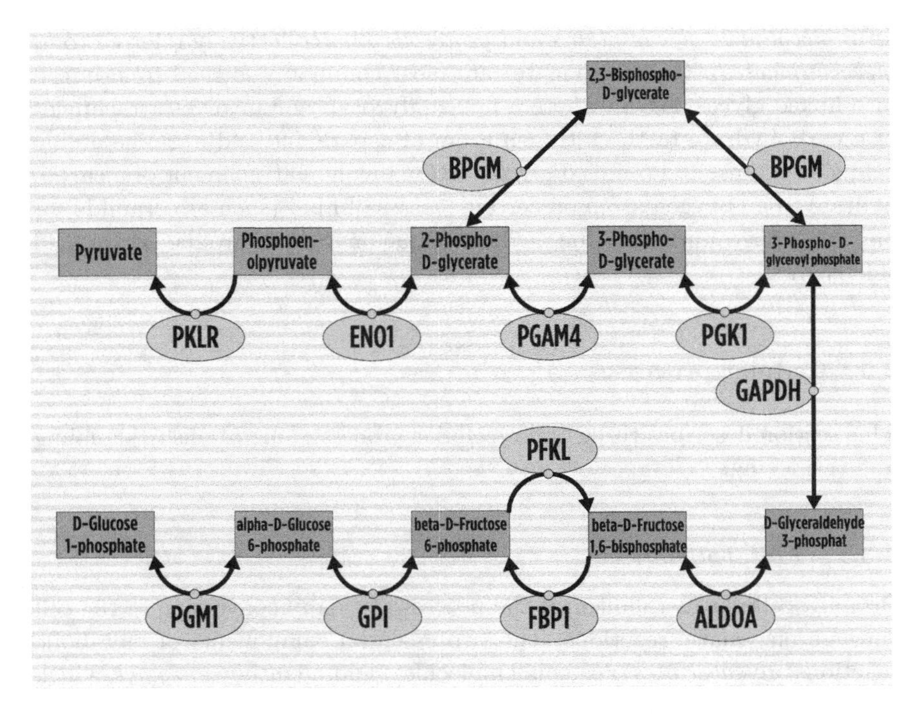

Figure 2.8 Glucose metabolism subpathway, illustrating the conversion of D-glucose 1-phosphate to pyruvate.

1,6-bisphosphate is an catabolic reaction, consuming one ATP and releasing one ADP in the process, whereas the conversion of phosphoenolpyruvate to pyruvate is anabolic, consuming two ADP and producing two ATP.

2.5.2 *Cellular signaling*

Cellular signaling works as a communication system within and between cells, and is essential for maintaining cellular response to changing environmental circumstances. The essential components of a signaling pathway include an extracellular signaling molecule (i.e. ligand), a cell surface receptor and one or more (often many) intracellular components that change behavior as a result. The majority of the intracellular components of a signaling cascade are proteins that exchange information through *phosphorylation* (i.e., addition of a phosphate group to a protein, in order to either activate or deactivate it). Similar to cellular metabolism, signaling

Autocrine, Paracrine, Endocrine

Signaling is generally divided into three classes, depending on the proximity of the source signaling molecule and target cell. In *autocrine signaling*, the source molecule originates from the target cell itself, whereas in *paracrine signaling* the source molecule targets cells in the immediate vicinity. *Endocrine signaling* involves production of signaling molecules from endocrine glands, and transport of these molecules through the bloodstream to distant target cells throughout the organism.

can be represented either as pathways or networks, depending on the preferred comprehensiveness. Figure 2.9 depicts a subset of the toll-like receptor signaling pathway, illustrating the binding of an extracellular signaling molecule to the cell surface receptor toll-like receptor 4 (TLR4) and subsequent activation of inflammatory cytokine gene transcription via a multiprotein signaling cascade. TLR4 is a critical part of the *innate immune system*, and can be activated by a variety of ligands, including lipopolysaccharide (a component of bacterial membranes) and saturated fatty acids.

There are two predominant classes of cell surface receptor related to cell signaling, which differ in the way they interact with the extracellular ligands and intracellular molecules in their respective signaling pathways. *G-protein coupled receptors* (GPCR) are cell surface receptors that are associated with an internal protein called a *g-protein* (short for guanosine nucleotide-binding protein). All GPCRs possess seven transmembrane domains (i.e., pass through the cellular membrane seven times) but beyond this similarity are a diverse family of receptors, with over 800 known unique GPCRs in the human genome. Upon activation, a GPCR undergoes a conformational changes which allows it to activate its associated g-protein by exchanging its bound guanosine diphosphate (GDP) for guanosine triphosphate (GTP). The internal GTP-bound g-protein then initiates further intracellular signaling or interacts directly with the target molecule, depending on the specific signaling pathway.

The second major type of cell surface receptor, *enzyme-linked receptors*, carry out catalytic activity that is activated on binding of the extracellular ligand. TLR4 is an example of this class of receptor. Enzyme-linked receptors may be themselves enzymes, or may be directly associated with an intracellular enzyme. Like GPCRs, enzyme-linked receptors are transmembrane proteins; however, they only contain a single transmembrane

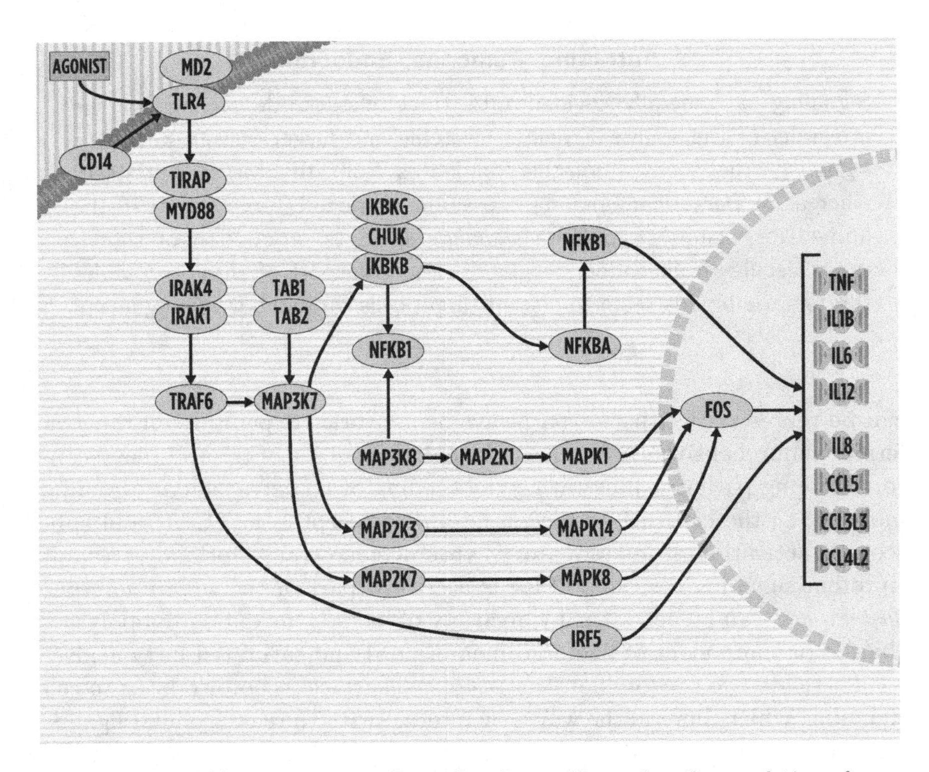

Figure 2.9 Toll-like receptor signaling subpathway, illustrating the regulation of gene expression by an inflammatory signaling cascade initiated by activation of toll-like receptor 4.

domain (although an enzyme-linked receptor may possess multiple subunits, each with a single transmembrane domain). There are six known classes of enzyme-linked receptors (receptor tyrosine kinases, tyrosine-kinase-associated receptors, receptor-like tyrosine phosphatases, receptor serine/threonine kinases, receptor guanylyl cyclases and histidine kinase associated receptors) however, the most common form is receptor tyrosine kinases (RTK). This subclass of receptor typically exists as *monomers* (i.e., single receptor units), that join together in pairs to form *dimers* on activation by binding of extracellular ligands. Each RTK monomer contains an intracellular domain containing multiple tyrosine amino acids. Upon ligand binding and dimerization, these tyrosine residues become phosphorylated (via enzymatic conversion of ATP $\rightarrow$ ADP+P and addition of the produced

P to the tyrosine residue). Once all tyrosine residues are phosphorylated, the RTK dimer is activated, and can now be recognized by intracellular proteins called *relay proteins*, which interact with the RTK and activate further signaling cascades within the cell.

2.5.3 *Trafficking and translocation*

Nearly half of the proteins in mammalian cells must be delivered to specific locations in order for the cell to function properly. The process of translocating proteins is also called *protein targeting* or *protein sorting.* Protein sorting occurs primarily through *secretory pathways* or through *non-secretory pathways.*

The secretory pathway (see Fig. 2.10; first and second row) targets secretory proteins (either during or post-translation) to the endoplasmic reticulum (ER) membrane, which are then translocated into the ER lumen. Once protein translation is complete, these proteins are further delivered to the lumen of Golgi and lysosomes, or to plasma membranes via vesicles. A uniform principle drives this type of protein sorting: proteins are moved from one membrane-bounded compartment to another membrane-bounded compartment, mediated by *transport vesicles* (third row). During vesicle-mediated transport, membrane cargo-receptors recruit the proteins to be transported. Then the membrane forms a vesicle with a Rab+GTP receptor for docking the vesicle to the target membrane. Once the Rab effector on the target membrane is identified and bound, the two SNARE counterparts tightly bind together, allowing the fusion of the vesicle with the target membrane. The whole content is delivered in the compartment bound by the target membrane. The translocation enablers are the v-SNARE and t-SNARE proteins. The non-secretory pathway sorts proteins to the ER, mitochondria (last row), peroxisomes and the nucleus.

Nuclear transport occurs through the *nuclear pore complexes* (NPC), which form holes on the nuclear membrane. Macromolecules equipped with *nuclear localization signals* (NLS) bind to *importin/exportin* proteins, which drive the macromolecules through the nuclear pores. The importin component of the complex interacts with Ran-GTP to release the cargo protein in the nucleoplasm and migrate back to the cytoplasm. Similarly, cargo proteins can migrate from the nucleoplasm to the cytoplasm by forming a complex with an exportin and a Ran-GTP.

CO-TRANSLATIONAL TRANSLOCATION TO THE ER

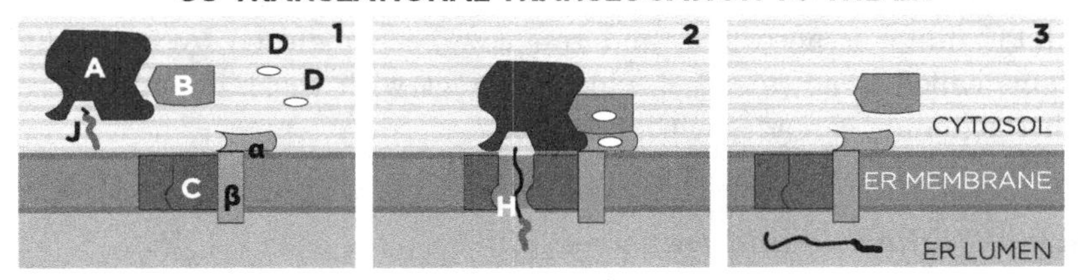

POST-TRANSLATIONAL TRANSLOCATION TO THE ER

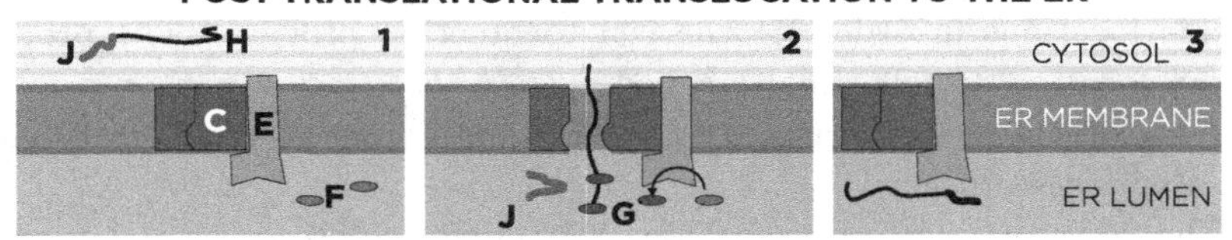

A Ribosome • **B** Signal-recognition particle (SRP) • **C** Translocon (Sec61) • **α, β** SRP receptor • **D** GDP molecules in 1 and GTP molecules in 2 • **E** Sec63 complex • **F** Molecular chaperone BiP of the Hsc70 family bound to ATP • **G** BiP bound to ADP • **H** Protein polypeptide • **J** ER signal sequence

VESICULAR TRAFFIC: DOCKING AND FUSION OF TRANSPORT VESICLE

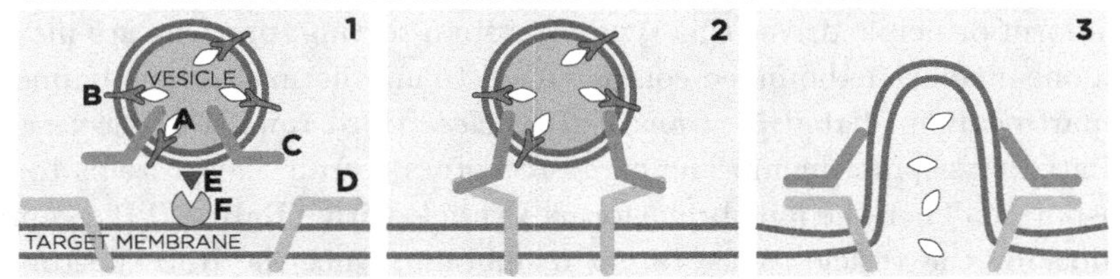

A Soluble cargo proteins **B** Membrane cargo-receptor proteins **C** v-SNARE proteins **D** t-SNARE proteins **E** Rab + GTP **F** Rab effector

POST-TRANSLATIONAL SORTING TO MITOCHONDRIA

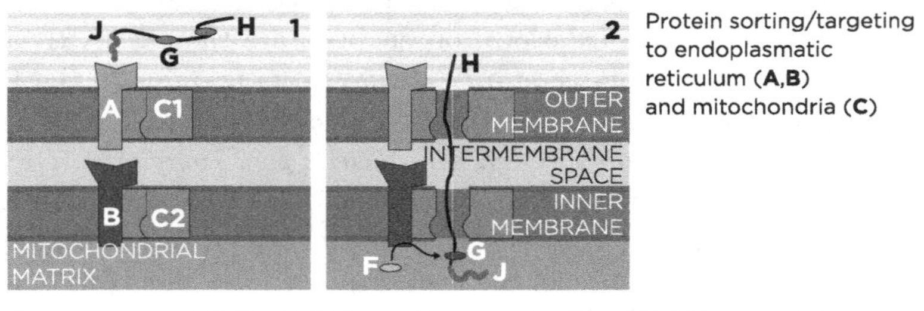

A Import receptor **B** Tim44 **C1** General import pore (Tom40) **C2** Inner membrane pore (Tim23/17) **F** Hsc70bound to ATP **G** Hsc70 (cytosolic in 1 and matrix in 2) bound to ADP **H** Protein polypeptide **J** Mitochondrial matrix signal sequence

Figure 2.10 Modes of cellular transport.

2.6 Experimental methods

Molecular biology is driven by quantities such as gradients, temperature, concentrations of elements, pH, etc. Since these quantities are important for understanding the mechanistic behavior of systems, a technological effort has been carried out in recent years to develop tools to measure these quantities of interest with ever-increasing precision.

With these rapid advances in biotechnology, biological data are increasingly being measured on a high-throughput scale, producing what is referred to as *omics* data. The suffix -omics refers to any data type that is measured at a system-wide scale. This does not necessarily mean that every object in a system is measured (due to technological and practical limitations); however, omics datasets carry the common feature of measuring large sets of variables that form biological systems. The most pervasive omics technologies (and those which will be addressed in this section) include genomics, epigenomics, transcriptomics, proteomics and metabolomics. However, with continuing technological advancements and expansion of *systems thinking*, the term ome/omics is being applied to a growing range of systems and data types. For instance, the physiome (physiology), exposome (environmental exposures), glycome (glycans) and secretome (secreted proteins) are all examples of data types that are being measured at a system level. As with any high-throughput data type, however, measurement of the system is only a starting point for understanding how the system functions. The consequent challenge lies in defining and applying data analytical strategies, often across multiple data types, in order to identify the key variables and processes that define a given biological system. Here we will briefly describe a selection of the most common high-throughput technologies that are used to produce omics data in systems biology.

2.6.1 *Microarray technology*

Defining the sequences and quantities of nucleic acid species inside the cell can reveal a great deal of information about the state of that cellular system. DNA microarrays measure cellular nucleic acids by taking advantage of the natural base pairing affinities of nucleotides. Although microarrays vary in their specific manufacture and application, all microarrays contain a very large number of single-stranded *oligonucleotide* (i.e., multi-nucleotide fragment) probes, fixed to the surface of the microarray. These probes are specifically designed to be complementary to the nucleic acid sequences of the target species. This feature of microarrays may be limiting in some study systems, since complete knowledge of the target sample genetic sequence is required in order to design the oligonucleotide probes. Microarray manufacturers have produced a wide range of microarray *platforms*, each specific to the application and target species of interest. For instance, if we are interested in comprehensively quantifying RNA from human tissue samples, we would use a human expression profiling microarray platform that contains complementary single-stranded DNA probes for the majority of known transcripts in the human genome.

DNA amplification

DNA amplification is a fundamental laboratory procedure in molecular biology that produces thousands or millions of copies of a DNA sample. Briefly, the process involves creating a mixture of the DNA sample and all molecular elements required for DNA replication, such as free nucleotides, oligonucleotide *primers* (i.e., single-stranded oligonucleotides) for initialization of replication and DNA polymerase. The mixture is then subjected to an period of temperature cycling wherein the template DNA is denatured, then primers are annealed, then the single strands are replicated. The temperature cycling is repeated on the growing number of newly replicated template DNA strands to produce the final amplified sample.

In order to produce high-throughput gene expression data from such a platform, it is first necessary to extract RNA from the target sample of interest. This extracted RNA may represent total RNA or mRNA only, depending on the extraction protocol. The purified RNA is then *reverse transcribed* to produce *complementary DNA* (cDNA), and the cDNA fragments are then *amplified* and labeled with a *fluorophore* (a fluorescent chemical that emits light following excitation). The type of labeling used here depends on whether *single-color* or *two-color* microarrays are being used. For two-color microarrays, two comparable samples (e.g., healthy vs diseased, tissue A vs tissue B, etc.) are labeled with different-colored fluorophores (commonly, red and green) then combined and applied to the same microarray. During the *hybridization* step, the labeled cDNA is first *denatured* (i.e., separated into single strands) then added to the microarray platform. The single stranded cDNA strands in the sample will then naturally hybridize (i.e., bind) to their complementary probes on the platform, in proportion to their abundance in the sample. Following hybridization, the fluorophores are excited with a laser and the emitted color intensities are quantified. This final step allows for assessment of the relative expression abundance in the two samples. For instance, if sample A is labeled with a red fluorophore and sample B with a green fluorophore, then we will observe a stronger red fluorescence signal from the probes corresponding to genes that are more highly expressed in sample A. The protocol for single-color microarrays is similar, except that only one sample is hybridized per array following labeling with a single fluorophore. Then, the overall light intensity for a given probe is compared between arrays.

While the protocol described here refers to gene expression profiling, the same general protocol can be applied to a range of study systems depending on sample preparation and probes used on the array. The box *Microarray applications* describes a number of the most commonly used applications of microarray technology.

2.6.2 *DNA sequencing*

Microarray technology (although still widely in use) possesses several inherent limitations including reliance on pre-defined platforms, relatively limited dynamic range ($\sim$200 fold) and presence of non-specific hybridization between probes and non-complementary target sequences. As an alternative technology that is not bound by these limitations, high-throughput sequencing (HTS; also called next generation sequencing, or NGS) is rapidly gaining use as a means for nucleic acid data generation in systems biology. Like microarrays, HTS can be used for a range of applications, such as gene expression profiling, genetic sequencing, and transcription factor binding site identification.

The technology of HTS largely evolved from the classical approach of capillary electrophoresis (CE) sequencing, also called Sanger sequencing after Frederick Sanger, who developed the original technique with colleagues in 1977. CE sequencing requires template DNA, short oligonucleotide primers (typically 18–24 nucleotides) that serve to initiate the DNA strand replication, DNA polymerase, *deoxynucleotides* (dNTP: dATP, dGTP, dCTP and dTTP) and *dideoxynucleotides* (ddNTP: ddATP, ddGTP, ddCTP and ddTTP). This latter form of nucleic acid is integral to the sequencing process, as they can be added to a growing strand during DNA replication, but their lack of a 3′-OH group prevents the formation of a phosphodiester bond between a following nucleotide, and thus strand elongation is terminated with the addition of a ddNTP (see Fig. 2.11A). The original

Microarray applications

The first column is the technology, the second one the application and the last column provides a brief description.

Gene expression profiling array	RNA abundance	Simultaneous assessment of high-throughput gene expression in parallel, using oligonucleotide probes that are complementary to known genes in the target species.
SNP array	DNA sequence polymorphism	Identification of genetic polymorphism through the use of probes matching known alternative alleles in a population.
Comparative genomic hybridization	Copy number variation	Identification of variation in the number of copies of DNA segments between two DNA samples, using probes designed to match the target genome sequences of interest. Similar in principle to RNA abundance assessment, the sample with higher copy number at a given locus will yield higher fluorescence in the corresponding probes on the microarray.
Chromatin immunoprecipitation	Transcription factor binding	Assessment of protein-DNA binding by cross-linking a transcription factor of interest to its bound DNA sites, then isolating the resultant protein-bound double stranded DNA fragments. The DNA fragments are then amplified and denatured, and hybridized to a microarray containing oligonucleotide probes complementary to the genomic region of interest.

protocol of Sanger sequencing involves running four parallel reactions, each containing amplified template DNA, primers, DNA polymerase, all dNTPs and one of the four ddNTPs. The DNA replication reaction contains many copies of the template DNA being elongated in parallel, and each elongation step may incorporate a dNTP or its corresponding ddNTP. The result of

Microarray applications — continued

The first column is the technology, the second one the application and the last column is a brief description.

Exon array	Alternative splicing	Similar to a gene expression microarray, but containing oligonucleotide probes specific to exons rather than genes. Some exon arrays use probes that are specific to the junctions between two adjacent exons (known as splice sites).
Tiling array	High resolution assessment	Tiling arrays are unique in that they contain probes that comprehensively cover an entire region of interest in the genome, rather than covering specific genes or polymorphic sites. Tiling arrays may be used for numerous applications (such as high-resolution gene expression or comparative genomic hybridization assessment) depending on sample preparation and platform design.
Methylation array	DNA methylation	Although there are a number of protocols to produce DNA methylation array data, one commonly used approach uses a modified DNA amplification procedure that converts unmethylated, but not methylated cytosine residues into thiamine. The resulting DNA product is hybridized to a microarray with probes designed specifically to match the modified sequences and thus differentiate methylated from unmethylated DNA.

each reaction will contain a mixture of fragment lengths, each terminating with the chosen ddNTP (as seen at left in Fig. 2.11B). The sequenced DNA fragments are labeled (either with a radioactive molecule or a dye that is visible under UV light) and, according to earlier protocols, separated by

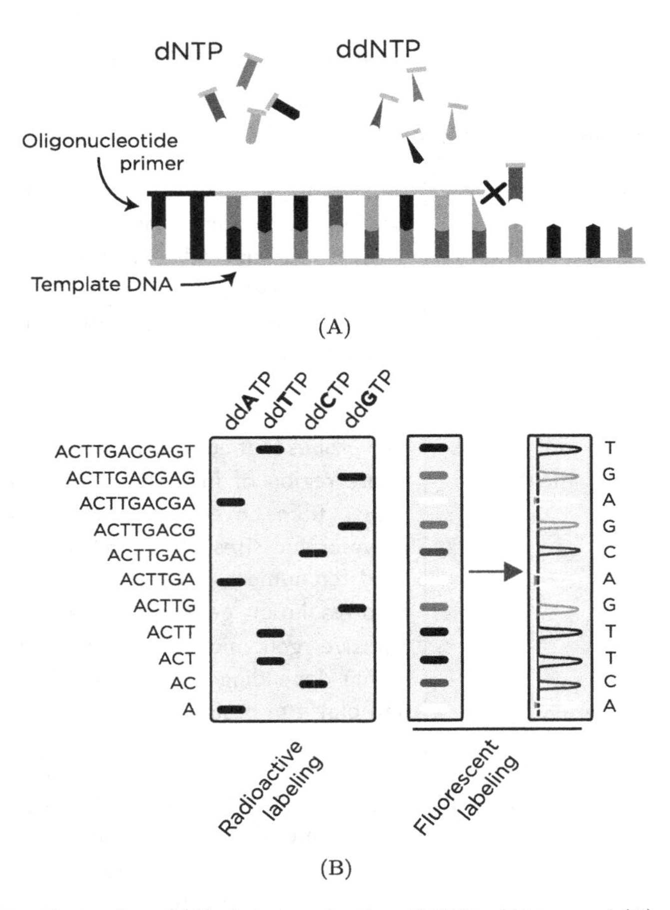

Figure 2.11 Illustration of (A) chain terminating ddNTP addition, and (B) sequencing by separation of ddNTP-terminated DNA via gel electrophoresis, either with radioactive or nucleotide-specific fluorescent labeling.

means of *gel electrophoresis*. Gel electrophoresis is a very common method in molecular biology, which involves loading a nucleic acid sample (or protein sample, with a modified protocol) into a viscous gel matrix then applying an electric current, thus forcing the fragments (which are negatively charged) to migrate toward the positive current (i.e., the bottom of the depicted gel at left in Fig. 2.11B). Shorter fragments will move more quickly than longer ones, resulting in a spectrum of fragment lengths, with the largest near the top and the smallest near the bottom. While the fragments are not naturally visible in the gel, they can be visualized through autoradiography (as was used in the original Sanger protocol) or with UV light. When the results of four parallel sequencing reactions (each containing one of the four ddNTPs) are loaded simultaneously, it is possible to determine the overall sequence by reading from the bottom to the top (Fig. 2.11B). Later modifications of the Sanger protocol involved labeling each ddNTP with a different fluorescent dye, allowing all four reactions to be performed in the same mixture. With this approach, each of the four dyes emits a unique wavelength that can be read with a chromatogram (as illustrated at right in Fig. 2.11B), thus automating a procedure that was originally labor-intensive and potentially error-prone, particularly for longer sequences. Another major modification of the original Sanger protocol involved replacing traditional gel electrophoresis with CE, which uses the same electrokinetic principles but involves loading the samples into a capillaries containing a polymer solution rather than a physical gel. This modification streamlined the laboratory protocol, and offered benefits in terms of scalability and precision in fragment length resolution.

CE sequencing is still in use in part for its wide accessibility and strength in sequencing long fragments of DNA. In fact, the first sequencing of the human genome was performed using CE sequencing — an effort that took ten years with a total project cost of nearly $3 billion US. More recently, the technology of HTS has rapidly gained popularity in analyses of nucleic acids. The fundamental innovation behind HTS is the massive parallelization of the sequencing reaction, allowing millions of short- to medium-sized fragments to be sequenced simultaneously. With this innovation, as of 2014 a human genome can be sequenced in ~30 hours for about $1000 US. Although the specific protocol for producing HTS data differs depending on the manufacturer of the sequencing machine being used, the general protocol entails: starting with a DNA (or cDNA) sample to be sequenced, the sample is fragmented into small segments (from ~30–1000bp, depending on the platform) and synthetic DNA adapters are added to each fragment end with DNA ligase, producing a *DNA library.* These adapters serve to attach the fragments to the physical sequencing surface (either a bead or a microfluidic channel). The anchored fragments are then amplified on the surface of the sequencing platform, producing dense clusters of template fragments that are then sequenced using an approach specific to the manufacturer of the platform. One common protocol for HTS was developed by the biotechnology company Solexa (later acquired by Illumina), and uses *reversible dye terminators* during the sequencing reaction, as shown in Fig. 2.12. Briefly, this approach involves addition of all four dNTPs to the

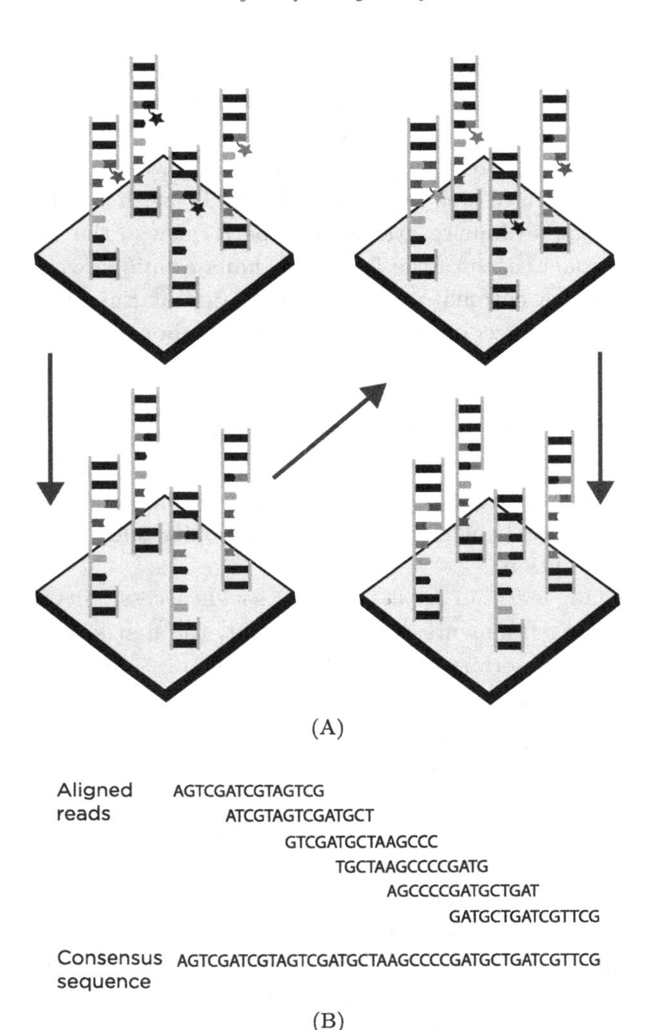

Figure 2.12 Illustration of (A) HTS with reversible dye terminators, and (B) alignment of multiple short reads to produce a consensus sequence.

platform, each dNTP being labeled with a different fluorescent dye. These dNTPs contain blocking groups (i.e., terminators) at the 3′-OH position of the ribose sugar, which prevents the addition of more than a single dNTP to a given strand (similar in concept to ddNTPs). Following addition of a single dNTP to each template strand on the platform, additional dNTPs are washed away and the fluorescence signal from each strand is recorded

(Fig. 2.12A, top-left panel). Therefore, whereas the Sanger/CE protocol involves two separate phases for DNA replication and sequence identification by gel/capillary separation, HTS integrates these two steps by performing sequence identification at each nucleotide addition. The fluorescent dyes and 3′ blocking agents are then removed from the fragments (Fig. 2.12A, bottom-left panel), and the procedure is repeated with a new set of labeled dNTPs.

The result of HTS is a dataset with a large number of short overlapping sequenced fragments, called *reads*. Various bioinformatic algorithms have been developed to assemble these reads accurately into longer contiguous sequences (Fig. 2.12B), either using only the fragments themselves (called *de novo assembly*) or using a reference genome as a guide against which to map the sequences. The latter case is very useful in the presence of higher genomic complexity, such as repetitive sequences and gene families. This is the case for the human genome, which contains ~45% repetitive content, and therefore a reference human genome (determined from low-throughput approaches) is often used to aid the sequence alignment. In the case of genome sequencing it is simply the unambiguous sequence of the genome that is the target of the analysis. However, HTS may also be used for other applications such as gene expression profiling, in which case RNA is used as template sample, and the number of reads mapping to the position of a given gene in the genome is equivalent to the expression level of that gene.

2.6.3 *Mass spectrometry*

Whereas the above technologies aim to clarify and quantify nucleic acids, proteins and metabolites do not posses the convenient complementary base pairing property, and thus require dedicated technologies. *Mass spectrometry* (MS) is a widely used approach for characterizing proteins and metabolites by measuring their mass-to-charge ratios (m/z). A typical MS experiment begins with vaporizing the protein/metabolite sample of interest then applying an electron beam, which causes electrons to be ejected from the atoms in the sample (thus producing positively charged ions). This process is called ionization, and may also be accomplished by chemical ionization for gas/vapor phase analytes, and by electrospray ionization and matrix-assisted laser desorption/ionization for solid or liquid samples. Regardless of the ionization process, many of the molecules in the sample will fragment during ionization, producing a mixture of charged and uncharged components of the original molecules. This mixture is then sprayed toward a detector with a magnetic field between the spray source and detector. The level of deflection of a given molecule by the magnetic field is proportional to its m/z, and thus serves to differentiate a complex mixture of molecules. Figure 2.13A illustrates an example mass spectrometry experiment on pure CO_2. Ionization and fragmentation of CO_2 results in a mixture of CO_2^+, CO^+, O^+, and C^+ (O_2^+ will not be formed, since the oxygen atoms are not directly linked in the native CO_2, molecule). When sprayed toward the detector, the magnetic field will deflect most strongly the lightest ion, resulting in a spectrum shown in Fig. 2.13B. The resulting mass spectra provide information on the m/z of the fragments in the CO_2 sample (x-axis), and the relative abundance of each (y-axis). Naturally, mass spectra may easily become very complicated for complex samples containing many distinct molecules. Generation of mass spectrometry data from complex mixtures is therefore often preceded by a preliminary fractionation step, in which the sample is divided into simpler subfractions and each is ionized and analyzed separately. Furthermore, mass spectrometry experiments are often run in tandem (called MS/MS) wherein a single m/z range is selected from the first mass spectrum and run on a second to produce more detailed information. In either case, mass spectra are typically analyzed by dedicated software that compare obtained spectra to known spectra, in order to identify and quantify the components of the sample.

2.6.4 *Nuclear magnetic resonance spectroscopy*

Nuclear magnetic resonance (NMR) *spectroscopy* can be used to characterize samples of proteins/metabolites by exploiting the atomic property of nuclear magnetism. The principle of NMR is based on the observation that some atomic nuclei (such as ^{1}H, ^{13}C, ^{15}N and ^{19}F) possess magnetic property called *magnetic moment*, causing them to have an angular momentum called *spin*. Under normal circumstances, the nuclear spin of a collection of atoms will be in random directions; however, the application of an external magnetic field will cause all of the nuclei with magnetic moments to adopt an orientation aligned either with or against the direction of the magnetic field. As a result, the nuclei aligned with the magnetic field will possess a lower energy, while those aligned against the field will possess higher energy. Radio frequency energy can then be applied to the sample in pulses, typically ranging from 40–800 Mhz. At a certain frequency, the energy will be absorbed and the nuclei aligned with the magnetic field will

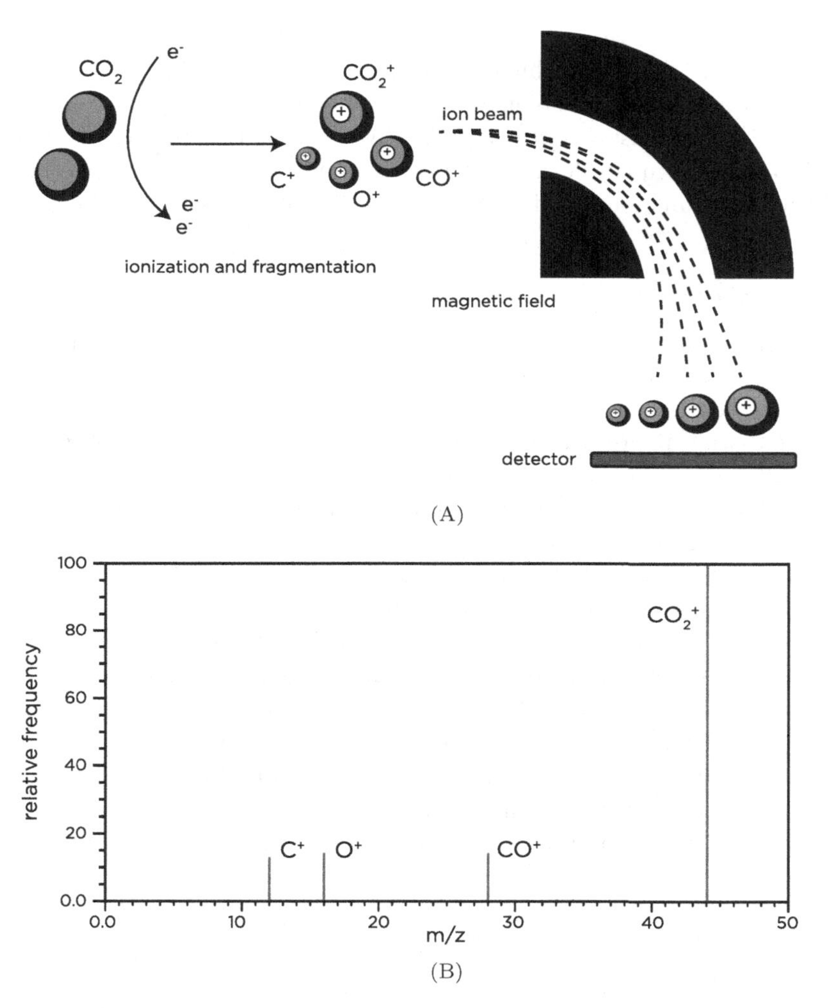

Figure 2.13 Schematic for the process of (A) ionization and ion separation to produce (B) mass spectra of a pure CO_2 sample.

flip, thus adopting the higher energy orientation. The specific frequency at which a nucleus absorbs energy is related to its precise chemical and physical environment, and thus can be used as a means to characterize complex biomolecular samples.

In comparison to MS, NMR is relatively less sensitive, but generates highly reproducible data. And in the case of both MS and NMR, the quantity of data that can be produced is considerably less than nucleic acid analysis, in terms of the number of uniquely identifiable compounds (typically on the order of hundreds of variables, for MS and NMR). However, technologies for proteomic and metabolomic analysis are constantly improving in precision and comprehensiveness. And despite the technical limitations, the quantification of proteins and metabolites is nonetheless essential for functional understanding of a complex biological system.

2.6.5 *Environmental data*

All of the above approaches aim to measure molecular level data. However, in many cases, the system being studied extends far beyond the scope of molecular data — for instance, a cohort of humans at varying stages of a multifactorial disease cannot be comprehensively characterized by molecular data alone. For such systems, it is important to appreciate that the molecular system under study is not a closed system, but rather is affected by a vast array of external factors. In reality, it is of course rarely possible to measure or even to have *a priori* knowledge all of the external factors that may affect the system, but it is substantially useful from a data analytical point of view to measure as much as is feasible. Apart from simple measures such as age and sex, other factors including background diet, activity level, medical history and chemical exposures form a sort of metadata for each individual, uniquely characterizing their environment. Particularly in human studies where the environmental background of each participant is often uncontrolled, such metadata can make the critical difference in shedding light on otherwise inexplicable patterns of variance in the data.

2.7 Summary

In this chapter we have introduced the major components and processes of the cell. Understanding the fundamental behavior of cells is substantially useful in analysis of biological data, but of course, the majority of modern biological research projects will be more highly focused on a specific biological phenomenon of interest. For instance, one might be interested in the onset and progression of subcutaneous adipose tissue inflammation following a high-fat dietary challenge in the diabetic state. In this case it is always advisable for the analyst to seek out a deeper understanding of this biological phenomenon, even before initiating analysis of the data. While it may be tempting to view a biological dataset as simply a set of numbers, in reality there are precise biological principles that uniquely characterize each dataset. Keeping these in close consideration will save the analyst from investigating infeasible patterns in the data (or, on the other hand, patterns that are already common knowledge) and will enable them to establish a more realistic and insightful plan for analysis.

2.8 Further reading

Extensive introduction to biochemistry and molecular biology can be found in [Alberts (2008)] and [Lehninger *et al.* (2008)], including expanded detail on each of the biological molecules and processes presented here. Introductory material to systems biology can be found in [Kitano (2002b)] and discussion on the difference with classic bioinformatics is in [Cassman *et al.* (2005)]. The interpretation of biological systems as computing systems is well exploited in [Hood and Galas (2003)].

Chapter 3

Systems and models

This chapter introduces the notions of systems and models, and also discusses their main properties.

Systems

A *system* is a set of integrated and interacting *components* or *entities* that form a whole with definite boundaries and surrounding environment.
A system has a goal to achieve by performing one or more *functions* or *tasks*.
Systems can be *aggregated* into a hierarchy. A system at a given level of detail can be a component at an higher level of detail.
A *complex (non-linear) system* is a system that does not satisfy the principle of superposition, i.e. the behavior of the system cannot be inferred from the behavior of its components.
A *dynamical system* is a system where fixed rules define the time dependencies of the system in a geometrical space.

3.1 Systems

A model is strictly connected to the *system* for which it is built. We will first introduce the notion of system and then we will move on to the model concept. We use the term "system" to refer to real-world objects that we want to study through analysis techniques. A system is therefore a collection of interacting entities that exhibit an overall behavior. Systems are studied by many disciplines such as systems theory, cybernetics, dynamical systems, thermodynamics, complex systems and computer science.

A system is characterized by boundaries within which we distinguish a *structure* defining the *components* (elements) of the system (e.g., the list of proteins and enzymes that we are considering in a specific biochemical signaling mechanism). Components have *attributes* defining their peculiar characteristics that may be perceived and measured (e.g., size, quantity, shape). The *behavior* of a system defines its processing features (e.g., how the signal is transmitted, amplified or reduced within a system; how an input is transformed into an output). A notion strictly connected to behavior is that of dynamical systems that evolve over time and space (e.g., the swing of a pendulum or the floating of proteins in the cytoplasm of a cell). Systems also have an *interconnection structure* (sometimes referred to as a *network*) that determines how the components can interact (e.g., the affinity of interaction between the proteins in the signaling cascade determines which proteins are compatible for interaction). The network represents the relationships between the components. We will sometimes use the term *subsystem* to refer to a system that acts as a component of a larger system.

3.1.1 *Classification of systems*

There are many kinds of systems (see Fig. 3.1). There are *open* and *closed* systems. Open systems interact with their external environment by exchanging input and output through the boundaries. Input and output are usually *energy* or *matter*, but they can also be *information* or *signals*. Changes in the environment may cause changes in the system and

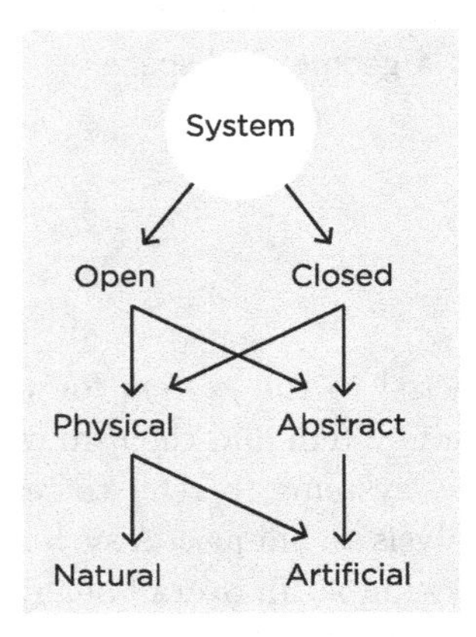

Figure 3.1 Classification of systems according to their characteristics.

vice versa. A closed system has no interaction with the environment outside its boundaries, except for the transfer of energy. Changes in the environment do not affect the behavior of a closed system (e.g., the Earth is usually considered a closed system). Closed systems that do not transfer even energy with the environment are called *isolated* systems (e.g., laboratory-controlled experiments or artificially built ecosystems).

Open and closed systems can be either *physical* tangible systems (e.g., a computer, an aircraft, a cell, an organ, a population, an ecosystem) or *abstract* intangible systems that are often conceptualizations of physical systems (e.g., the model of the atom, Newton's laws of motion, the equations representing chemical reactions, a cultural system defined by the interaction of cultural elements). Abstract systems are the result of a cognitive activity and most of the time are models for physical systems. Physical systems can be either *natural* (a cell, an organ, a population, an ecosystem, the interception of solar radiation by the Earth) or *artificial* (a computer, an aircraft). Abstract systems are artificial.

A distinct classification of systems is driven by our understanding of the system's structure and behavior (see Fig. 3.2). A *morphological* system is one that we understand through the relationships between its components and their attributes, but we do not exactly understand the mechanisms of interaction between the components through their connections (e.g., we statistically know that modifications in a set of genes cause a disease, but we do not know the actual mechanistic interactions that make the disease emerge). A *cascading* system is one for which we know the mechanistic

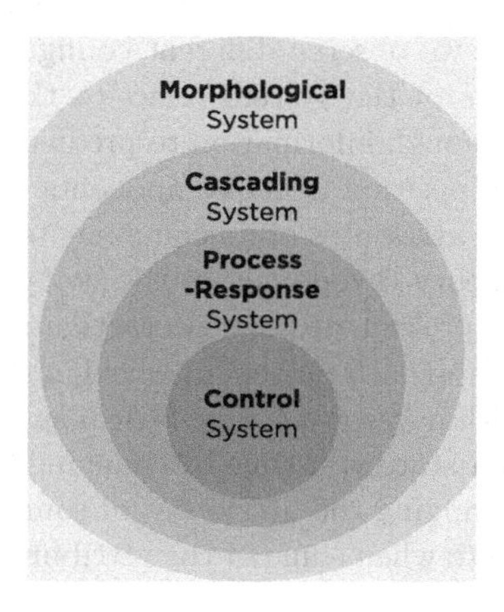

Figure 3.2 Classification of systems according to our understanding.

relationships between its components, but we do not fully understand the quantitative relationships driving the behavior. We are usually interested in the flow of energy/material/information from one component of the system to another (e.g., how a signal represented by a receptor binding on a cell membrane is transferred to DNA into the nucleus). A *process-response* system is one of which we understand all the qualitative and quantitative details. Finally, a *control* system is one that we can manipulate to tune its behavior (e.g., remote control devices).

The characterization of the system we are interested in according to the previous classifications is a preliminary and fundamental step that can help us select the best methods and technologies to model it. For instance, in this book we are interested in the analysis of biological systems that are open, physical, natural and dynamical systems that interact through modulation of energy and matter. The outcome of our modeling activity will be an abstract, artificial system that evolves through information exchange (in place of energy and matter) and that is a model of the biological system we consider. We will mainly address morphological and cascading systems due to the lack of knowledge of the mechanistic behavior of biological systems or the difficulty/impossibility of having precise quantitative measures of their parameters governing their behavior.

3.1.1.1 *Dynamical systems*

Dynamical systems have a space and time dimension because they change their characteristics over time. Therefore if we pick snapshots of the system at different time points we observe different configurations of the system. A *configuration* or *state* of the system refers to the current condition of the system and stores enough information to predict its next move. A state is characterized by the position of its components in a geometrical space and by the values of the attributes of its components (the precise definition according to system theory is reported in the box on the next page). For instance the state of a protein–protein interaction system is given by the concentration or number of each protein involved in the interaction network and by the location within its physical or logical space.

Systems change their state over time by changing the location of some of their components or changing the attributes of some of their components. We refer to a *steady state* when some of the attributes of the system are no longer changing in the future. During the time needed to reach the steady state, the system is in a *transient state*. The set of the states through which a

State, State space

A *state* of a dynamical system is the additional information to the input required at time t to uniquely determine the output at time $t' > t$. A *steady state* with respect to an attribute A is one in which A is constant independently of the ongoing processes in the system. A *transient state* is one in which the system is warming up to reach a steady state.
The *state space* of a system is the set N of all possible states of the system. The state space is *discrete* if N is finite or infinite enumerable (there is a bijection between N and $\mathbb{N}$, i.e., we can assign a different natural number to each element of N). The state space is *continuous* if N is infinite and not enumerable.

system can transition is called the *state space* of the system. A *discrete* state space is represented as a graph whose nodes N are the states through which the system can transition and the edges E represent the transitions. For instance the state space of a computing machine is a structure (N, E, S, G) where (N, E) is a graph, $S \subseteq N$ is the set of *starting* states of the computation and $G \subseteq N$ is a set of *goal* or *final* states where the computation stops. We use discrete state spaces when the dynamic of the system is discrete (e.g., the number of individuals in a population are always a natural number) or our experimental setting requires us to approximate a continuous value by sampling it at discrete time points (e.g., the amount of water in a recipient that is filled by a pipe observed every three seconds).

A *continuous* state space can be obtained by the motion of a particle in a geometric space. The state of the particle is a vector that has its origin in the current position of the particle and that defines the direction and velocity of the particle.

The state change is determined by a transition function that governs the dynamics of the system. Transition functions can be either local or global. Depending on the kind of system, components interact with each other only for some specific tasks (e.g., consider a set of companies working in the same market that interact sporadically only through the market). In this case the behavior of each component can be specified locally to the component and interaction rules determine how components exchange information. This is the essential property of distributed systems in which no centralized point of control exists. An example of a canonical distributed system is the Internet.

Graph

A *graph* is a pair (N, E), where N is a set of nodes and $E \subseteq N \times N$ are the edges connecting nodes.
A graph is *directed* if pairs in E are ordered (i.e. $(n_i, n_j) \neq (n_j, n_i)$) and it is *undirected* if pairs in E are unordered (i.e. $(n_i, n_j) = (n_j, n_i)$).
The *order* of a graph is $|N|$, the *size* of a graph is $|E|$ and the *degree* of a node is the number of edges connecting to it (double counting self-loops). The *in-degree* and the *out-degree* of a node in a directed graph is the number of edges entering and exiting the node.
A *weighted graph* is a graph with a weighting function $w : E \rightarrow \mathbb{R}$ that labels edges with numbers. A *labeled graph* is one where a set of labels $\mathcal{L}$ replaces $\mathbb{R}$ in the weighting function.
A *multigraph* is a graph where multiple copies of edges can connect the same pair of nodes.
A *path* from a node n_0 to a node n_1 in a graph is a sequence of edges connecting n_0 (the *source* of the path) to n_1 (the *target* of the path). The path is *directed* if the graph is directed and each edge leads to the next one in the path. The number of edges in the path is the *length* of the path. A path can be infinite. A path of length 1 is a *transition*. A path can be alternatively defined as the sequence of nodes connected by the edges.
An undirected graph is *connected* if for each pair of nodes there is a path between them.
A directed graph is *connected* if for each pair of nodes there is a directed path between them. It is *weakly connected* if the graph obtained by removing the orientation of edges is connected.

Biological components, when interpreted as information and computational devices, give rise to distributed systems. In fact in a biological system there are millions of simultaneously active computational threads with partial knowledge of the interconnection structure (e.g., metabolic networks, gene regulatory networks, signaling pathways) that dynamically vary over time. Furthermore, each component has only a limited view of the whole system (e.g., a gene or a protein has a very limited view of the cell) and interactions can only occur if components are correctly located (e.g., they are close enough or they are not divided by membranes) as occurs in distributed systems with administrative domains. For instance, each protein in

Distributed system

Distributed systems were introduced in the 1960s to manage processes within operating systems and communication networks. Distributed systems are made up of components physically or logically distributed in space, that compete for resources and operate to reach a system goal. There is no precise definition of distributed system, but there is a set of properties that these systems share:

1. the components of the system are autonomous computational entities and have a local state and a local memory;
2. the components may communicate with each other;
3. each component has only a limited view of the system and each component may know only one part of the input;
4. the interconnection structure of the system (network topology, network latency, number of nodes) is not known in advance.

a cell has a local view of the whole and the behavior of the cell is determined by local interactions between its components.

3.1.1.2 *Determinism vs stochasticity*

Systems also differ in the way in which they react to external stimuli (input) according to our observations. *Deterministic* systems always react in the same way to the same set of stimuli. These systems are completely determined by the initial state and the input set. For instance, the motion of an object on the Earth is completely determined by the basic parameters of Newton's laws of motion (being that the speed of the object is considerably smaller than the speed of light) or that an electric circuit always behaves the same way when it is provided with the same input. A sequential computing program always produces the same result for the same input. The essence of deterministic systems is that each event is causally related to previous events and choices are always resolved in the same way in the same context.

When a system generates multiple outcomes from the same input in different observations, and hence we cannot predict the output from the input, we say that the system is *nondeterministic* (i.e., we cannot make any assumption on the output of the system and on the way in which it resolves the choices that lead to different outcomes). For instance, we usually assume that distributed computing systems are nondeterministic

Causality, Determinism

Causality is the relationship between an event (the cause) and a subsequent event (the effect) interpreted as a consequence of the first. A causal chain implies the temporal ordering on its events, but temporal ordering does not imply causal relation (independent events may be ordered in a temporal sequence).
Determinism conjectures that every event is causally determined by previous events and that choices are uniquely resolved in the same context. A deterministic system always produces the same outcome for the same initial conditions.

because we do not know the relative speed of concurrent processes, the scheduling algorithms of the operating systems or the routing strategies of the interconnection network and therefore it is impossible to hypothesize on the order in which independent events are observed. The lack of assumptions on the architecture of the system allows us to prove properties that are hardware independent and hence can be related to the essence of the system we are studying.

Nondeterminism, Stochasticity

Nondeterminism is the quality of producing multiple outcomes from the same state with the same input in different observations. *Stochasticity* is the quality of lacking any predictable order or plan. A stochastic system may produce different outcomes for the same initial conditions in different observations, although we can associate probabilistic distributions to predict speed and alternatives.

We can relax the nondeterministic assumption when we have more contextual information. For instance, in biochemical reactions we know that there are probabilistic distributions governing the choices between different enabled reactions. The kinetics of reactions is then driven by the affinity between reactants and the concentration of reactants. If the quantities of components are limited in the reaction volume, the system is driven by probabilistic distributions that determine the order in which we observe the events. For instance, the expression of genes in a cell under pre-defined conditions is a *stochastic* system.

It is possible to transform a nondeterministic system into a stochastic one by attaching probabilities to the selection points so that we turn nondeterministic choices into probabilistic choices. We can observe how the system behaves in the selection points on repeated experiments and compute an approximated probability from the observed frequencies of the actions performed. Note that this approach can only be applied if we can experiment on the real system, i.e., if we have the system and experimenting on it is neither too dangerous nor expensive.

3.1.2 *Properties of systems*

Systems expose many different properties depending on how they are observed or studied. Many properties of systems are expressed in formal languages to facilitate their verification and assessment. First-order logics is the ground for such formalisms. The behaviors that a system can exhibit are characterized by a set of possibly infinite sequences of states through which the system can pass. A *property* of the system can also be defined as a set of sequences of states. A property holds for a system if the set of sequences of states that the system can generate is contained in the property.

There are two main classes of properties called *safety* and *liveness* properties. *Safety* means to be in a safe state for static systems and that bad things do not happen in the operation of the system. A property P is a *safety property* if each prefix of the sequences generated by the system and not in P cannot be extended to a sequence in P. *Liveness* means that good things eventually do happen in the operation of a system. A property P is a *liveness property* if each finite sequence of states can be extended to a sequence in P. An important liveness property is *progress*: an action enabled will eventually be executed, thus letting the system progress. The opposite of progress is called *starvation*, a situation in which an action is never executed. Another fundamental liveness property is *fairness*: if the system performs a choice infinitely often, then each alternative will be picked infinitely often as well (e.g., if we are tossing a coin infinitely often, we expect infinitely often head and infinitely often tail).

There is a theorem ensuring that each property P can be expressed as $P = P_1 \wedge P_2$, with P_1 a safety property and P_2 a liveness property. It is also useful to know that safety and liveness properties are closed under Boolean operations, but they are not closed under complement. Safety properties are closed under union and intersection, while liveness properties are closed only under union.

Safety, Liveness

Let σ, possibly indexed, be a sequence of states of the system, and let P be a property of the system represented as a set of sequences of states.
P is a *safety* property if $\forall \sigma_0 \notin P$, $\exists \sigma_0^i, i \geq 0$, such that $\forall \sigma_1, \sigma_0^i \sigma_1 \notin P$, where σ_0^i is the prefix of σ_0 made up of the first i states.
P is a *liveness* property if $\forall \sigma_0$ is finite, $\exists \sigma_1$ such that $\sigma_0 \sigma_1 \in P$.

Example 3.1. [Safety and liveness properties]
Consider a single-lane bridge over a river that can be passed in both directions alternately. A safety property states that there is no situation in which two cars moving in opposite directions are simultaneously on the bridge. A liveness property states that a car waiting to enter the bridge in one direction will eventually enter it, thus ensuring progress of the cars in each direction. This also ensures that the controller of the bridge that manages the switch from one direction to the other is operating fair choices.

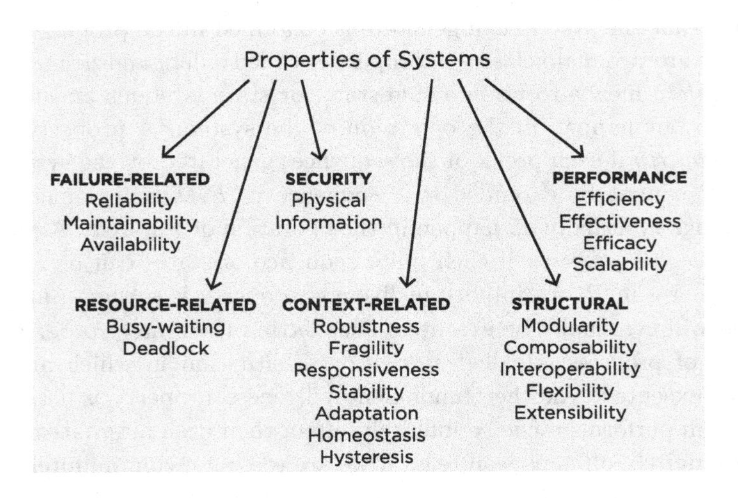

Figure 3.3 Classification of properties of systems.

Figure 3.3 shows a hierarchical classification of properties. Failure-related properties deal with the proper functioning of systems and are usually quantitative properties. The way in which the system uses its resources introduces some properties that may deteriorate the intended behavior of the system. Security is another class of properties related to the accessibility of a system's components from external agents. Systems operate in different contexts that can influence a system's internal behavior. There is a class of properties relating systems to their context. Structural properties define the characteristics of a system with respect to its interaction with other systems and possible modification of the system itself. An evaluation of the overall performance of a system allows us to compare different systems that are addressing the same problem.

3.1.2.1 *Failure-related properties*

Systems may fail to operate correctly by experiencing *failures* due to random errors, design mistakes or deliberate attacks by adversaries (e.g., an elevator may stop working due to a software error or loss of electric power, an automatic pilot of a plane may stop working according to specifications, a shut-down procedure of a nuclear reactor may fail to properly shut down the reactor, a heart may beat improperly, a railway signaling system may function badly, a virus invasion may make a cell experience a disease). All the examples listed so far refer to systems whose malfunctioning may cause severe damages and are usually identified as *safety-critical systems.* Some failures do not cause loss of life-critical resources. For instance, failures in accounting systems may cause loss of money and damage to the organization running the system, but do not destroy equipment, environment or people. This is an example of a *business-critical system.* The failure of the versioning system in a software company may alter the software production which is the mission of the company. This is a *mission-critical system.*

Critical systems

A *life-critical* or *safety-critical* system is a system whose failure or malfunctioning may cause death of people or loss of equipment or environmental damage.
A *mission-critical* or *business-critical* system is a system whose failure or malfunctioning may cause damage to the mission or to the business of an organization.

Systems can react differently to failures, ranging from no effect on the operation of the system to the complete crash of the system. In other words, systems have different degrees of reliability. *Reliability* is the ability

Reliability, Maintenability, Availability

Reliability is the probability that a system will not fail.
Maintenability is the probability that a system successfully recover from a failure.
Availability is the proability that a system is operational at time t.

of a system to perform correctly in normal and hostile conditions. A *fail-operational system* continues to operate in presence of failures of its control system (e.g., an organism with a non-vital damage continues to maintain its basic life functions, an elevator is a fail-operational system because failures to the power system of an operating elevator cause the elevator to be safely driven to level 0 followed by the opening of the elevator door). A *fail-safe system* is a safe system when it cannot operate (e.g., railway signaling systems are designed to be fail-safe). A *fail-passive system* continues to operate in the event of a system failure (e.g., an aircraft autopilot that fails allows the pilot to take over the control).

A *fault-tolerant system* maintains a reasonable level of services even when faults are experienced by the system (e.g., error-repairing pathways that recover from minor errors in DNA copying, redundancy of software or hardware components, different routing paths to a node of a network). A *fail-secure system* ensures maximum security when it cannot operate (e.g., intrusion-detection systems or alarm systems may block doors or functions of the controlled system in the presence of failures).

Systems that are 100% reliable do not exist. Therefore, we need a measure that tells us how easy it is to recover from a failure. Such a measure is the *maintenability* of a system. The qualitative counterpart of maintenability is sometimes called *resilience*: the ability of a system to recover from a fault or a damage. A more general property is *survivability* that describes the ability of a system to survive (being totally or partially operational) in the presence of a natural or artificial disturbance (e.g., an organism under virus attack can operate and survive during the immune system response to the pathogen invasion, the nuclear reactor of a power plant should perform the shut-down procedure correctly even in the presence of external disturbances).

There are timeframes in which a system is operational and timeframes in which it is not because it failed and is under repair. The *availability* of a

Busy-waiting, Deadlock

The *busy-waiting* of a component of a system is the continuous checking for an event that enables the further activities of the component without carrying out any useful job.

Deadlock is a state of a system in which no action can be performed because two components are waiting reciprocally for the completion of an activity that enables the other to proceed.

system measures how often the system is operational. Availability accounts for reliability and maintenability at the same time.

3.1.2.2 *Resources-related properties*

It is good practice to minimize the waste of resources when designing systems. Ensuring liveness properties of some operational patterns of systems (e.g., allocation of processes to processors in multiprocessor hardware so that a job scheduled will eventually start execution) is a way of stating that there is no infinite busy-waiting and no deadlock in a system for the chosen operational pattern.

Example 3.2. [Deadlock]
Components A and B both need resources R_1 and R_2 available in unique copies to complete their behavior. A acquires R_1 and B simultaneously acquires R_2. If there is no mechanism for which A or B releases its resource, no component will be able to acquire the other resource to proceed and the system is blocked, i.e., the system is in a deadlock state.

An abstract biological example of deadlock could be the need by two entities (e.g., proteins) A_1 and A_2 for two activators P_1 and P_2 to become active. If A_1 binds to P_1 and A_2 binds to P_2, neither A_1 nor A_2 can become active because each of them is missing one of the two activators. Nature prevents deadlock states by the reversibility of reactions (after binding of an A with a P, we can also observe the unbinding) that is an implementation of releasing resources.

Busy-waiting is an active waiting in which the resources of the system are consumed to check for an event (e.g., a person continuously calling a busy line until the connection is established). A situation worse than busy-waiting is *deadlock*. A system is in a deadlock state when two competing actions are simultaneously waiting for each other to allow the system to proceed.

Security

Security states that each component of a system can only access the resources for which it owns the access rights.
Information security states that confidential information cannot be accessed by nonauthorized components, be it legally or maliciously.

3.1.2.3 *Security properties*

A property that could seem similar to safety is *security* (cf. fail-safe and fail-secure in the degrees of reliability). Security concerns access rights to resources that can be either physical or logical, spatial or temporal. A great deal of research is related to information security, i.e., access rights for digital content in computing systems. Enforcing security policies to avoid forbidden accesses to resources is a central matter in the design of digital systems as well as in the definition of strategies for homeland security.

Example 3.3. [Security]
Molecular systems implement security policies for access rights by controlling biochemical interactions through the affinity between the reactants (a type of system with a notion of compatibility on top of the system's components). Protein A can access a resource R only if there is compatibility between them or there is a third element (a chaperone or a small molecule) working as an adaptor.

Space is also controlled in molecular systems by the mechanisms of active transport through cytoskeleton elements or through transporter molecules. Active transport is a way of avoiding unwanted access to locations by wrong components.

Finally, the whole immune system is continuously performing intrusion detection tasks and policy enforcement.

3.1.2.4 *Context-related properties*

The behavior of most systems is determined by environmental conditions (e.g., a biological system may behave differently at different temperatures, level of acidity, presence or absence of nutrients, etc.). The ability of a system to maintain its characteristic behavior under different conditions (i.e., under different parameter values) is sometimes referred to as the *robustness* of the system. The complementary property to robustness is *fragility*: how easy it is to break the normal behavior of a system. Most of the times fragility emerges from the strong coupling of systems components. Physical fragility refers instead to the easiness of destroying a system.

Robustness, Fragility

Robustness is the ability of a system to operate despite abnormal input.
Fragility is the quality of a system of being easily broken or destroyed (both physically and logically).

There are systems that have a bounded time to react to external stimuli (e.g., the time needed by the air-bag system of a car to be operational after an accident, the time needed by an intrusion detection system to signal attacks and block the door of a vault, the time needed by the immune system to react to external invasion). The ability of a system to react properly to the stimulus within a given amount of time is called *responsiveness*.

Responsiveness, Stability

Responsiveness is the ability of a system to perform a task or to react to an external stimulus within a given time.
Stability is the ability of a system to produce bounded outputs for every bounded input.

A property of systems that is related to responsiveness is *stability*. Stability relates a system to its response to inputs or disturbances. A system is stable if it is in a steady state unless affected by an external action, and it returns to a constant state when the action is completed. Stability is sometimes also related to safety (e.g., instability in the control system of a nuclear reactor violates safety conditions). An *adaptive system* may change its behavior depending on the environmental conditions and its internal state. Adaptive systems populate ecosystems, biology, human communities and organizations (e.g., the immune system works differently depending on the invaders, the gene regulatory network modulates gene expression depending on the environment and the state of the cell, investments depend on the status of the market).

A main mechanism for adaptation is the ability of a system to produce loops of cause–effect events. In this way an event of the system can influence

Adaptation, Events chain

Adaptation is the ability of a system to react to environmental changes or changes in its interacting components.
A *feed-back loop* is a chain of causally related events that forms a loop. Feed-back loops can be either *positive* (magnify a stimulus to enhance the reaction) or *negative* (reduce a stimulus to inhibit a reaction).
A *feed-forward chain* is a chain of causally related events that transfers a signal from a source to a target.

the present or future behavior of the system itself. Such loops of cause–effect events are commonly called *feed-back loops.* Molecular machineries contain many feed-back loops and *feed-forward chains* to modulate and tune their reactions. In systems that are characterized by input/output relationships, feed-back loops are sometimes implemented by feeding back to the system part of its output. In feed-forward systems an event acts on more than one component on a downstream chain of causally related events. Most of the time feed-forward chains form nonoriented loops (see Example 3.4 and also Fig. 3.4).

An important property of many biological systems is *homeostasis* (e.g., organisms tend to keep their pH or temperature constant, as well as the amount of glucose in the blood). Homeostasis is also used in ecology to explain nutrient recycling and population dynamics.

We conclude by introducing *hysteresis* — that is, the dependence of a system on its current and past environment and state. This dependence arises because the system can be in more than one internal state and its history must be known to decide the next state.

3.1.2.5 *Structural properties*

Systems are built by connecting components. Sometimes it is useful to think of a system as a component of a larger system, or it could be the case that different systems work together to accomplish a complex task. Three properties allow us to understand how easy it is to connect different systems together: *modularity, composability* and *interoperability.* Modularity is the ability of representing a system as a collection of subsystems or components with well-defined functionalities. Modularity allows the recombination of modules to obtain functionalities different from those of the

Example 3.4. [Interaction patterns in molecular biology]
The topology of the interaction networks in molecular biology (gene regulatory, metabolic and signaling networks) presents recurrent patterns that can be interpreted as modules of the system and that can be associated with typical interaction patterns among components. Some authors call these biological modules *network motifs*. The main classes of biological network motifs/interaction patterns are: (see Fig. 3.4)

(1) *Autoregulation*. Negative autoregulation occurs when a component represses its enabler or generator (e.g., a transcription factor represses its own gene or a protein inhibits its own activity by *autophosphorilation*). Positive autoregulation occurs when a component increases its production rate.

(2) *Cascades*. 'Component A activates B that activates C' is a positive cascade, while 'component A inhibits B that inhibits C' is a negative cascade.

(3) *Positive feed-back loops*. A *double-negative loop* is when two components repress each other, while a *double-positive loop* is when two components enhance each other. Double-positive loops are *regulated* if a third component is regulating both components in the loop; positive autoregulation of the loop components and of the regulating component is frequent. The pattern 'gene A produces a protein that activates gene B that causes, through its product protein, faster or slower promotion of transcription of gene A' is recurrent in many organisms.

(4) *Negative feed-back loops*. A component A slowly (quickly) regulates a component B that quickly (slowly) regulates A; the two regulations must be opposite, i.e. one positive and the other negative. A characterizing feature here is the difference of speed of the reactions forming the loop.

(5) *Feed-forward loops*. A component A regulates B that regulates C together with A. Regulations can be either positive or negative. The possible combinations originate eight feed-forward loops. Furthermore, the regulation of C can be by a simultaneous signal from A and B (AND gate) or by alternative signals from A or B (OR gate). The pattern gene A activates gene B, and then the two genes A and B activate a third gene C is recurrent in yeast and *E. coli*.

(6) *Single-input module*. A component A regulates a set of components $B_1 \dots B_n$.

The basic motifs identified above can be combined to give rise to many different behaviors in the cell.

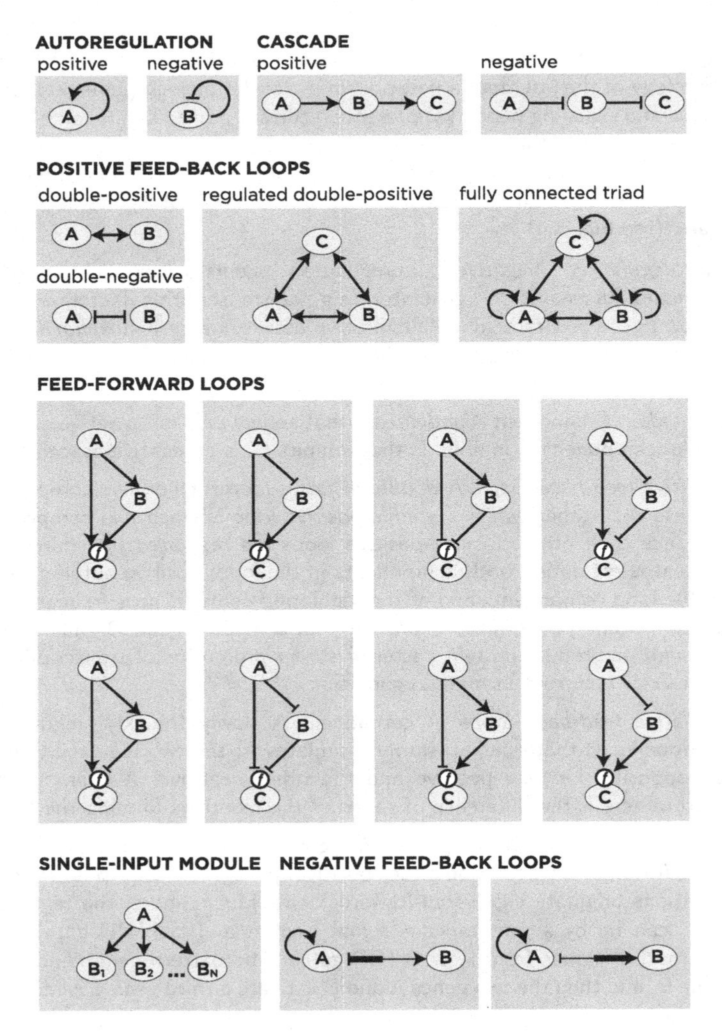

Figure 3.4 Network motifs. Autoregulation can be either positive or negative. Cascades are sequences of interactions of the same type (either all positive or negative). Positive feed-back loops originate from bidirectional interactions of the same sort (the figure exemplifies the cases of two and three nodes). Feed-forward loops originate when different paths target the same node. The figure illustrates the case in which two interactions target the same node C. The alternatives are history-dependent, because the type of initial interactions also differentiate the behavior of the pattern. The f circle in the picture is needed to show how the different signals affect the target node C, being the simplest cases $f \in \{and, or\}$ (simultaneous or alternative presence). Single-input modules originates when a node can interact with a set of different nodes. Finally, negative feed-back loops originate when bidirectional interactions have complementary sorts and different speeds combined with a self-activation (the thickness of the arrows in the picture denotes the speed, where thicker corresponding to faster).

Homeostasis, Hysteresis

Homeostasis is the ability of a system to maintain a steady state.
Hysteresis is the history dependence of a system to select its next state.

system in hand. Modules are identified as building blocks of system dynamics (e.g., gene regulatory networks and metabolic or signaling pathways in biological systems can be interpreted as modules that are coupled to obtain cell functioning). The recombination of modules introduces the property of composability. The composition of components or modules to build a system must obey the compatibility properties of these blocks (e.g., LEGO® systems are made up of components that can be assembled together if they satisfy a compatibility relationship).

Composability is a high-level feature of modules and usually refers to conceptual descriptions of systems and their behavior. Sometimes composability is interpreted as a strong form of interoperability (see below). In this case we have *dynamic composability*, obtained when the modules under coupling share the understanding of each other's state change, and *conceptual composability*, obtained when the understanding of reality (conceptual level) of the modules under coupling is the same.

Modularity, Composability, Interoperability

Modularity refers to the (logical) partitioning of a system in subsystems or set of components called *modules* that can be differently recombined.
Composability is the ability of a component or a module to be assembled in various combinations to form a new system.
Interoperability is the ability of a group of systems or modules to work together. *Syntactic interoperability* is obtained when common data format and resources are used by the systems under integration. *Semantic interoperability* is obtained when the systems under integration share the meaning of the information and of the resources used. *Pragmatic interoperability* is obtained when the systems under integration use the same methods to manipulate information and resources.
Integrability is the ability of a group of systems or modules to be technically or physically coupled.

Interoperability is a weaker notion of composability and it is obtained by letting the systems under integration *dialogue* (exchange information or resources) to cooperate on a complex task. This property does not affect the identity of each system involved. There are different levels of interoperability usually referred to as *syntactic*, *semantic* and *pragmatic*. At lower levels we usually consider *integrability*, i.e., the ability of two systems to physically or technically interact (e.g., different hardware components in a computer).

Other properties that characterize the structure of a system are *flexibility* and *extensibility*. Flexibility refers to how open a system is to changes. Note that flexibility and *adaptability* define the same concept in two different contexts. Adaptability is related to functionalities of the system (e.g., an autoradio can automatically re-tune a station depending on the location of the car to maintain the quality of the signal), while flexibility refers to the structure of a system (e.g., a bag can be slightly compressed to fit a pre-determined space). Extensibility refers to how easy it is to extend the functionality of a system (e.g., adding a new feature to a software program as an extension of the previous version of the software system).

Flexibility, Extensibility

Flexibility is the ability of a system to adapt effectively to internal and external changes. In biology this property is called *adaptability*.
Extensibility is the ability of a system to accept future functional expansions.

3.1.2.6 *Performance properties*

The properties mentioned so far highlight specific aspects or attributes of systems. It is useful to have a more general measure of how good a system is. Such a measure is called the *performance* of the system and it is meaningful with respect to a specific target. It can be determined as a weighted aggregate of some of the properties introduced below. We measure how well a system behaves with respect to the intended purpose and goal. Each system has to use and consume resources to pursue its goals. The best allocation of resources to a system is the one in which all and only the needed resources are assigned and are all consumed to achieve the goal. Resources are sometimes not continuous in nature and therefore we can only rely on integer multiples of a basic block of a given resource. In this case we may consume only a part of the assigned resources. The percentage of resource consumption by a system is the measure of its *efficiency* in managing resources.

Efficiency, Effectiveness, Efficacy, Scalability

Efficiency is a measure of how much a system uses its resources for the intended purpose.
Effectiveness is a measure of the degree to which a system produces the intended effect.
Efficacy is a measure of the success of a system in achieving a pre-determined goal.
Scalability is the capability of a system to smoothly adapt to increased workload.

The measure of how close the job performed by a system is to the goal is the *effectiveness* of the system. Note that we can have a very efficient system (usage of resources allocated) that is not effective at all (it performs a task different from the expected one).

A measure combining efficiency and effectiveness is the *efficacy* of a system, i.e. the ability of a system to achieve exactly the pre-defined goal with a minimal consumption of resources. A system can be effective because it reaches the pre-determined goal, but it could have low efficacy because it takes too much time to reach the goal or it consumes too many resources.

A further measure in evaluating a system is *scalability* (e.g., an operating system that smoothly increases its ability to manage computing resources as the number of users grows over a threshold is scalable).

3.1.3 *Complexity*

Complexity arises when interacting components self-organize to form evolving structures that exhibit a hierarchy of emergent system properties (e.g., a cell is complex because DNA strands and proteins interact to regulate gene expression and to produce new proteins that mediate intracellular signals to let the cell react properly to external stimuli). Many definitions have been given for complexity in the last 80 years. All of them fail to capture some aspects that are recurrent in complex systems. For this reason we leave the characterization of complex systems to intuition.

We consider the notion of complexity and emergence of behavior to be tightly coupled. We consider all those systems whose behavior cannot be predicted or defined by merging the behavior of their component in some way to be complex. *Emergent* behavior is usually originated by a collection of components that interact in the absence of a centralized point of control to produce something that has not been designed or programmed in the system construction or evolution. The Internet, ant colonies, intelligence and consciousness are all examples of emergent behavior: we cannot understand these phenomena just by looking at the components of the systems.

An evaluation scale of complexity is related to the difficulty of predicting (behavioral) properties, assuming knowledge of the properties of the constituent components. According to this intuition, linear systems are not complex at all because the degree of emergence of their properties is zero.

Computational complexity

Computational complexity is the amount of resources, measured as a function of the size of the input, needed to execute an algorithm. *Computational space complexity* refers to the amount of memory needed, while *computational time complexity* refers to the number of instructions to be executed.

Computational complexity is strictly connected to the notion of complexity. The performance of the technological solutions in analyzing biological systems on digital computers can be measured through an estimate of their computational complexity (see eponymous box, above). Minimizing the computational complexity of a solution, still remaining effective, makes efficacy grow. The same measure can be applied to all systems that can be interpreted as information processing systems and for which we have a mechanistic description of their behavior (an algorithm). Biological systems are commonly interpreted in systems biology as information processing systems whose mechanistic behavior is determined by the interaction of the components (biochemical reactions) that store (e.g., DNA) and transfer information (e.g., activation and de-activation of proteins via phosphorilation and dephosphorilation) from one another.

When dealing with technological and software solutions, it is important to understand whether the proposed solution is feasible for a given problem or not. We can distinguish problems (in our context, systems to be modeled, simulated and analyzed) into three macro classes (see Fig. 3.5). There are problems for which there exists no algorithm to solve their general

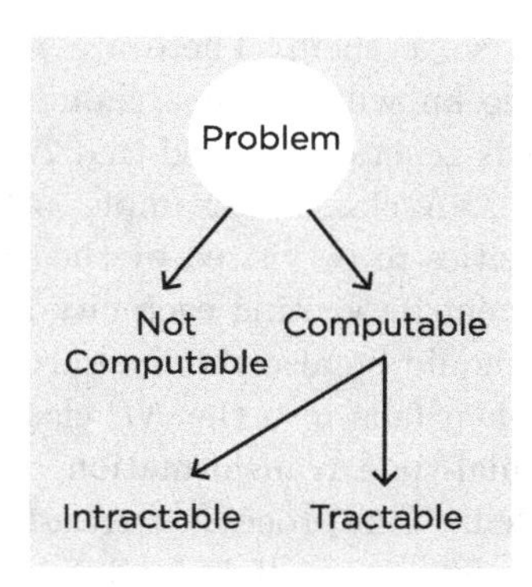

Figure 3.5 Classification of problems with respect to their computational complexity.

formulation. The most famous *not computable* problem in computing is the halting problem: given a program and its initial input, determine whether the program, when executed on this input, ever halts (completes). The alternative is that it runs forever without halting. According to Turing, it is not possible to solve the halting problem effectively through an algorithm for all possible program–input pairs.

Computability

Computability is the ability to effectively solve a problem by providing an algorithm for it.

Hints on tractability

growth	tractability
$n!$	20
2^n	40
n^2	100
$n \log n$	10^9
n	10^9
$\log n$	every size

The class of *computable* problems is further partitioned into two sub-classes depending on the computing effort needed to solve a problem. There are problems that need at most polynomial time (a polynomial number of instructions with respect to the size of the input) to solve them exactly (e.g., string matching algorithms) or even less in the case of logarithmic time (e.g., searching on a binary tree). These problems are also called *tractable* or *P* problems because there exist exact, polynomial-time algorithms in the size of the input to solve them. There are *intractable* or *NP* problems for which there is no known exact, polynomial-time algorithm. Even if not proved formally, it is generally belived that *NP* problems do not have polynomial-time solutions. A classical example is the traveling salesman problem. Given a set of cities to be visited by the salesman, determine the shortest path between them by visiting each city at most once except for the starting city, which should be also the final arriving city.

To show that a problem falls into the *NP* class, it is enough to show that there is a polynomial-time transformation of the problem in hand into a known *NP* problem. The process of transforming a problem into another is called *problem reduction*. It is fundamental to determining the complexity class of a problem and then using this knowledge to address the problem with the right tools for analysis. This is even more important when addressing biological problems due to the size of the input of genome-wide systems. Practical hints to decide whether a system is tractable are reported in the box entitled *Hints on tractability* on the previous page where we associate the growth of computational complexity with respect to the size N of the input with a threshold value for tractable input.

3.1.4 *Hierarchical systems*

Large and complex systems are usually hierarchical, multilevel systems. Many artificial and natural systems (ICT, biology, transport, epidemics, meteorology, energy management, climate, financial markets, urban planning, social simulation) differ greatly in structure and organization. In order to understand and eventually control these systems, we have to manage large amounts of heterogeneous data gathered at various scales. In general we may assume that a system is characterized by a hierarchy of *layers* (see Fig. 3.6 (A)) $L_0, \ldots, L_{i-1}, L_i, L_{i+1}, \ldots$ Figure 3.6 (B) and (C) are layered representations of a computing system and of a cell. A couple of examples reflecting the hierarchy are reported below.

Each layer is characterized by a set of *resources* $R_1, \ldots, R_n$ and a set of actions or *primitives* $A_1, \ldots, A_m$. Moving upwards in the layers we perform an abstraction, while moving downwards we perform a concretization. The most detailed layer of a system is L_0 (e.g., L_0 is the physical level of the wires in the hardware of computing systems and it could be the physical structure of genes in chromosomes in a cell).

Since each layer is an abstraction of the resources and primitives available at lower layers, a primitive or a resource at layer i is implemented

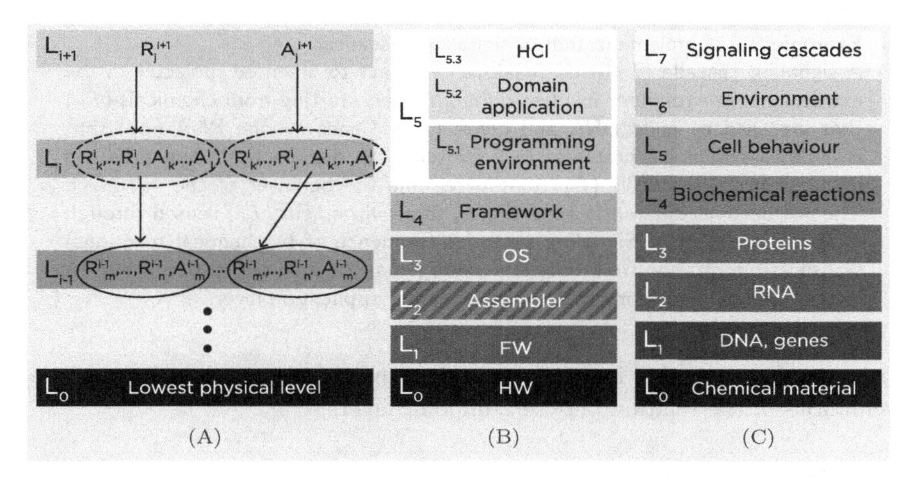

Figure 3.6 Layered characterization of systems (A) and its application to computing systems (B) and molecular systems (C). It is $l - k > 0$ and $n - m > 0$ including their′ variants. Detailed discussion of (A) is in the text. In (B) the Assembler level (L_2) may exist or may not, while the Application layer can be sublayered into three different abstraction levels, human-computer interaction being the highest abstraction (see also Example 3.5). A detailed description of (C) is given in Example 3.6.

Example 3.5. [Implementation of programming primitives]
A primitive of a programming language (application layer) may require many assembler/firmware instructions to be implemented. Similarly, each assembler/firmware primitive may require more than one clock cycle at the hardware level. Furthermore, the programming language primitive may also need OS routines to manage access to the computing resources. The OS routines are also implemented by sets of assembler/firmware primitives.

by a set of primitives and resources at layer $i - 1$ (e.g., the duplication of a cell can be considered a single step at a layer i of abstraction, but its implementation at the molecular layer $j < i$ needs a lot of protein–protein interactions).

The same process performed by the system may involve many different layers (e.g., metabolic networks, signaling networks, gene regulatory networks) and can be studied at different levels of abstraction (i.e., taking a specific layer as main reference). As a consequence, there are many different views of the same system that must be coherent. The mapping between a

Example 3.6. [Implementation of signaling cascades]
A signaling cascade (L_7) that makes a cell react to a sensed molecule in the extracellular environment involves different layers starting from chemicals (L_0) that are used to build DNA and genes (L_1). Genes encode RNA (L_2) that is used to synthesize proteins (L_3). Proteins are the reactants and products of biochemical reactions (L_4) that determine the behavior of the cell (L_5). The cell behaves differently according to the environment (L_6) sensed through membrane receptors. Signaling cascades (sequence of biochemical reactions) transport information from different layers (from the environment to the DNA in the nucleus) and represent primitives at the application layer.

primitive/resource and its implementation and the reverse mapping are the semantics of the relation between different layers.

Example 3.7. [Tissue formation]
A tissue is a network of interacting cells that together form a structure with proper identity and functions (e.g., an organ, a bone, the skin). The tissue formation can be studied through a model whose resources are cells and whose primitives are the binding/unbinding of cells, their duplication or apoptosis, messaging through hormones, etc. A view of the same system at a lower level considers each cell as a system that describes the cell-cycle machinery for duplication. Here resources are chemicals and proteins that drive and control the cell cycle. The main interactions of these chemicals and proteins are the primitives that let each cell transition in the phases of the cell cycle (G_1, S, G_2, M). The tissue behavior must be coherent with the internal machinery of each cell. We could also go further ahead by including the circadian clock machinery in each cell to roughly synchronize the cell cycles so that the tissue has a uniform behavior (in fact, the uniform behavior is the reason that we can abstract from the internal processes of each cell when modeling the tissue).

3.2 Model

The theoretical grounds and development of all scientific disciplines is model building and validation of these models through experiments. A model can also be used as a guide to build a real artifact (e.g., the reduced-scale models of buildings used by architects or designs done by painters before delivering the final artwork). *Representation* and *abstraction* (see box below) are two concepts that are fundamental to understanding what a model is. A model of a system is an abstract representation suitable to investigating a particular set of the system's properties. Different sets of properties characterize different experimental frames for the model. Note that the same system can have many different models and each model generates different experimental frames depending on the properties of the system that we are investigating.

Representation, Abstraction

A *representation* is a set of symbols used to convey information and knowledge about a system. It is either physical as a cell or an ecosystem, or artificial as a computer network or an economic market.
An *abstraction* is a representation that ignores some aspects of a system which are not of interest for the current investigation.

The entire field of *computational systems biology* is devoted to building models of biological systems and to their validation and analysis with many different techniques, including statistical analysis and simulation.

Models are needed because the system may not be accessible to experimenting with it (e.g., a marine ecosystem from which you cannot remove all the whales to observe what happens; astronomy and astrophysics cannot experiment directly on planets and stars) or it could be dangerous to experiment with it (e.g., the effect of a new drug on humans). There are also cases in which the real system does not exist yet (e.g., the performance and properties of a new aircraft under construction) or it is unacceptable to experiment with it (e.g., the effect of radioactive exposition on human cells or disater recovery in environmental studies).

Models, when coupled with statistical analysis and simulation, help us investigate the correlation-based and mechanistic properties of a system in a

Model

A *model* is an abstraction of a system. A model has its own interacting components that are characterized by the *attributes* that we want to observe. The set of all the attributes in a model is the *experimental frame*.
A *dynamic model* aims at predicting the behavior of the system in time/space through *what if* analysis. What if analysis investigates how a change in some attributes affects the behavior of the modeled system.
A *computational model* is a model that can be manipulated by a computer to observe properties of the corresponding system.

faster (e.g., many cell cycles of mammalian cells can be studied in seconds on a computer, while it takes days in a lab) and cheaper way (e.g., performing crash tests on a computer avoids destroying physical objects or analyzing the effect of perturbation on biological systems may save animals).

Models, as opposed to computational ones, exist like physical models (e.g., a prototype of a car) or animal models (e.g., mice in labs are used to experiment on drugs to be used for humans).

A model used as a prototype to drive the building of an artificial system precedes in time the system that it is modeling. A dynamic model used to investigate the behavior of a natural phenomenon (e.g., the mechanistic behavior of an organ or the effect of an extinction in an ecosystem) follows in time the system that it is modeling. The models that precede the system in time are used for the *synthesis* of the system (models as prototypes). The models that follow the system in time are used for the *analysis* of the system (models as theories).

Synthesis, Analysis

Synthesis: M is a prototype (model) for the system S that is built from M by inheriting its properties.
Analysis: M is a model for the system S and there is a mapping from the components of M to the components of S.

A model for a system is an abstraction and cannot answer all the questions over the system, but only the ones for which it has been specifically built. A model ignores the details of a system that are not relevant for the investigation that one wants to perform and of course ignores the details of the system that are not known. A model is the result of a process of abstraction on the knowledge we have concerning the system guided by the class of properties we are interested in, plus the management of lack of knowledge concerning the system.

According to Rosen, a model is an encoding of a piece of nature. The model is used to infer the relationships between its parts that mimick the causal relationships between real events in nature. The decoding process allows the modeler to reflect the properties studied on the model to natural phenomena (see Fig. 3.7). The sucessful outcome of this process can be a better understanding or an augmented knowledge of natural phenomena.

A good model represents interesting phenomena (systematization of knowledge) and predicts the behavior of the system (a tool to experiment

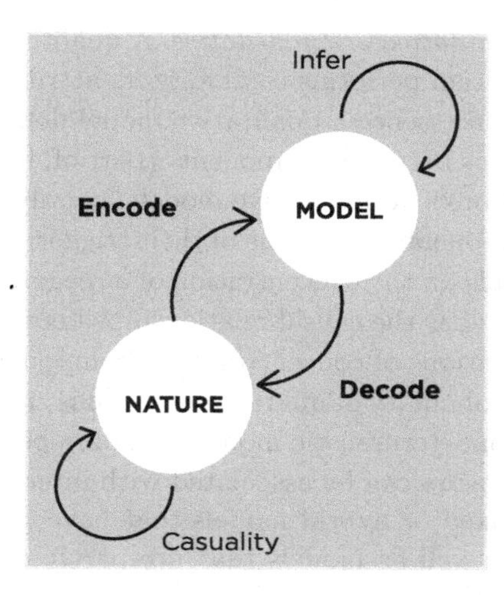

Figure 3.7 Relationship between nature and models.

on the system and predict the outcome of controlled perturbations). This is needed to make educated decisions and to plan experiments or activities on the basis of foreseen properties of the system.

A model is a tool (like a microscope or a sequencer) for carrying out experiments (that scientists have to creatively design and soundly perform). Therefore a model is not an answer *per se* to a problem, but most of the time is a generator of new questions.

3.2.1 *Classification of models*

Models, like systems, can be static and dynamic. Static models represent mainly structures (e.g., the DNA sequence, the folding of a protein, the interaction capabilities of a set of proteins and their relationship network, or the architecture of a building) and do not include the time dimension among their attributes. Dynamic models depend on the flow of time for understanding how the values of the attributes vary.

Models can be classified as deterministic, nondeterministic or stochastic. A model is deterministic if it always produces the same (analysis or simulation) outcome once the parameters and the input are fixed. A model that produces different outcomes with fixed parameters and input is either nondeterministic (if we cannot associate probabilities with different outcomes) or stochastic (if we can associate probabilities with different outcomes).

Models can be *qualitative* or *quantitative*. A qualitative model describes the steps that the system performs to modify its attributes without resorting to any quantity. In essence a qualitative model defines an algorithm or structural relationships between components (part of, interact with, needed for, etc. kind of relations). A qualitative model can only highlight the interactions between components with some of their triggering conditions (e.g., a condition that depends on the concentration of a reagent cannot be qualitatively captured) as well as the causal relationships between events. If we are interested in the variations of concentration of components in a solution or in the speed or the probability of interactions we must resort to quantitative models. Note that nondeterministic models can only be qualitative because no distinguishing measure can be associated with interactions. Finally, it is possible to create mixed or hybrid models that have qualitative parts and quantitative parts as well as models that are partly deterministic, partly nondeterministic and partly stochastic.

Mathematical modeling

Mathematical modeling can be traced back to the 1960s and even earlier to the Schrödinger. It has its roots in Newton's physics and it has already been used in the early 20th century to model rhythmic biological phenomena by Volterra.

Models can also be classified according to the modeling technology (see Fig. 3.8 and the respective box). We stick here to computational models and we attempt to identify the main technologies for modeling biological systems. We briefly mention classes of technologies here and we will deal with them in detail in the chapters on static (Chapter 4) and dynamic (Chapter 5) modeling technologies. Historically, mathematical modeling based on equations has been the first approach adopted to define the behavior of biological systems. Ordinary differential equations (ODEs) and difference equations are the main representatives of this class of technologies.

Another successful approach is *network-based modeling*. It represents the interconnection structure of the components of a system and identifies basic characteristics of the system by investigating the topological properties of the interconnection network or the enrichment of their components with respect to some experimental measure. Networks are mainly related to graph structures, but finite-state automata are also used as behavioral models of computing machines abstracted as a set of states connected by

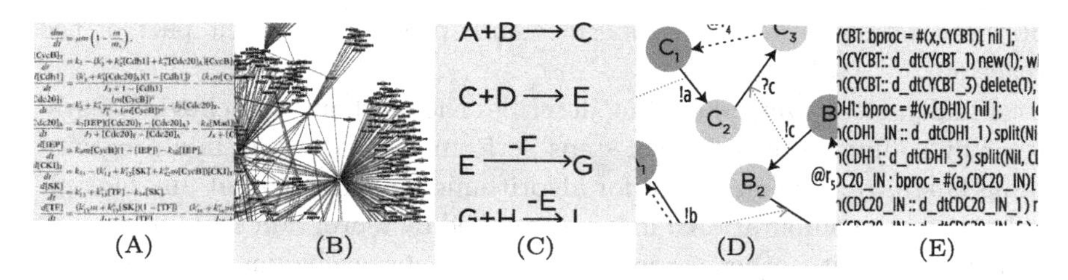

Figure 3.8 Dynamic modeling technologies. Ordinary differential equations (A), protein–protein interaction network (B), chemical reactions (C), finite-state automata (D), programming-based models (E).

Modeling technologies

Equation-based modeling represents system attributes by variables and describes the system behavior with a set of equations relating these variables over time.

Network-based modeling represents the interconnections among the components of a system and studies the topological and correlation properties of the obtained graph.

Automata-based modeling represents the dynamics of systems by a set of states connected by arcs that represent the transitions between states and that are triggered by input signals. Input traces or transition functions determine travel among the states, identifying the final state as the result of the computation.

Rewriting modeling represents the dynamics of a system by replacing at each step some part of the initial representation with a new one until no more replacements are possible or a terminating condition is met by the configuration reached. There are either *term-rewriting* (for textual representations) or *graph-rewriting* (for graphical representations).

Algorithmic modeling describes the algorithms executed by the system through programming language technologies and the dynamics is described by the execution of the algorithms.

arcs representing the changes. They encode the basic steps that an executor performs in response to input values. The state abstracts all the specific details of the real machine represented. Other formalisms we consider in this class are Petri nets and Boolean networks.

Rewriting modeling defines rules to rewrite configurations of systems into new configurations (e.g., chemical reactions allow reactants — left part of the reaction — to be transformed into products — right part of the reaction).

Finally, *algorithmic* modeling describes the mechanistic behavior of systems by coding the sequence of steps performed by a system in a programming language. The metaphor for algorithmic modeling is that any biological entity is a computational unit represented by a program that cooperates concurrently with other computational units by exchanging information. The capability of interaction between biological elements is provided in the representation through typing compatibilities between program interfaces, and actual interaction is modeled by message passing, shared memory or scoping rules (see Fig. 6.1). Algorithms abstract from the executor exactly as automata do, so that a collection of transistors or a soup of proteins and genes make no difference at this level of abstraction. We include in this class of models natural constrained languages, both textual and graphical programming or specification languages. Figure 3.9 reports the classification of models according to the aspects discussed so far.

3.2.2 *Properties of models*

The main class of properties of a model are depicted in Fig. 3.10. The structural and performance properties of systems can be applied to models as they stand. Modularity, composability, interoperability and integrability

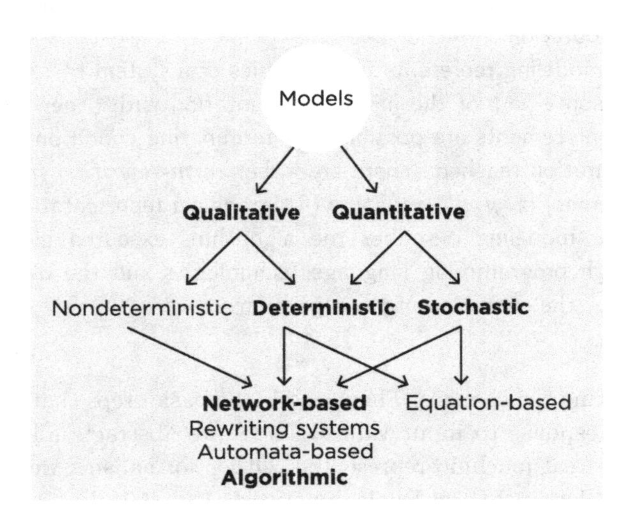

Figure 3.9 Classification of models. The words in bold identify the main classes of model dealt with in this book.

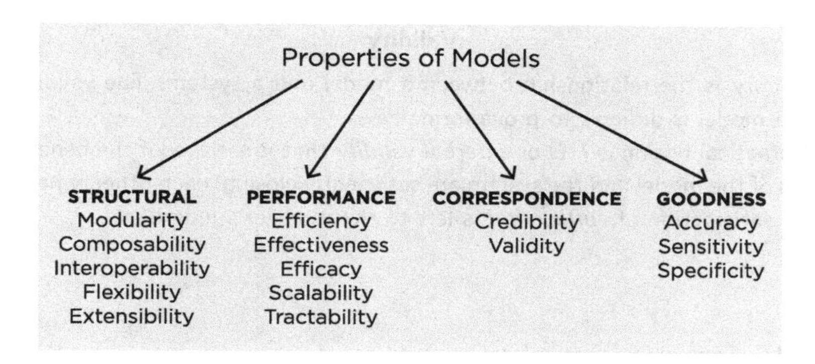

Figure 3.10 Classification of properties of models.

of models refer to the model-building process and to the characteristics of the technology adopted. For instance, it is not easy to compose ODE models without heavily reworking them, while it is easier to compose programming-based models or network-based models. Similarly, *flexibility* and *extensibility* refer to the capacity of maintaining a model over time.

A model is extremely similar to a software artifact (actually it is, in the case of programming-based modeling) and hence all the properties of software systems apply to models as well, including performance measures. Therefore, efficiency, effectiveness and efficacy refer to the ability of a model to perform the intended task with a minimum amount of resources, but still meeting the pre-defined goals. *Scalability* is the ability of a model to operate smoothly when the input data or the number of components grows. We add *tractability* to the performance properties of models. Tractability is related to computational complexity and it is independent of the characteristics of the available hardware. A model is computationally tractable if its computational complexity grows at most polynomially with the number of the states of the model components (see Sect. 3.1.3).

Tractability

A model is tractable if its computational complexity increases at most polynomially with the number of attributes of interest.

Validity

Validity is the relationship between a model and a system. The validity of a model is difficult to prove formally.
A practical notion is *I/O* or *external validity* that is satisfied if the behaviors of the model and the system are sufficiently close to each other, where the satisfaction of *sufficiently* is left to the modeler's judgment.

3.2.2.1 *Correspondence-related properties*

Other properties of models refer to the correspondence between the model and the system and to the relationship between the model and its users. The model must be trusted by its users, i.e., the model must have credibility. *Credibility* is the relationship between the model and its target users. Users of a model must be an active part in the model-building process to make the model credible.

Validity is a fundamental property of models and witnesses the capacity of a model of making good predictions because the model soundly captures the aspects we are interested in. We need to assess the validity of a model before using it to predict the behavior of a system. Assume that M is a model for the system S and $\mathcal{M}$ is the modeling process. Let $s(t)$ and $m(t)$ be the state of the system and of the model at time t, and f_s and f_m the state transition functions of the system and of the model, respectively. Finally, let $I_s(t)$ and $O_s(t)$ be the input and output of the system at time t. Similarly, we write $I_m(t)$ and $O_m(t)$ for the model.

The input and output are here generalized concepts: input can be any perturbation of the system or of the model and output can be any observable property causally related to the input. A model M is valid for a system S if $f_m(\mathcal{M}(s(t_0))) = \mathcal{M}(f_s(s(t_0))) = m(t_1)$ (see Fig. 3.11A). When dealing with nature, we cannot validate models according to the above definition simply because we do not know f_s. We can only observe the output of the system caused by a perturbing input and compare this output to the one produced by the model on the same input. Consequently, we can only falsify models (because they have different behavior on some input with respect to the system), but we cannot validate them, as well expressed by Karl Popper. We still need, however, to acquire confidence on the fidelity of the model with respect to the system and we resort to I/O or external validity. I/O validity is based on the comparison of the output of the system and of

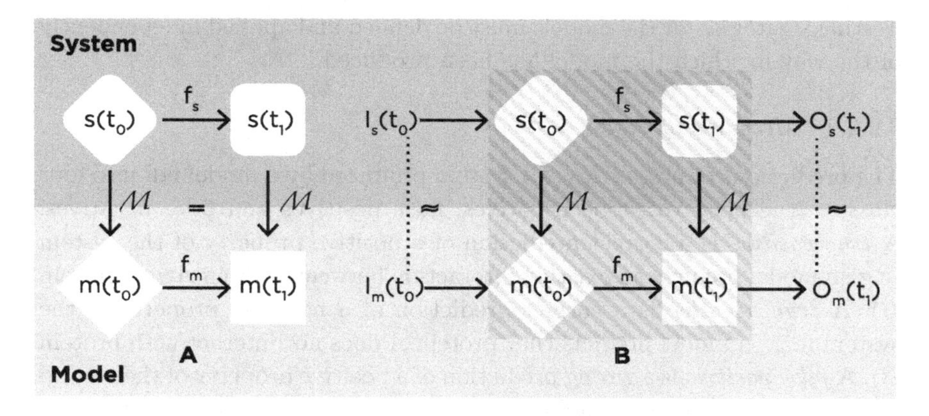

Figure 3.11 Validity (a) and I/O validity (b) of a model.

the model and on an interpretation of the closeness of the results observed from the system and the model. Pictorially, I/O validity amounts to saying that the diagrams in Fig. 3.11B commutes i.e., all the paths leading from $I_s(t_0)$ to $O_m(t_1)$ are equivalent.

I/O validity can be checked by using data sets produced by the model and observed and measured on the system. An issue in this comparison process is *overfitting*, i.e., a model is extermely well tuned with respect to a specific data set used to build the model, but poorly related to other data sets that the model must manage when used. Overfitting is more frequent for complex systems. A strategy to limit the problem is *cross-validation.*

Overfitting, Cross-validation

Overfitting of a model is when the model is tuned only with the data set used to build it.

Cross-validation allows the modeler to check overfitting by testing the model on data sets different from the ones used to build and calibrate or train the model.

The relationship between model building and software development introduces the software development practices also used in modeling. Verification of a model is needed to remove bugs from the preliminary versions.

Testing strategies for the models must be defined and applied independently of the way in which the model has been produced.

3.2.2.2 *Measures of goodness*

The predictions of properties of a system produced by a model fall into four classes: true positives, true negatives, false positives and false negatives. A *true positive* is a correct prediction of a positive property of the system (e.g., a model predicts an existing interaction between protein A and protein B). A *true negative* is a correct prediction of a negative property of the system (e.g., a model predicts that protein A does not interact with protein B). A *false positive* is a wrong prediction of a positive property of the system (e.g., a model predicts an interaction between protein A and protein B but A and B do not interact). Finally, a *false negative* is a wrong prediction of a negative property of the system (e.g., a model predicts that protein A and protein B do not interact when they do).

Example 3.8. [Diagnostic test]
Consider a medical test based on a biological marker. A cutoff value for the marker is determined such that values under the threshold are normal, while values above the threshold denote a disease state (see Fig. 3.12). The distributions of health and disease patients overlap and hence the test can correctly predict a disease (true positive — TP in the figure) or the absence of the disease (true negative — TN in the figure), but also wrongly predict a disease (false positive — FP in the figure) or the absence of a disease (false negative — FN in the figure). The choice of the cutoff value highly affects the performance of the test.

According to the classification above, there are three *measures of goodness*: accuracy, sensitivity and specificity. The *accuracy* of a model is the percentage of true positives and negatives, i.e.,

$$\text{accuracy} = \frac{|\text{true positives}| + |\text{true negatives}|}{|\text{total predictions}|}.$$

Accuracy, Sensitivity, Specificity

Accuracy is the fraction of correct predictions made by the model.
Sensitivity is the frequency of correct identifications made by the model.
Specificity is the frequency of identifications that should be made by the model.

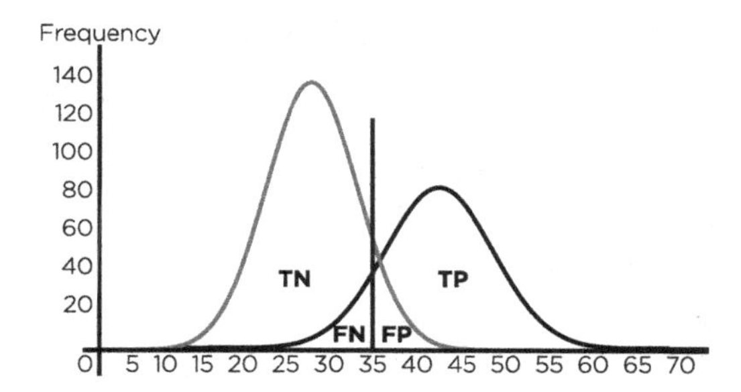

Figure 3.12 The gray distribution denotes health patients and the black distribution denotes disease patients. The vertical line is the cutoff value of the medical test. The areas under the curve labeled TN, TP, FN, FP are the true negative, true positive, false negative and false positive outcomes of the test.

When we are looking for a rare event in a system, we can get a reasonably accurate model by simply saying that the event never occurs. If the event we are looking for occurs once over hundreds of experiments we can easily get accuracy around 99%. The prediction of the rare event could be of uttermost importance and hence we need a different measure to evaluate the validity of a model that correctly identifies the event when it occurs (true positive) and that minimizes the situation in which the model does not report an event that actually occurs in the system (false negatives). *Sensitivity* is a measure that captures how often a model identifies what it is supposed to identify (in our example the rare event) and it is

$$\text{sensitivity} = \frac{|\text{true positives}|}{|\text{true positives}| + |\text{false negatives}|}.$$

Another measure that can account for the rare events better than accuracy is *specificity*. Here we want to limit false positives, i.e. the situation in which the model reports an event that does not occur in the system. Specificity is

$$\text{specificity} = \frac{|\text{true positives}|}{|\text{true positives}| + |\text{false positives}|}.$$

Models may perform differently depending on the measure of goodness selected. Sensitivity and specificity account respectively for the true positives and false positives (see also Example 3.8). A compound measure is the receiver operating characteristic (ROC) curve that plots true positives rate vs false positives rate for a test performed on a model (see Fig. 3.13). The ROC analysis comes from signal theory and was developed to analyze radar images during the Second World War. The diagonal ROC

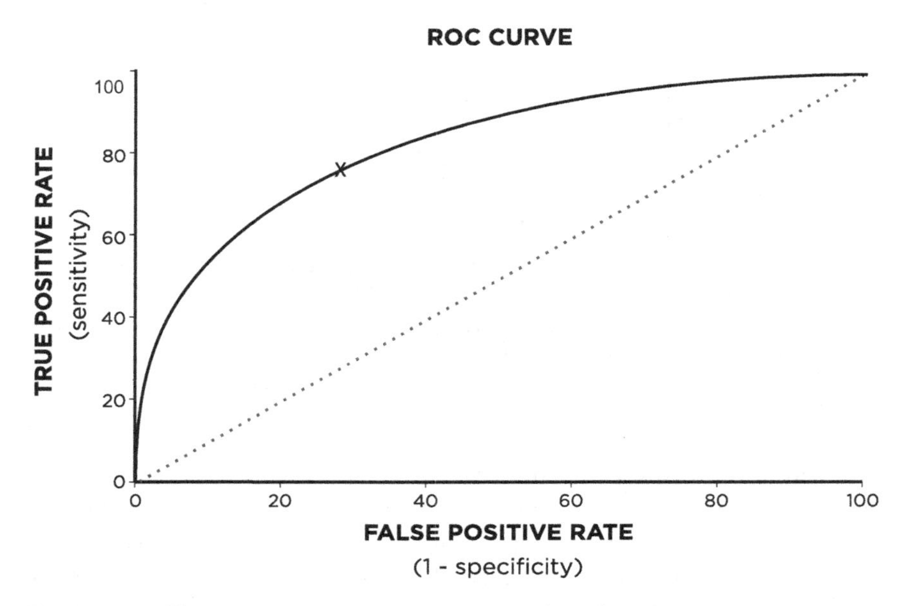

Figure 3.13 The receiver operating characteristic (ROC) curve. The x on the curve represents the cut-point of the test. The slope of the tangent line at x gives the likelihood ratio for that value of the test. The area under the curve is the accuracy of the test.

curve represents pure-chance performance (i.e., no predictive value). The closer the curve is to the y-axis and to the top border of the diagram, the more accurate the test is (indeed the accuracy is the area under the curve). An area of 1 represents a perfect test, while an area of 0.5 (the area under the diagonal) is a pure-chance test. The area represents the ability of the test to discriminate between healthy and diseased patients correctly.

3.2.3 *Complexity*

The main concern when developing a model is its computational complexity. The ability to determine the computational complexity of a model is a fundamental tool to decide both the modeling approach to be adopted and the algorithms to be used. If a problem is not tractable, approximate solutions can at least provide hints on the behavior of the system.

If approximation is not acceptable or it does not allow us to gain in efficiency, we must rethink the model by changing the basic abstraction chosen to represent the system.

Example 3.9. [The right abstraction]
A great example here is the sequencing of DNA. Figure 3.14 shows representations at different levels of abstraction and complexity. It has been possible to complete the human genome project when the right abstraction for representing DNA has been found: a string over a four-letter alphabet, i.e., DNA $\in \{A, G, T, C\}^*$. Then string algorithms and pattern matching applied smoothly to genome sequences led to our current understanding and development of genome studies.

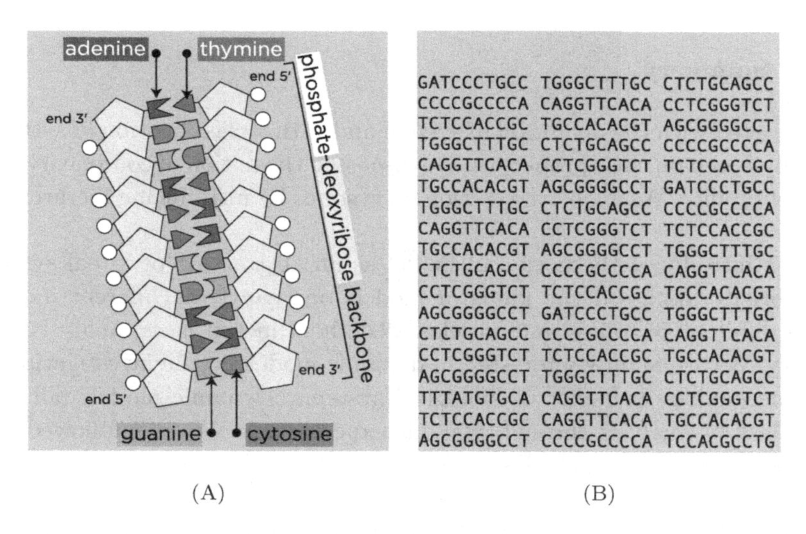

(A) (B)

Figure 3.14 DNA representations. Besides those in Fig. 2.2, there is the abstraction of nucleotides through names (A) and that of a string over the alphabet of the names of nucleotides (B).

3.2.4 *Hierarchical models*

When considering multilevel systems, many coherent views of the same system must coexist to study different attributes or properties of the system. A layered logical organization of a model can help manage different views of a layered system. Therefore, the layered organization depicted in Fig. 3.6 applies to models as well. The very same discussion of layered systems still works by replacing system for model.

Interfaces between different layers are fundamental in model definition and development. In fact, they define the kind of information that can be exchanged between resources/primitives and their implementations at lower levels. Most of the structural properties of models (e.g., modularity, composability, flexibility, extensibility) depends on non-ambiguous definitions of these interfaces.

The interfaces between different layers in the hierarchy are also needed to carefully represent *upward* and *downward causation.* An event at a layer may cause events at higher (upward) or lower (downward) layers in the hierarchy (e.g., the molecular machinery within a cell causes the cell — a component at the higher level — to duplicate, while the state of the cell controls the rate of the molecular reactions inside the cell).

3.3 Summary

We started with the definition of system and with a classification of systems mainly based on their properties. We discussed the notion of complexity and computability. We ended the section on systems by introducing hierarchical systems.

We introduced the notion of model as an abstraction of a real system and briefly discussed the kinds of models one can use. Different models have different purposes and we classified them in synthesis, analysis and theory. We are mainly interested in dynamic models in the investigation of the mechanistic behavior of biological systems. Dynamic models fall into the class of analysis through what if experiments. We also followed the structure of the presentation of systems to classify models according to their properties.

We defined the notion of validity of a model and we discovered that the formal definition does not apply to natural systems, for which we must rely on a notion of I/O validity. This led us to introduce some measures of the goodness of models (accuracy, sensitivity and specificity) in order to compare different models of the same system. Strictly coupled with the notion of validity is the overfitting of a model, i.e., a model is extremely good for the data set used to build the model, but it is poorly related to other data sets it will manage later on.

Computational complexity and tractability of models are important concepts for discriminating among different models of the same system and among different modeling technologies. We ended the section on models by dealing with hierarchical models and the main issues to be dealt with in this setting.

3.4 Further reading

General concepts from systems theory are in [Sibani and Jensen (2013); Fuchs (2012); Auyang (1999)]. The general introduction to models and modeling is derived from [Raczynski (2006)] and [Fishwick (2007b)]. Further discussion on Rosen's view of modeling and its relationship to nature is in [Rosen (1991)]. The definition of safety and liveness properties is introduced in [Alpern and Schneider (1985)]. The section on validity of models is mainly referring to [Schwartz (2008)]. Sensitivity and specificity are presented in [Spitalnic (2004)]. Deeper treatment on usefulness of models by George Box is in [Box and Darper (1987)]. Network motifs, modularity and interaction patterns can be further investigated in [Alon (2007); Shoval and Alon (2010)]. A historical paper on complexity is [Weaver (1948)]. A survey of the different definitions of complexity can be found in [Edmonds (1996)]. Background and deeper analysis of computability and computational complexity is available in [Papadimitriou (1994); Sipser (2005)]. Hierarchical modeling in biology is studied in [Maus *et al.* (2008)].

Chapter 4

Static modeling technologies

In this chapter we present static data analytical approaches for systems biology. The goal of this chapter is to provide the reader with a practical introduction to a variety of statistical and bioinformatic approaches to analysis of biological data. Some of the methods presented are specifically designed for high-throughput data; however, many of the approaches are broadly applicable to smaller datasets as well. To contain this material within a single chapter, we do not present a deep theoretical treatment of the presented methods. Moreover, we are not able to discuss every method in statistics and bioinformatics. We instead provide a practical introduction to linear regression, dimensionality reduction and pathway/network analysis, as we feel that the methods presented in these sections will allow the reader to approach analysis of a dataset from a number of angles, and will also provide the reader with a foundation for understanding more specialized or advanced techniques.

4.1 Preliminary assessment

Analysis of biological data frequently involves assessment of the relationship between two or more continuous or discrete variables. In the case of data from clinical or nutritional studies, these variables may represent blood measurements of clinically important physiological biomarkers, or in the case of omics data, the variables may be transcript, protein or metabolite levels in a given tissue. Before beginning any statistical analysis of a dataset, an important first step is building an understanding of the biological characteristics of the data being analyzed. For instance, for glucose levels one should appreciate the physiological significance of blood glucose, or in the case of transcriptomic assessment before and after a given treatment, one should aim to understand what the general anticipated effect of this treatment is, as well as the basic mechanics of DNA transcription. Such a background understanding allows for a more knowledge-driven approach to the data analysis, and facilitates identification of samples with potential technical problems. Further to this, it is critical to examine basic statistical qualities of the dataset, such as mean, variance and distribution across variables/samples, and presence of outliers. Such an assessment builds a global picture of the dataset being analyzed, and is an important part of the data's quality assessment.

Example 4.1. [Pima Indians — summary statistics]
We consider a subset of a simple dataset representing diabetes-related phenotypic characteristics from a cohort of adult female Pima Indians (available at http://archive.ics.uci.edu/ml/datasets.html). We first examine a set of summary statistics about the variables in the dataset.

Pregnancies	**Glucose** *(mg/dL)*	**Diastolic** *(mm Hg)*	**Triceps skin** *(mm)*
Min.: 0.000	Min.: 0.0	Min.: 0.00	Min.: 0.00
1st Qu.: 1.000	1st Qu.: 99.0	1st Qu.: 62.00	1st Qu.: 0.00
Median: 3.000	Median : 117.0	Median: 72.00	Median: 23.00
Mean: 3.845	Mean : 120.9	Mean: 69.11	Mean: 20.54
3rd Qu.: 6.000	3rd Qu.: 140.2	3rd Qu.: 80.00	3rd Qu.: 32.00
Max.: 17.000	Max.: 199.0	Max.: 122.00	Max.: 99.00

The table above describes the minimum, 1st quartile (1st Qu.), median, mean, 3rd quartile (3rd Qu.) and maximum values across four variables: number of previous pregnancies, fasting blood glucose, diastolic blood pressure and triceps skin thickness (see Appendix B.3.1 for a description of quartiles).

We can understand important aspects of the study sample by examining these summary statistics. For instance, we can see that some individuals in the cohort have clinically high blood pressure, and that the population minimum glucose measurement is 0, which suggests at least one sample of questionable quality, as this is not a physiologically reasonable value. We may therefore decide to remove any low-quality samples before continuing the analysis.

It is always important to identify *outliers* (i.e., samples that strongly deviate from the rest of the sample). While there may be a reasonable explanation for outliers in a dataset, such observations can strongly influence the results of certain statistical modeling algorithms.

Further assessment of the variables in the dataset in Example 4.1 can be achieved with summary plots such as histograms, boxplots and scatterplots

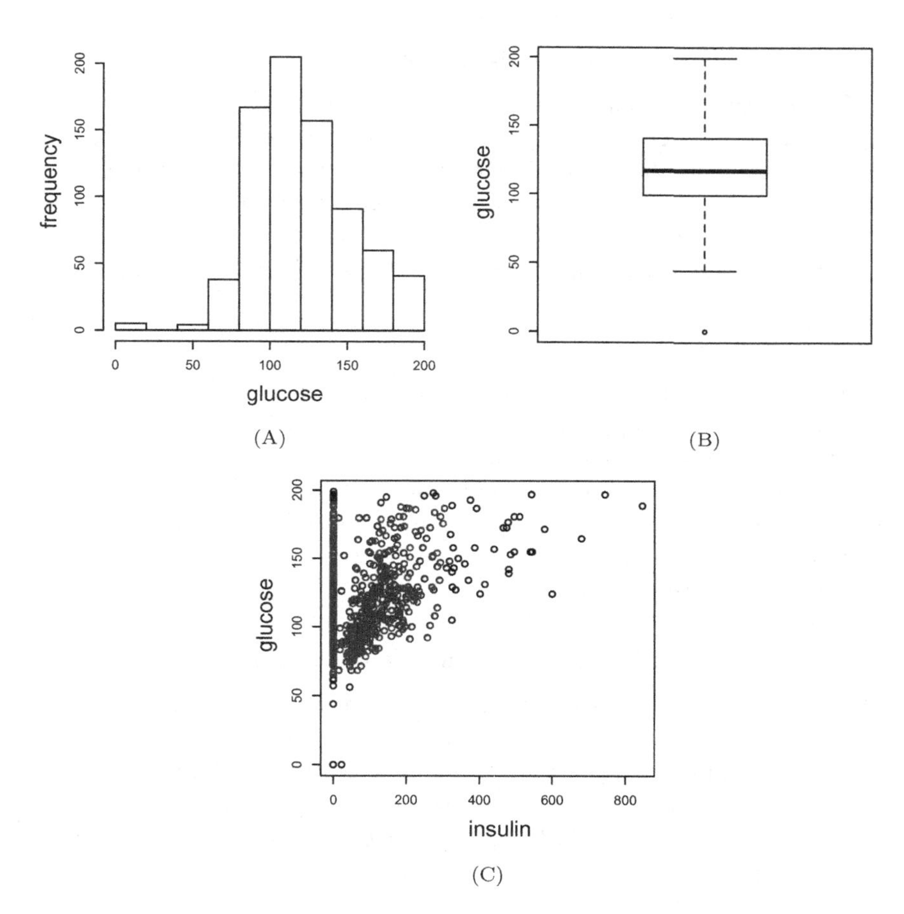

Figure 4.1 Summary plots. (A) Histogram and (B) Boxplot showing distribution of glucose measurements; (C) Scatterplot showing correlation between glucose and insulin measurements.

(Fig. 4.1). The first two are especially useful for visualizing the *distribution* of a variable. Taking note of whether the data are *normally* distributed, *skewed*, or display a *bimodal* distribution is essential for correct application of many statistical algorithms. From these plots we can see that the blood glucose variable generally exhibits a normal distribution (Fig. 4.1A and B) and that blood glucose and insulin display some degree of correlation (Fig. 4.1).

Once the dataset has been quality checked, the next step is precisely specifying a statistical question (i.e., a question that can be directly

Distribution

The *frequency distribution* of a variable describes its spread in a sample. This representation is formed by dividing the variable into classes/bins, and determining the relative frequency of the values that occur in each bin. The resultant distribution may then be described in terms of its similarity to known *probability distributions* (see Appendix B.2.1 and B.2.4), such as the normal distribution, Poisson distribution or uniform distribution.

answered with statistical analysis). For instance, if we are generally interested in the relationship between blood glucose and insulin, an appropriate starting statistical question may be "can we define a statistical model that reliably predicts glucose levels from insulin levels?". This question may be addressed using *linear regression*, a fundamental supervised statistical approach for investigating the relationship between one or more explanatory variables and a response variable.

Supervised, unsupervised learning

Supervised statistical methods are those in which we utilize known class labels/responses, whereas *unsupervised* methods do not assume this predictor/response paradigm, and focus instead on the overall relationships among the variables in a dataset. *Regression analysis* is a classic example of supervised learning, whereas *clustering analysis* is an unsupervised approach.

4.2 Linear regression

A linear model describes a relationship between one or more predictor variables and a response variable, where the response is assumed to be a linear function of the predictors. Linear regression refers to the technical approach for estimating and evaluating the coefficients in a linear model. *Simple linear regression* (or univariate linear regression) describes the case where the number of predictor variables $j = 1$, whereas *multiple linear regression* indicates a model with more than one predictor variable (i.e., $j > 1$). Regression analysis in general can be seen as a *supervised* learning approach.

Linear regression — assumptions

The fundamental assumptions of the linear model are that

(1) there is a linear relationship between the predictors and response;
(2) errors are homoscedastic (i.e., constant variance); and
(3) errors are independent and normally distributed.

Linear regression

A linear model fitted to an $n \times 1$ response variable $\mathbf{y}$ and predictor variable(s) $X_1, .., X_p$ takes the following form:

$$\mathbf{y} = \beta_0 + \sum_{j=1}^{p} X_j \beta_j + \epsilon$$

where β_0 represents the intercept term (i.e., the modeled response when all predictor variables are set to zero), β_j represents the estimated coefficient for variable j, and ϵ represents the error (i.e., the remaining variance in $\mathbf{y}$ that is not explained by the predictor variables) where $\epsilon \sim N(0, \sigma^2)$. We may also use vector-matrix notation to represent this model

$$\mathbf{y} = \mathbf{X}\beta + \epsilon$$

where $\mathbf{X}$ contains the $n \times p$ matrix of predictor variables, and the vector β contains the estimated $\beta_0, .., \beta_p$ parameters in the model (the first column of $\mathbf{X}$ is a column of ones, which allows estimation of β_0).

If the assumptions of linear regression (see box above) are violated, the resulting model will likely produce inaccurate predictions, and other model classes should be considered. The most common approach to estimating the β values in a linear model is *ordinary least squares* (OLS), wherein we choose the values that minimize the sum of squared residuals (RSS)

$$\sum_{i=1}^{n} (y_i - f(x_i))^2,$$

where y_i represents the observed response and $f(x_i)$ represents the estimated response for a given set of β values. The OLS estimate of β can be calculated by

$$(\mathbf{X}^T\mathbf{X})^{-1}\mathbf{X}^T\mathbf{y}.$$

Although the OLS fit may be calculated by hand, it is rare to do so since such models can easily be fitted using a variety of statistical software.

Estimation of the reliability of a fitted linear model, in statistical terms, is performed by determining whether a given coefficient β_j is *significantly* different than zero. Our confidence in this difference from zero is quantified with a parameter called a p value (see Appendix B.3.3 for a general introduction to hypothesis testing, p values and assessment of significance in statistical inference). Naturally, if a coefficient β_j is not significantly greater or lesser than from zero, then the corresponding variable X_j does not carry a statistically significant predictive effect on the response variable in our model. Typically, we consider a coefficient to be statistically significant if its associated p value is less than 0.05 (although this threshold may vary depending on the application).

Example 4.2. [Pima Indians — predicting glucose levels from insulin]
Considering the dataset in Example 4.1, we may estimate the coefficients in a linear model with insulin as predictor and glucose as response

$$glucose \sim \beta_0 + \beta_1 * insulin.$$

Note that this is a physiologically meaningful model, as blood glucose levels are maintained in part through molecular signaling by insulin.
For this model comparing glucose and insulin in the Pima Indian cohort, the OLS estimate of the coefficients is

$$\sim 99.07 + 0.15 * insulin.$$

According to this model, the estimated glucose level is 99.07 when insulin is 0, and each unit increase in insulin results in a 0.15 unit increase in glucose. We may plot the line representing this fit onto a scatterplot of our observed glucose and insulin measurements, as shown in Fig. 4.2. Naturally, since there is not a perfect 1:1 relationship between glucose and insulin, there is not a perfect fit of the model to our data. And it is possible that a more complex model (e.g., a non-linear model, or a model including age or BMI as additional predictors) may provide a better fit; however, the equation determined from simple linear regression still provides a reasonable estimate of average glucose at a given level of insulin.

In Example 4.2, we were dealing with a relatively simplistic dataset and fitting a simple linear model. However, modern approaches to data generation in systems biology routinely produce datasets with thousands, or even millions of variables. For instance, common microarray platforms

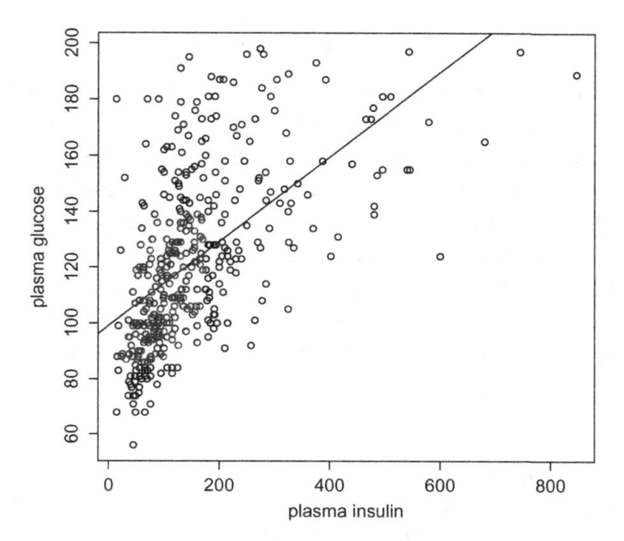

Figure 4.2 Observed relationship between glucose and insulin, with line of best fit from linear regression.

produce expression measurements for more than 20,000 unique transcripts (see Sect. 2.6). Given such a dataset, we may be interested in defining a model between these transcripts and some outcome variable, such that we identify the subset of transcripts that is most strongly predictive of our outcome variable.

The standard linear model breaks down when we have more predictor variables than samples (i.e., $p > n$), because the maximum likelihood estimation of the vector of β coefficients requires the inversion of the predictor matrix $\mathbf{X}$. In the case where $p > n$, the matrix $\mathbf{X}$ is singular and thus non-invertible, and therefore the coefficients in the model cannot be uniquely estimated.

Furthermore, even in the case where p is large but not larger than n, least-squares fit often suffers from high variance and low interpretability (due to many predictors with non-zero coefficients). A simple approach to regression when $p > n$ is conducting serialized simple linear regressions (i.e., fitting a unique model for each predictor variable p in the dataset). This is a very common approach in high-throughput analysis and is often sufficient for the interests of the researcher; however, one major limitation is that these repeated statistical tests produce a phenomenon called the *multiple testing problem.* This refers to the inflation of overall statistical uncertainty when a given statistical test is performed many times (see Example 4.3).

Bias–variance tradeoff

Bias and *variance* describe common, complementary sources of error in supervised learning. Model bias describes the difference between modeled responses and observed responses. Variance in a model is the sensitivity of the model to new training samples (i.e., the samples on which we estimate the coefficients in our model). A model with high variance will therefore change dramatically upon re-estimation of the model coefficients with a new training set. A model with high bias may not change with a new training set, but will be systematically incorrect in its predictions. In practice, there is generally a tradeoff between bias and variance.

Example 4.3. [Multiple testing]
Suppose we perform a simple linear regression using 0.05 as a p value threshold (often referred to as α) to define a significant relationship between a predictor and response variable. At this threshold, there is a 5% chance of rejecting the null hypothesis (i.e., the hypothesis that there is no true relationship between the predictor and response) if the null hypothesis is in fact true. If we then perform 100 such tests (e.g., with 100 different genes as predictors) then we expect to incorrectly reject the null hypothesis in five of these cases.

In the case of performing univariate regression for each of 20,000 genes on a microarray and aiming to maintain an overall α of 0.05, then an individual regression p value would have to be $< 2.5e-06$ in order to be considered significant after Bonferroni correction (see below for definition of Bonferroni correction).

The concept of multiple testing is prevalent, and sometimes contentious, in high-throughput data analysis. Various approaches have been proposed to correct for multiple testing, two of the most prevalent in high-throughput data analysis being *Bonferroni* correction, and the Benjamini and Hochberg method.

Bonferroni correction is simple, and is also the most conservative method. If m is the number of tests performed, and α is the chosen significance threshold, then Bonferroni correction states that accepting only $p < \alpha/m$ will keep the *overall* proportion of incorrectly rejected null hypotheses at a level below α. In practice, this multiple testing correction is very stringent when applied to high-throughput data (see Example 4.3).

A second approach to multiple testing correction was proposed by Benjamini and Hochberg (often called *BH procedure*). Given a ranked set of p values $P_i, \ldots, P_m$ corresponding to $H_i, \ldots, H_m$ hypothesis tests, the BH procedure constrains the *false discovery rate* (i.e., the rate of non-significant discoveries that are falsely declared significant) to α, by identifying the largest value of k such that $Pr_{(k)} \leq \alpha(k/m)$. Following the BH procedure, all tests H_i for $i = 1, ..., k$ are declared to be false positives. A range of other algorithms for multiple testing correction have been proposed, particularly for the case where the m tests are non-independent (as is often the case in genetic association analysis, for instance).

Another practical limitation of performing serialized univariate regression is that this approach will produce a set of univariate models that cannot be collectively used to make a single prediction of the response value of interest. Because of these reasons, an alternate class of methods, sometimes called *regression shrinkage methods* or *penalized regression*, has therefore been developed for the specific case of large p and $p > n$ in regression analysis.

4.2.1 *Penalized regression*

Shrinkage methods seek to shrink the estimated coefficients in a fitted regression model in order to reduce variance of models with many predictor variables. *Ridge regression* (also called *Tikhonov regularization*) achieves shrinkage by imposing a limit on the l_2 norm of the coefficient vector β.

Vector norm

A *norm* is a function that assigns a length to a given vector. While there are multiple ways to define a norm on a vector x, common norms used in penalized regression are the l_1 norm ($\|x\|_1$) and l_2 norm ($\|x\|_2$) defined as

$$\|x\|_1 = \sum_{i=1}^{n} |x|_i \qquad \|x\|_2 = \sqrt{\sum_{i=1}^{n} x_i^2}.$$

Comparing our least-squares fit from standard linear regression, we can see that the $\widehat{\beta}_{ridge}$ estimate is a standard RSS with the additional penalty term $\lambda \sum_{j=1}^{p} \beta_j^2$. Therefore, as λ approaches 0, $\widehat{\beta}_{ridge}$ approaches the least squares β.

Large values of λ impose a stronger penalty on the l_2 norm of the β coefficient vector, thus favoring solutions with smaller coefficients. For a

Ridge regression

The ridge regression estimate $\widehat{\beta}_{ridge}$ is the value β that minimizes the equation

$$\sum_{i=1}^{N}(y_i - \beta_0 - \sum_{j=1}^{p} x_{ij}\beta_j)^2 + \lambda \sum_{j=1}^{p} \beta_j^2.$$

chosen value of λ, it can be shown that $\widehat{\beta}_{ridge}$ can be calculated by

$$(\mathbf{X}^T\mathbf{X} + \lambda\mathbf{I})^{-1}\mathbf{X}^T y$$

where $\mathbf{I}$ is a $p \times p$ *identity matrix* (i.e., a matrix with ones on the main diagonal, and zeroes elsewhere). While ridge regression works well in reducing variance of models with a large number of multicollinear (i.e., correlated) predictor variables, models with a large number of predictors will still suffer from low interpretability as is the case in standard linear regression, due to many predictors with non-zero coefficients in the fitted model.

Lasso regression provides an alternative approach to penalized regression that is similar to ridge regression, but with the important difference that the lasso penalty may constrain some coefficients in the model to zero.

Lasso regression

The lasso estimate $\widehat{\beta}_{lasso}$ is the value β that minimizes the equation

$$\sum_{i=1}^{N}(y_i - \beta_0 - \sum_{j=1}^{p} x_{ij}\beta_j)^2 + \lambda \sum_{j=1}^{p} |\beta_j|.$$

We can see β_{lasso} like the $\widehat{\beta}_{ridge}$ estimate, except with a penalty term $\lambda \sum_{j=1}^{p} |\beta_j|$. $\widehat{\beta}_{lasso}$ thus restricts the l_1 norm of the β coefficient vector, and in practice produces increasingly sparse coefficient vectors (i.e., with more zero entries) as the value of λ increases. For both lasso and ridge regression, it is clear that the selection of λ has a major impact on the

k-fold cross-validation

Cross-validation is a procedure for estimating the quality of a model fit, by determining how well it performs on an independent set of samples. In *k-fold cross-validation*, the sample set is divided into k equal groups. One of the k groups is retained as a testing set, and the remaining $k-1$ groups are used to parameterize the model. The fitted model is then used to predict the response of group k, and model quality is assessed by determining the difference between predicted and observed response (i.e., model error). Each of the k groups is used one at a time as a testing set, yielding information on both mean model error and error variance.

Example 4.4. [Ridge and lasso regression]
While ridge and lasso regression work well in the $p \gg n$ context, for simplicity we will use an example dataset of 72 scaled variables (anthropometric characteristics, plasma lipids, amino acids and other metabolites) measured across 117 individuals (available from www.ncbi.nlm.nih.gov/pubmed/19357637), and fit ridge and lasso regression models with blood glucose as response and all other variables as predictors. Figure 4.3 illustrates the fitted coefficients for both modeling approaches, across a range of values of λ. In the case of ridge regression, we see that increasing λ parameter (thereby increasing the stringency of the penalty) decreases the estimated coefficients. An important point to note is that at more stringent penalty levels, lasso shrinkage forces the coefficients of many variables to zero, and thus acts as an effective variable selection procedure. Examining the fitted models using ten-fold cross-validation, we see that at the optimal log-λ value of -2.3 the lasso fit contains only 14 non-zero coefficients, while the ridge fit maintains the full 72.

estimated coefficients. A common approach to the selection of optimal λ is k-fold cross-validation, wherein the value of λ that yields the minimum mean cross-validated error is selected.

When extending lasso regression to very high-dimensional datasets, it is easy to see how this variable selection may be a useful feature. However, a known limitation to lasso regression is that in the case of strong multicollinearity (i.e., correlation among predictor variables), lasso models tend to retain at random a single variable from a group of correlated variables. To address this limitation, a third type of penalty, the *elastic net* penalty, has been proposed.

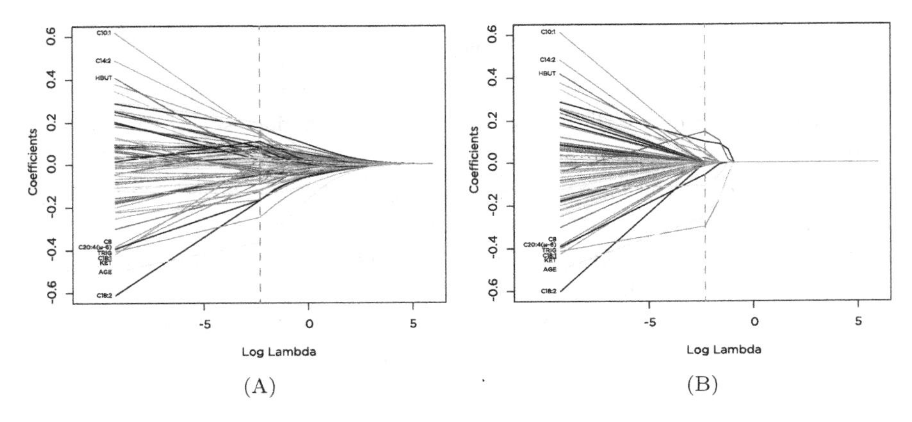

Figure 4.3 Comparison of coefficient estimates for (A) ridge and (B) lasso regression across a range of lambda coefficients. The predictor variables with the largest ten coefficients in the least-constrained models are labeled in each plot. The vertical dashed lines indicate the optimal lambda value, as selected by cross-validation. Variable abbreviations are as follows: C10:1: Decenoyl carnitine, C14:2: Tetradecadienoyl carnitine, HBUT: beta hydroxybutyrate, C8: Octanoyl carnitine, C20:4(ω-6): arachidonic acid, TRIG: Triglycerides, C18:1: Oleyl carnitine, KET: Ketones, C18:2: Linoleyl carnitine.

Elastic net penalty

The *elastic net penalty* is a combination of ridge and lasso penalties, and is defined as

$$\sum_{i=1}^{N}\left(y_i - \beta_0 - \sum_{j=1}^{p} x_{ij}\beta_j\right)^2 + \lambda\sum_{j=1}^{p}(\alpha\beta_j^2 + (1-\alpha)\,|\beta_j|)$$

which we can see combines ridge and lasso penalties in proportion controlled by the α coefficient. That is, $\alpha = 0$ reduces to lasso regression, $\alpha = 1$ reduces to ridge regression and $0 < \alpha < 1$ combines the two. The elastic net therefore acts as a variable selection method due to the lasso penalty, and shrinks correlated variables together as a group due to the ridge penalty.

4.3 Dimensionality reduction methods

Although regression analysis is straightforward and easily interpretable, it does not explicitly model the inherent correlation structure within a dataset, which is often of primary interest in biological data analysis. For instance, we may measure the abundance of 1000 proteins and compare these levels to disease state, but we may also be interested in *covariance* patterns among the proteins, or among the samples in the study. Multivariate analysis incorporates a range of approaches for exploring this correlation structure within and between multivariate datasets. Dimensionality reduction — i.e., identifying informative low-dimensional representations of high-dimensional datasets — is a common theme across many multivariate approaches.

Covariance, Correlation

Covariance is a measure of the extent to which two variables change together. Given two vectors $\mathbf{x}$ and $\mathbf{y}$ of dimension n, their covariance is defined as

$$cov(x,y) = \sum_{i=1}^{n} \frac{(x_i - \overline{x})(y_i - \overline{y})}{n}$$

where $\overline{x}$ and $\overline{y}$ are the means of x and y. Covariance is thus closely related to *correlation*, which is defined as

$$cor(x,y) = \frac{cov(x,y)}{\sigma_x \sigma_y}$$

where σ_x and σ_y are the standard deviations of x and y.

4.3.1 *Principal component analysis*

Principal component analysis (PCA) is a widely used multivariate approach that identifies linear combinations of the variables in a multivariate dataset that maximally represent the variance structure in a smaller number of uncorrelated variables (often called latent variables). These latent variables are the principal components of the original variables. To illustrate, consider a dataset containing abundance measurements for a set of molecules across a cohort of individuals. It may be the case that a subset of these molecules displays a strong correlation with each other, and thus could be represented by a composite (latent) variable that captures a large amount of the variance in these variables. The *orthogonal* transformation imposed by PCA is such that the first latent vector (first principal component) accounts for the largest amount of variability in the dataset, with each successive component accounting for less variability and being uncorrelated to

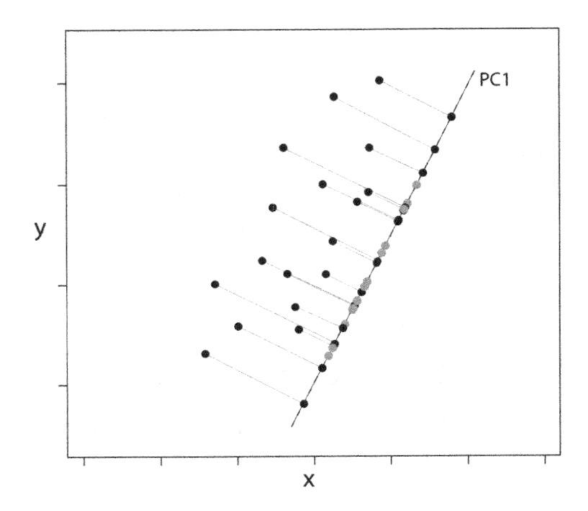

Figure 4.4 Projection of two-dimensional data onto its first principal component.

the rest (i.e., due to the orthogonality constraint). Figure 4.4 illustrates the concept of projecting variables onto a new axis. In this example, we can see that the variables x and y are correlated, and thus there is a clear vector on which the variance in the dataset is greatest. This vector is defined as the *first principal component* (PC1), and all successive components are perpendicular (i.e., orthogonal) to the first.

When PCA is applied to a dataset with many variables, plotting the samples in this low-dimensional space (on the first two principal components, for example) may reveal informative grouping structure across the samples, such as grouping by dietary treatment, genotype or gender. Of course, the goal is to balance variance explained and visual clarity, and so it is commonplace to present the dataset in terms of two or three principal components. Plotting the bar chart of variances of each principal component (known as a *scree plot*) is a common approach to selecting the number of components that are deemed adequate to describe the overall variance in the dataset. Note that there is not a strict rule on deciding the number of components; rather, it is common to look for a steep drop in the variance from one component to the next, and to consider the overall variance accounted for by each principal component.

Typically, in PCA and other dimensionality reduction procedures, we assume that the data are standardized such that each variable has mean zero and standard deviation one. In order to calculate the principal components given a set of multivariate data, we rely on principles from linear

Single value decomposition, Eigendecomposition

Both *single value decomposition* (SVD) and *eigendecomposition* denote factorizations of a matrix $\mathbf{X}$ into a product of matrices. Specifically, the SVD of an $n \times p$ matrix $\mathbf{X}$ is

$$\mathbf{U}\Sigma\mathbf{V}^T$$

where $\mathbf{U}$ is a unitary $n \times n$ matrix (i.e., $\mathbf{U}^T\mathbf{U}$ produces a square matrix with ones on the diagonal and zeroes elsewhere), Σ is an $n \times p$ matrix with the diagonal entries representing the *singular values* of $\mathbf{X}$ and $\mathbf{V}^T$ is a $p \times p$ unitary matrix ($\mathbf{V}^T$ may be written as the conjugate transpose $\mathbf{V}*$, if $\mathbf{V}$ does not consist entirely of real numbers). The n columns of $\mathbf{U}$ are called the *left-singular vectors* (LSV) of $\mathbf{X}$, and the p columns of $\mathbf{V}$ are the *right-singular vectors* (RSV) of $\mathbf{X}$.
Eigendecomposition is instead carried out on a square matrix, in this case the $p \times p$ matrix $\mathbf{A}$, defined as $\mathbf{X}^T\mathbf{X}$. The eigendecomposition of $\mathbf{A}$ is

$$\mathbf{PDP}^{-1}$$

where $\mathbf{P}$ is a $p \times p$ matrix with each column representing an *eigenvector* of $\mathbf{A}$, and $\mathbf{D}$ is a diagonal $p \times p$ matrix with each diagonal entry representing an *eigenvalue* of $\mathbf{A}$.
SVD and eigendecomposition are closely linked, in that the LSV of $\mathbf{X}$ are the eigenvectors of $\mathbf{X}^T\mathbf{X}$, and the diagonal entries of Σ are the square roots of the eigenvalues of $\mathbf{X}^T\mathbf{X}$. There are a number of algorithms for SVD and eigendecomposition (such as the QR algorithm, LAPACK subroutine, and Jacobi eigenvalue algorithm) and a variety of statistical software packages implementing these algorithms.

algebra. While there are multiple approaches to the calculation of principal components of a matrix $\mathbf{X}$, one common approach is to calculate the single value decomposition (SVD) of $\mathbf{X}$ (i.e., $\mathbf{U}\Sigma\mathbf{V}^T$; see box above). Note that this is equivalent to calculating the eigendecomposition of $\mathbf{X}^T\mathbf{X}$. The resulting columns of $\mathbf{U}\Sigma$ represent the principal components ($n \times p$), and $\mathbf{V}$ represents a $p \times p$ *loading matrix*, with each column defined as the correlation between the original variable and a given principal component. For instance, in Fig. 4.4, it is clear that among variables x and y, variable y is more strongly correlated with PC1. Therefore, y would have a larger entry in the first column of the loading matrix. It should be noted that PCA can be directly applied to datasets where $p \gg n$, although the interpretation

Example 4.5. [PCA — lipidomics]
Using a simple dataset of 20 lipids measured in the plasma of 72 individuals (a subset of data available from www.ncbi.nlm.nih.gov/pubmed/22257447), we can perform PCA and plot the first two principal components using a graph called a *biplot*, which plots both the principal components and loadings onto a common space. Plotting these variables illustrates a clear covariance pattern among the lipids in the dataset (Fig. 4.5A), as closely covarying variables will possess similar loading values and thus will plot in close vicinity. In addition, we can see that the first two dimensions in our PCA model account for about 55% of the summed variance of all principal components; including the third component increases this to 70% (Fig. 4.5B).

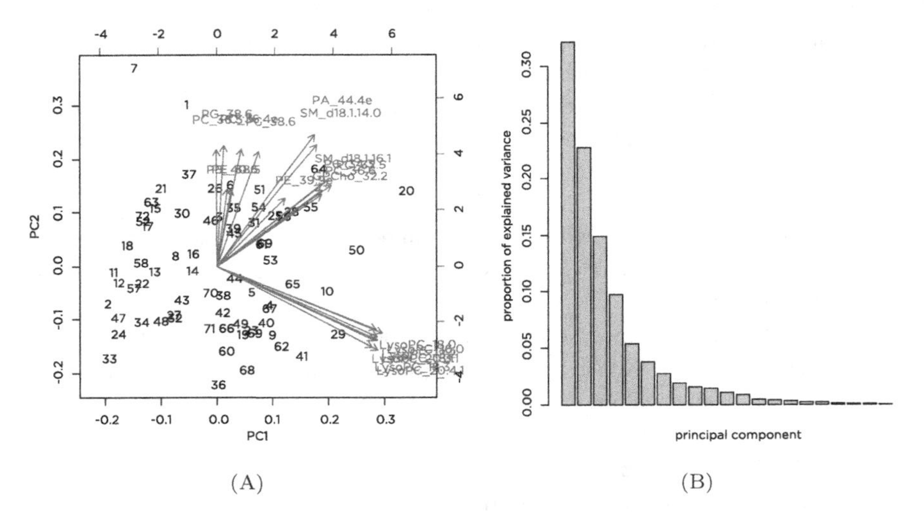

Figure 4.5 Principal component analysis of lipidomic dataset, displaying (A) a biplot of the first two principal components, and (B) a bar chart illustrating the proportion of variance captured by each component.

of the loadings may become challenging for matrices with very large p. To improve interpretability, algorithms for sparse PCA use elastic net or other penalties to calculate principle components with sparse loading vectors.

4.3.2 *Linear discriminant analysis*

Whereas principal component analysis defines the latent variable with maximal variance in the samples, *linear discriminant analysis* (LDA) seeks to define the latent variable that maximally discriminates between two or more

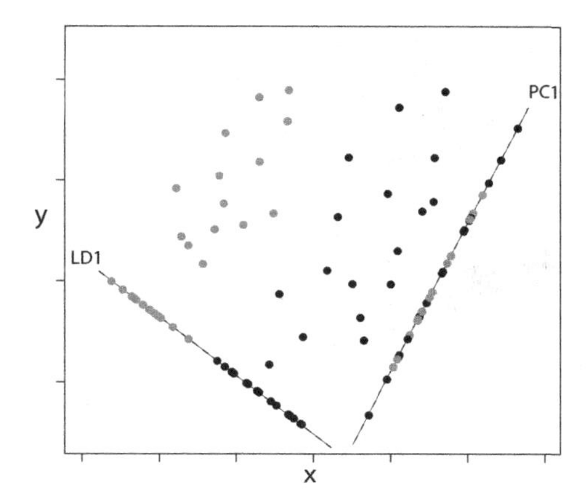

Figure 4.6 Illustration of dimensionality reduction via PCA and LDA.

Example 4.6. [LDA vs PCA — dimensionality reduction]
Figure 4.7 illustrates dimensionality reduction using a set of 100 genes measured in skeletal muscle tissue of 118 individuals from three known phenotype classes: normal, glucose intolerant, and diabetic (data available from www.ncbi.nlm.nih.gov/geo/query/acc.cgi?acc=GSE18732). It is clear from the illustration that LDA seeks a low-dimensional representation that maximizes separation of the groups, whereas PCA does not.

known classes in a dataset. Given a simple dataset with two variables, Fig. 4.6 illustrates the difference between projecting to a principal component axis and a linear discriminant axis. Although LDA is very similar to (and often used interchangeably with) a variant called Fisher's linear discriminant analysis, they are in fact slightly different in that Fisher's LDA does not make the same assumptions of within-class normal distribution and equal covariance. The box *Linear discriminant analysis* presents Fisher's approach to LDA. Note that similarly to PCA, LDA can be applied in the $p > n$ context, but benefits from sparse approaches for interpretability.

4.3.3 *Canonical correlation analysis*

With the growing accessibility of high-throughput data generation, as well as a growing appreciation of the multifaceted nature of biological systems, it is becoming increasingly common to measure more than one data type on a given set of samples, such as genotype and transcript, or protein and

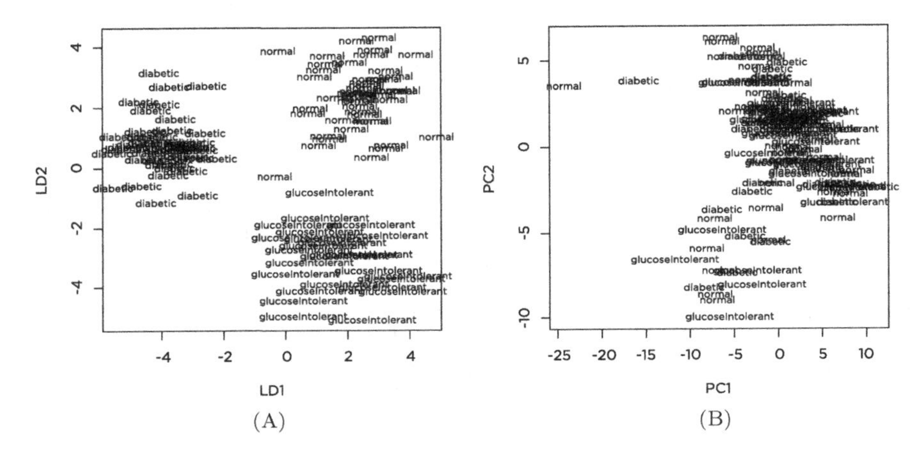

Figure 4.7 Illustration of dimensionality reduction via (A) LDA and (B) PCA.

Linear discriminant analysis

Given an $n \times p$ dataset $\mathbf{X}$ with C class labels for the n samples, LDA seeks to identify the $n \times C$ *transformation matrix* $\mathbf{W}$ such that the projection $\mathbf{Y} = \mathbf{W}^T\mathbf{X}$ has maximum difference between class means and minimum within class variance, thus maximizing class separation in the projection. In order to identify $\mathbf{W}$, we need to maximize the objective function

$$J(\mathbf{W}) = (|\mathbf{W}^T\mathbf{S}_B\mathbf{W}|)/(|\mathbf{W}^T\mathbf{S}_W\mathbf{W}|)$$

where |...| refers to the matrix determinant, and $\mathbf{S}_B/\mathbf{S}_W$ refer to the between-group/within-group scatter matrices (a measure of variance)

$$\mathbf{S}_W = \sum_{i=1}^{C} \sum_{x_k \in c_i} (x_k - \mu_i)(x_k - \mu_i)^t, \quad \mathbf{S}_B = \sum_{i=1}^{C} n_i(\mu_i - \mu)(\mu_i - \mu)^t$$

where n_i is the number of samples in class i, μ_i is the sample mean of class i, and μ is the mean of all samples. Maximizing $J(\mathbf{W})$ is equivalent to the generalized eigenvalue problem $\mathbf{S}_B\mathbf{w} = \lambda\mathbf{S}_W\mathbf{w}$ and thus, the projection vectors $\mathbf{w}$ (i.e., column vectors of $\mathbf{W}$) that maximize class separation are the eigenvectors corresponding to the largest eigenvalues of $\mathbf{S}_W^{-1}\mathbf{S}_B$.

metabolite. Methods for exploring correlation structure within and between two multivariate datasets (measured on the same samples) have therefore become commonplace in analysis of biological systems. *Canonical correlation analysis* (CCA) is one such method, using a latent variable approach to identify low-dimensional correlations between two matrices of continuous variables measured on the same samples. Whereas PCA seeks a decomposition of a single matrix to produce latent variables with maximal variance, CCA seeks a decomposition of a pair of matrices resulting in pairs of latent vectors with maximal correlation.

Canonical correlation analysis

Canonical correlation analysis (CCA) accepts as input a matrix $\mathbf{X}$ of dimension $n \times p$ and $\mathbf{Y}$ of dimension $n \times q$, both measured on the same n samples. The specific aim of CCA is to identify H pairs of loading vectors $\mathbf{a}_h$ and $\mathbf{b}_h$ of dimension p and q, respectively, that maximize the objective function

$$cor(\mathbf{Xa}_h, \mathbf{Yb}_h), h = 1...H$$

subject to $\mathbf{a'}_h\mathbf{a}_h = 1$ and $\mathbf{b'}_h\mathbf{b}_h = 1$, where h is the chosen dimension of the solution, H is the number of variables in the smaller of the two matrices and the n-dimensional latent variables $\mathbf{Xa}_h$ and $\mathbf{Yb}_h$ are the canonical variates.

Similar to PCA, the first canonical variate pair (i.e., first dimension) represents the most highly correlated canonical variates, with successive canonical variate pairs carrying successively weaker correlations.

CCA cannot be directly applied in cases where $p+q \gg n$, as the solution of the objective function requires calculation of the inverse of the covariance matrices $\mathbf{XX}^T$ and $\mathbf{YY}^T$, which are singular in the case of $p+q \gg n$ and thus lead to non-unique solutions. We therefore need to use penalized versions of CCA in the high-dimensional setting. One such approach uses an elastic net penalty (see subsection on penalized regression in Sect. 4.2), wherein the ridge regularization obtains non-singular covariance matrices $\mathbf{XX}^T$ and $\mathbf{YY}^T$ by imposing a weighting penalty on the diagonal of the covariance matrices, such that highly correlated variable pairs receive similar weights. Lasso shrinkage then imposes an additional penalty that shrinks variables with low explanatory power to zero, and thus can be viewed as a variable selection procedure.

Example 4.7. [CCA]
Using a dataset of 21 fatty acids and 120 genes measured in the livers of 40 mice (available from http://www.qfab.org/mixomics), we can apply CCA to determine the canonical variates with maximal correlation between the two sets of variables. Figure 4.8 depicts the correlation between each successive pair of canonical variates, illustrating that the first canonical variate pair is highly correlated, and each successive pair is less strongly correlated.

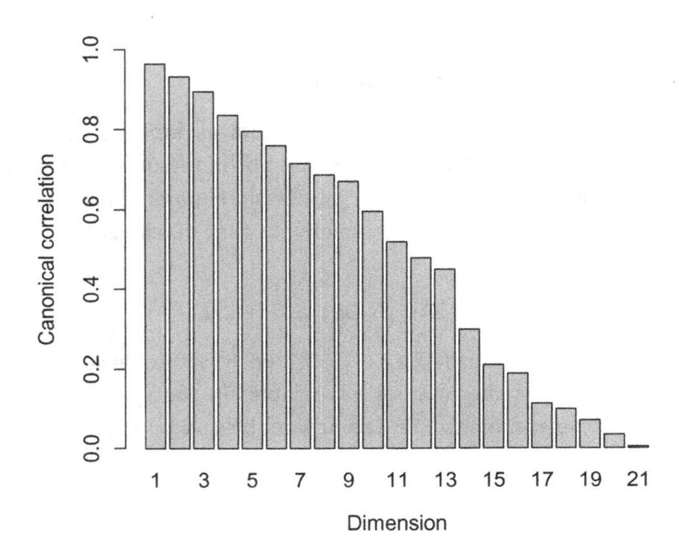

Figure 4.8 Illustration of correlation between canonical variate pairs resulting from CCA.

4.4 Clustering

Given a set of multivariate data, we may be interested in understanding the inherent grouping pattern of samples across all variables (or variables across all samples). Cluster analysis is a widely used method for accomplishing this task. Hierarchical and k-means clustering are the most commonly used approaches to clustering, and can each be considered as *unsupervised methods* as they assume no prior knowledge of sample or variable classes. However, clustering analysis is frequently performed in a blinded fashion on data with known classes, in order to determine whether distinctions between the classes naturally emerge. For simplicity, we will discuss clustering of samples across a set of variables; however, these same approaches are equally applicable to clustering of variables across a set of

Example 4.8. [Visualizing CCA]
In applying CCA, our primary interest is the correlation structure between the two datasets. Interpretation of CCA results is strongly aided by plotting *correlation circles*, which illustrate the correlation between each original variable (in both datasets) and selected canonical variate. Figure 4.9A depicts a correlation circle plot following canonical correlation analysis of the dataset in Example 4.7. In this plot, variables that plot further from the origin are more strongly correlated with at least one of the first two canonical variates (for simplicity, we only visualize the variables that are correlated with at least one of the first two canonical variates with a correlation coefficient of at least 0.65). From this illustration, we can examine the correlations among the variables within and between the datasets. For instance, in this plot we can see that the fatty acid C16:0 is plotted closely to the gene THIOL, and by plotting these two variables together in a scatterplot (Fig. 4.9B) we can see that they are in fact strongly correlated.

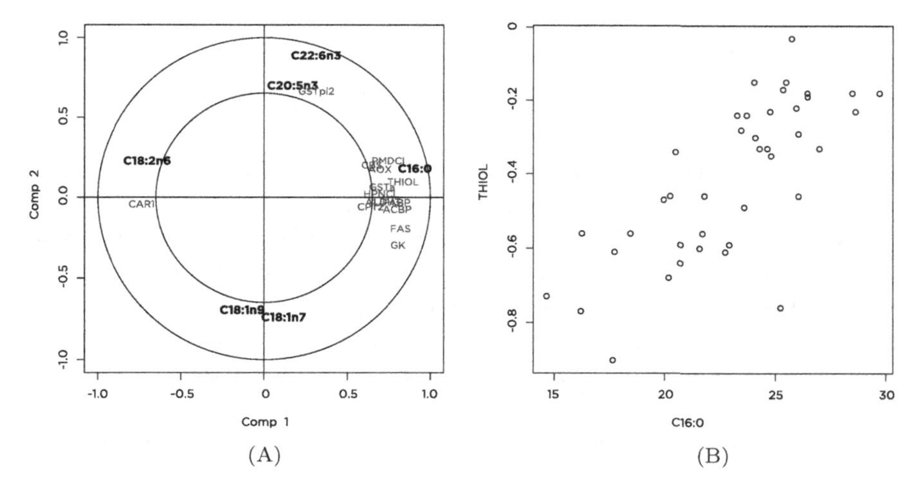

Figure 4.9 (A) Illustration of correlation circle from CCA. (B) shows an example plot of two variables from the correlation circle that are similarly correlated with their first and second canonical variates.

samples. Although there is no technical obstacle to applying hierarchical and k-means clustering to a dataset with $p \gg n$, it is important to keep in mind that clustering a set of samples based on their entire transcriptome/proteome/metabolome is a very broad clustering. And although this may certainly be informative, it is often also worthwhile to perform the clustering on subsets of the variables that represent more cohesive biological processes (e.g., clustering the samples across all genes involved in carbohydrate metabolism, or all proteins involved in immune response). In doing so, we can explore our data from multiple angles and may uncover unexpected patterns therein.

4.4.1 *Hierarchical clustering*

Hierarchical clustering does not assign samples to discrete clusters, but instead constructs a hierarchy of similarity between the samples such that the number of clusters ranges from 1 to n, the number of samples. Two general schemes for constructing this hierarchy are *agglomerative* (starting by grouping together the most similar samples and progressively adding more to build the hierarchy) and *divisive clustering* (starting with the complete dataset, and successively dividing it into subgroups). In either case, a *similarity metric* is used to define the similarity between samples. Common similarity metrics include Euclidean distance, Manhattan distance and Mahalanobis distance respectively defined as

$$\sqrt{\sum_{i=1}^{p}(x_i - y_i)^2}, \qquad \sum_{i=1}^{p}|x_i - y_i|, \qquad \sqrt{(\mathbf{x}-\mathbf{y})^T\mathbf{S}^{-1}(\mathbf{x}-\mathbf{y})}$$

where $\mathbf{x}$ and $\mathbf{y}$ represent two samples with p variables measured on each, and $\mathbf{S}$ is the matrix of covariances between $\mathbf{x}$ and $\mathbf{y}$.

Whereas the similarity metric measures the similarity of two observations, the *linkage criterion* defines the distance between clusters of observations. Common linkage criteria include *single linkage* (the closest two observations between two groups), *complete linkage* (the most distant two observations) and *average linkage* (the distance between the centroids of the two groups) as shown in Fig. 4.10. The general approach to agglomerative hierarchical clustering of n samples in an $n \times p$ dataset is defined in Algorithm 4.1.

4.4.2 *K-means clustering*

Unlike hierarchical clustering, *k-means clustering* results in a discrete assignment of observations to clusters (Algorithm 4.2). As this algorithm requires the selection of k by the user, it is typical to rerun the algorithm across a range of values of k and assess the fit of each clustering.

The fit may be assessed by calculating the *within-cluster sum of squares* (WCSS), which is the sum of squared distances between each data point and its respective cluster center. Naturally, a smaller WCSS indicates a better fit; however, the WCSS approaches zero as k approaches n. Therefore it

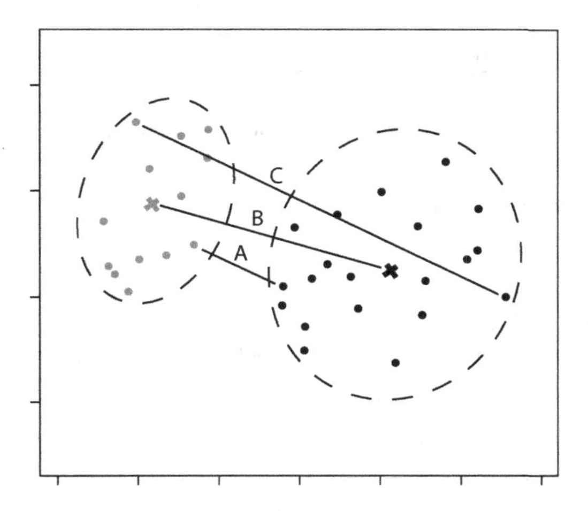

Figure 4.10 Distance between two clusters of observations, as defined by single linkage (A), average linkage (B) and complete linkage (C). The X marks the centroid of each cluster.

Algorithm 4.1: Hierarchical clustering

Input: An $n \times p$ dataset, and definition of similarity and distance metric.
Output: A hierarchy of observation similarities, ranging from n clusters to a single cluster.

begin
  assign each observation to its own cluster;
  repeat
    choose the two most similar clusters, based on the chosen similarity and distance metrics;
    merge them to form a single cluster;
  until *all observations in the dataset are merged into a single cluster*;
end

---

is common to plot the WCSS across a range of values of k, and choose the "elbow" point in the plot (i.e., the point at which the WCSS ceases to drop sharply as k increases; see Fig. 4.12A). Another method for selecting optimal k is called the *gap statistic*. This method involves first calculating the WCSS for a range of k, then performing the same procedure on a large number (typically 100 or more) of simulated reference datasets. These reference datasets match the dimension of the original dataset, and may be generated in multiple ways, such as by uniform sampling from the range of

Algorithm 4.2: k-means clustering

Input: Parameter k and a dataset.
Output: An assignment of each observation to a cluster.

begin
  select k initial cluster centers and assign each observation to its closest center;
  repeat
    calculate new cluster centers based on this assignment;
    assign each observation again to its closest cluster (some observations will change cluster while others will remain the same);
  until *no observation changes cluster*;
end

Example 4.9. [Hierarchical clustering — cell lines]
To illustrate hierarchical clustering, we can use a transcriptomic dataset comprising 54,675 transcripts measured across 59 distinct cell lines from nine distinct tissues (available from discover.nci.nih.gov/cellminer). Using Euclidean distance and complete linkage, we obtain the clustering dendrogram illustrated in Fig. 4.11. By labeling the dengrogram leaves with the tissue of origin for each sample, it is evident that the samples strongly cluster based on tissue.

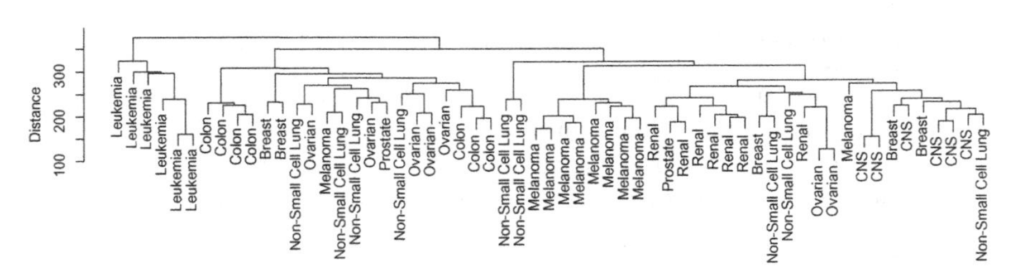

Figure 4.11 Hierarchical clustering of transcriptomes measured across nine distinct cell lines. The distance scale indicates the similarity between two clusters at each bisection, based on the chosen similarity metric and linkage criterion.

the original variables. The result of this procedure is an average reference WCSS for each value of k. Finally, for each k, the original log-WCSS is subtracted from the average reference log-WCSS, and the value of k yielding the largest positive difference (i.e., gap) is selected as optimal.

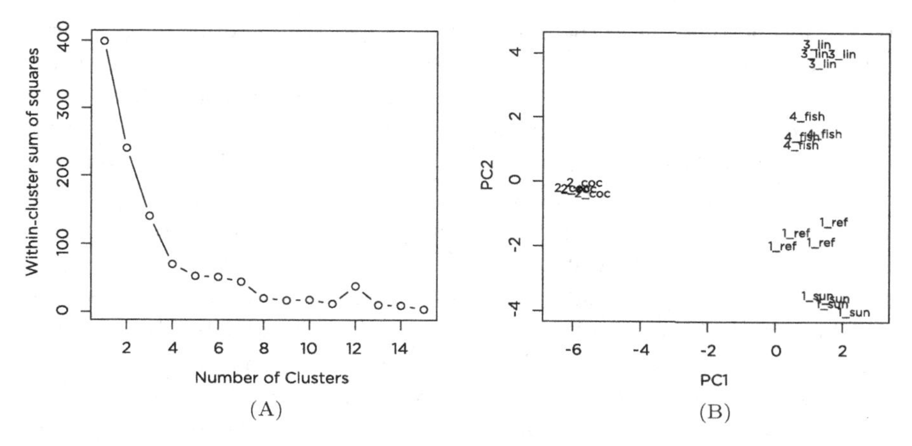

Figure 4.12 *K*-means clustering of 20 mice across 21 hepatic fatty acids, illustrating (A) WCSS at increasing values of k and (B) a PCA plot showing each of the samples on the first two principal components. In the PCA plot, samples are labeled with their assigned clusters from k=4, and their assigned dietary group, where ref: reference diet, coc: hydrogenated coconut oil, sun: sunflower oil, lin: linseed oil.

Example 4.10. [k-means clustering — liver fatty acids]
To illustrate k-means clustering, we will use a set of 21 fatty acids measured in the livers of 20 *wild type* mice (i.e., genetic variant found in the wild). These 20 mice were divided into five groups of four, each receiving a diet with distinct fatty acid composition (data available from http://www.qfab.org/mixomics). With these data, we apply k-means clustering across a range of k from 1 to 15. Figure 4.12A shows the reduction in WCSS with increasing k. From this figure, it appears that k=4 may be an appropriate choice for this dataset. This choice is largely supported by the PCA plot in Fig. 4.12B, which shows a clear division of the samples by dietary group, except with the reference and sunflower oil group clustering together according to k=4.

4.5 Gene set analysis

One of the primary challenges in working with high-throughput data is the interpretation of results obtained by primary statistical analyses. For instance, regression analysis of a transcriptomics dataset may highlight hundreds or thousands of transcripts that are strongly correlated with some outcome phenotype. Translating these results into a cohesive biological explanation is critical for understanding the biological system, and producing further hypotheses for follow-up work. *Gene set analysis*

(GSA) has become one of the most widely adopted method for the interpretation of high-throughput analysis results. It is most often applied to gene expression data, but is in principle applicable to all data types. The general premise of GSA is to first assign each gene in a high-throughput dataset to known functional categories, determined from publicly available annotation databases. Then each functional category is tested as a whole for significant correlation with the outcome of interest. By reducing gene-level information to the level of biological process, GSA has advantages both in terms of statistical power (by dramatically reducing the burden of multiple testing) and biological interpretability. The most commonly used functional categories (i.e., gene sets) are metabolic/signaling/regulatory pathway models (such as those in the Reactome database — www.reactome.org) or Gene Ontology terms (www.geneontology.org); however, in principle, any grouping could be used. Other examples of gene sets that have been successfully applied in this context include cytobands, genes regulated by common miRNA/transcription factors and pre-defined groups of disease genes, among others.

There are numerous algorithms that have been developed for GSA, and a wide range of implementations available through standalone software, libraries for programming languages such as R or python or via web servers. Despite the breadth of approaches to GSA, most methods can be broadly categorized as those based on *over-representation*, or abundance change scoring. Over-representation (OR) methods are the earliest examples of GSA, and are most broadly applicable across data types. The general approach to OR proceeds as follows:

(1) A list of significant genes (also called *hits*) is determined from the high-throughput data using t-test, linear regression, etc., at a defined significance threshold. This gene list may represent, for instance, genes significantly correlated with a phenotype of interest or a perturbation.
(2) The number of significant genes is counted both in the background list (typically the entire data platform, such as a microarray) and in each pre-defined gene set.
(3) Each gene set is tested for significant enrichment with the genes of interest (i.e., significantly more hits than expected by chance), typically using a test such as Fisher's exact, hypergeometric, chi-square or binomial test.

The simplicity of the OR approach can be a strength in some settings. Using only a list of significant genes as input (without the requirement for specific quantitative values), this approach can be directly applied to gene lists that do not possess abundance values (such as sets of genes containing SNPs that are statistically associated with a specific phenotype). However, this simple use of gene lists also implies that OR approaches consider marginally significant genes as identical to highly significant genes, thus ignoring an informative feature of the data. Furthermore, the chosen significance threshold for defining the initial gene list may exclude potentially informative genes that marginally miss this threshold.

Abundance change scoring (ACS) approaches inherently consider quantitative data on each variable in calculation of gene set enrichment, and do not impose pre-selection of a significant gene list. ACS approaches to GSA begin with:

(1) Calculation of differential expression scores for each gene in the dataset. Abundance change scores may be, for instance, log-fold change between two groups of samples, or β coefficients from a fitted linear model.
(2) These raw scores may be directly used, or they may be transformed to remove the distinction between up- and down-regulated genes, if preferred (e.g., absolute values, squared values, or rank statistics).
(3) Abundance change scores are then aggregated to gene set level statistics, such as by averaging across the gene set, then
(4) significance of each gene set is determined, often by a permutation test, either permuting the class labels or gene labels.

4.6 Analysis of biological networks

Molecular networks are fundamental constructs in systems biology as they carry the potential to represent all known molecular details of a biological system of interest, depending on the granularity and complexity chosen for the analysis. As such, it is becoming more commonplace to use global interaction networks as a frameworks for analysis of high throughput data. The most common network types used in bioinformatics are metabolic (enzymatic reactions between enzymes and metabolites), protein–protein (complex formation and signal transduction) and regulatory (interactions between transcription factors and target genes); however, it is important to appreciate that in reality, the elements of each network type are interlinked in a common system. For instance, certain products of metabolism activate signaling cascades, which in turn activate gene regulatory interactions.

Example 4.11. [OR — skeletal muscle transcriptome]
Using a set of 31,001 genes measured in the skeletal muscle tissue of 118 individuals, along with fasting blood glucose measurements (data available from www.ncbi.nlm.nih.gov/geo/query/acc.cgi?acc=GSE18732), we can use an OR approach to identify biological pathways with gene expression significantly correlated with fasting glucose. To carry out this analysis, we first need to identify the subset of genes that are significantly correlated with glucose levels. We fit a linear model to each gene in our dataset, using the model

$$glucose \sim \beta_0 + \beta_1 * expression + \beta_2 * gender.$$

We have included gender here, as it may be a confounding variable (i.e., a variable that correlates with both the predictor and response, and thus should be included to avoid identifying a spurious relationship between predictor and response). As discussed in the section on linear regression, our goal in this modeling is to determine if the β_1 coefficient is significantly different than zero. The result of this regression analysis is 898 significant genes, after correcting for multiple testing using the BH procedure and a significance threshold of 0.1 (as is the conventional threshold in bioinformatics when considering corrected p values).

Network models and pathway models are similar in that they both represent molecular interactions; however, an important difference is that global networks inherently consider the overlap and crosstalk between different canonical pathways. Analysis of omic data in the context of global networks is difficult, however, due to the considerable size of global interaction networks. For instance, the I2D database (http://ophid.utoronto.ca), which assembles protein interaction data from multiple databases such as BIND, IntAct, BioGrid and others, contained 183,524 human protein interactions as of the publication of this book (2015). While such a network could be used in an omics data analysis, it is important to be aware that the interactions in such databases are identified using both experimental and computational approaches (such as prediction of an interaction between molecules x and y in species A following observation of this interaction in species B). Furthermore, some of the technologies used to identify molecular interactions experimentally are prone to false positives, leading to an artificially inflated number of apparent interactions in interaction databases. Finally, public databases typically contain what are known as *global* interaction networks, such that they contain any reaction that could presumably occur in the species of interest. Therefore, they do not capture the tissue-, cellular

Example 4.11. *continued* [OR — continued]
The 898 genes can then be used in conjunction with the pathway models from the Reactome database (www.reactome.org) to identify those pathways that are significantly enriched with our genes of interest.

Description	Hits	Path size	P value	BH
Meiosis	11/386	50/4766	0.002	0.092
Antigen processing	25/386	177/4766	0.004	0.092
Class I MHC mediated antigen processing and presentation	28/386	213/4766	0.007	0.092
Signaling by Type 1 Insulin-like Growth Factor 1 Receptor (IGF1R)	12/386	70/4766	0.010	0.092
IGF1R signaling cascade	12/386	70/4766	0.010	0.092
Fatty acid, triacylglycerol and ketone body metabolism	17/386	115/4766	0.010	0.092
Translocation of GLUT4 to the Plasma Membrane	11/386	63/4766	0.011	0.092
Insulin receptor signaling cascade	12/386	72/4766	0.012	0.092
IRS-mediated signaling	11/386	65/4766	0.014	0.092
Adaptive Immune System	53/386	489/4766	0.014	0.092
IRS-related events triggered by IGF1R	11/386	67/4766	0.018	0.094
IRS-related events	11/386	67/4766	0.018	0.094

From these results, we can see that 12 pathways are significantly enriched with our 898 genes (386 of which occur in the Reactome pathway models), using a pathway-level BH-corrected p value threshold of 0.1. Although these pathways do not represent our entire set of 898 significant genes, they provide an informative starting point to interpreting our results. For instance, among these results is the pathway *translocation of GLUT4 to the plasma membrane*. GLUT4 is the glucose transporter gene in humans, and is therefore mechanistically involved in moving glucose from the blood stream into peripheral tissues, such as muscle tissue. Also physiologically relevant among these results are the *insulin receptor signaling cascade* and the *IRS (insulin receptor substrate)-linked pathways*. In addition to the sensible or expected results, it is common to see potentially unanticipated significant pathways, which may provide novel insight and new hypotheses on the biological system being studied. These pathways may also serve as starting points for mechanistic modeling, although it is important to consider that the boundaries of these pathways may not represent the true boundaries of the system of interest. It may therefore be worthwhile to examine how these pathways overlap and intersect, or to look to network models as a starting point for mechanistic modeling.

compartment-, and condition-specificity of molecular interactions that are critical aspects of biological systems. Identification of these specificities (as well as specification of quality scores for interactions) has been the subject of recent work in molecular network construction, and provides a stronger foundation for high-quality interaction networks that are specific to the biological system of interest.

4.6.1 *Network topology*

A network can be described as a formal mathematical object (sometimes called a *graph*) with distinct topological properties. The topology of a network is frequently investigated in network analysis, as topology is intrinsically linked to function. The typical notation for a network G with V nodes and E edges (i.e., links) is (V, E), and each edge may carry directionality (indicating that the interaction only occurs in a single direction). In a given network, *centrality* is a measure of importance of a given node, and can be defined in multiple ways. *Degree centrality* is the simplest measure of centrality, and measures the number of links of a given node v (i.e., $D(v) = deg(v)$). For networks with directed edges, degree may be defined as both *in-degree* and *out-degree*. The *degree distribution* of a network is often used to summarize the heterogeneity and spread of degree in a given network. For instance, a network may consist of nodes all carrying a similar degree, or alternatively may possess a few highly connected nodes, and many weakly connected nodes. This second example is more common in real-world networks, both in biology and in other fields. Specifically, this type of network topology is called *scale free*, and describes a network with a degree distribution that decays as a power law. A *scale free network* can be defined as $P(k)\ k^{-\lambda}$ where $P(k)$ is the probability of a node having degree k, and λ is a parameter that defines the sharpness of the decay. Important measures of centrality are defined in the box *Network measures of centrality*.

The measures of centrality described here refer to individual nodes, but each measure can be averaged across all nodes, or alternatively viewed in terms of their distribution, to obtain an overall view of network shape. The topological feature of *modularity*, on the other hand, is a network-level topological property. Modularity measures the degree to which a given system can be partitioned into semi-independent modules/subnetworks. In biological networks, modules are analogous to semi-independent biological functions, and are therefore particularly useful from a modeling standpoint.

Example 4.12. [Network topology]
Figure 4.13 shows a simple undirected example network, illustrating the node with highest degree in black ($D = 8$), and the node with highest betweenness and closeness in dark gray ($B = 98$; $C = 0.032$). The two nodes in light gray have a clustering coefficient of one, which is the highest in the network and indicates that all of their neighbors are directly connected to each other.

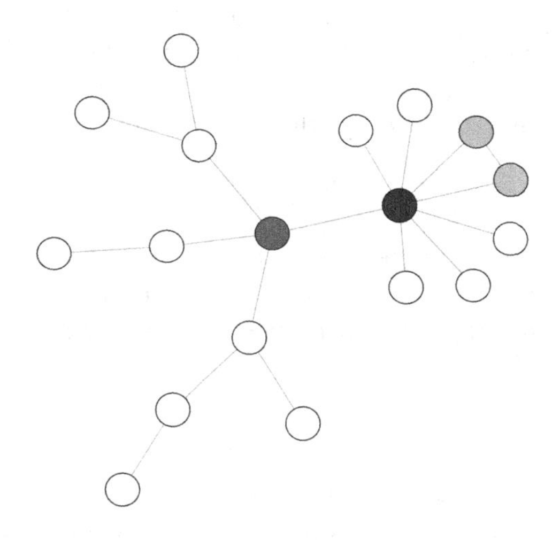

Figure 4.13 Simple undirected example network.

Although there are multiple variants on the definition of modularity, a commonly used simple definition is

$$\sum_{s=1}^{N_m} \left[\frac{l_s}{E} - \left(\frac{D_s}{2E} \right)^2 \right]$$

where N_m is the number of modules in the network, E is the number of edges in the network, l_s is the number of edges between nodes in module s and D_s is the summed degrees of all nodes in module s. A network partition with high modularity is therefore one with a high density of edges within modules, and a low density of edges between modules. Note that modularity must be calculated on a partition of a network. Therefore, a number of heuristic algorithms are available for identifying the maximally modular

Network measures of centrality

Betweenness centrality, $B(v)$, measures the centrality of nodes v in terms of the number of shortest paths in the network that pass through v, defined as

$$\sum_{s \neq t \neq v \in V} \frac{\omega_{st}(v)}{\omega_{st}}$$

where ω_{st} is the total number of shortest paths between nodes s and t, and $\omega_{st}(v)$ is the number of those that pass through node v. A node with high betweenness centrality can therefore be seen as an important mediator in a network. *Closeness centrality*, $C(v)$, measures the vicinity of a node v to all other nodes in the network, as

$$\sum_{v \neq t} \frac{1}{d(v,t)}$$

where $d(v,t)$ is the shortest distance between v and t. *Clustering coefficient*, $Cl(v)$, is a measure of the degree to which nodes in a network tend to cluster together into groups. In undirected and directed networks, $Cl(v)$ is defined respectively as

$$\frac{2\epsilon_v}{D_v(D_v - 1)}, \quad \frac{\epsilon_v}{D_v(D_v - 1)}$$

where D_v is the degree of v and ϵ_v is the number of connected pairs between all neighbors of v. In both cases, the clustering coefficient of a node defines the proportion of a node's neighbors that are themselves directly connected.

partition of a network. Figure 4.14A shows an example of a strongly modular partition of a network, while Fig. 4.14B shows a comparatively lower modularity partition.

4.6.2 *Identification of active subnetworks*

Analysis of high-throughput data in the context of pathway models is straightforward with the use of GSA methodology. Global interaction networks are more problematic, as they are exceedingly complex and therefore require dedicated algorithms to identify specific regions of the global network that are most strongly *active* in a given omics dataset. Identification

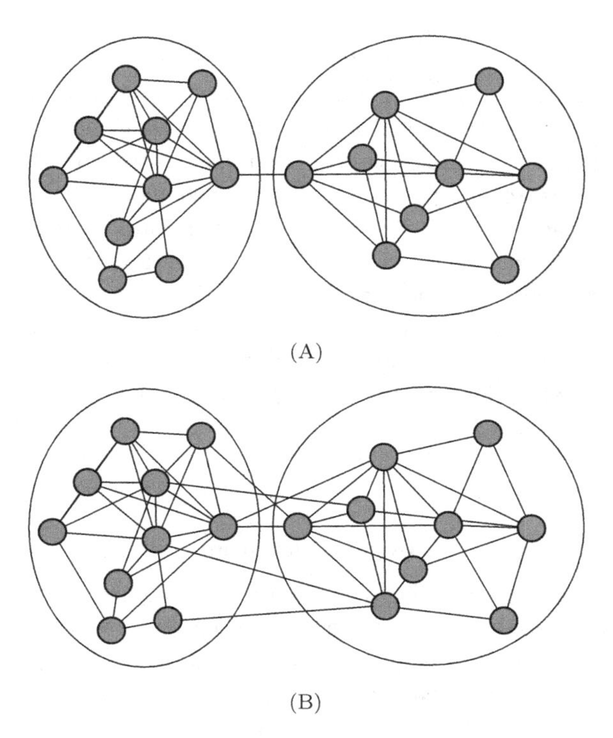

Figure 4.14 Example network, displaying (A) high modularity, and (B) lower modularity.

of activity in a network may be based on node values (e.g., raw expression values or fold-change values between two conditions), edge values (e.g., co-abundance between a pair of interacting proteins) or both.

Broadly, algorithms for identification of active subnetworks (also called modules) can be categorized into those based on: 1) *high-scoring subregions* (HCS), 2) *network clustering or partitioning* (NCP) and 3) *network information flow* (NIF), although there is considerable technical diversity within each. HCS approaches are the most common and most diverse class of method for active subnetwork identification. In HCS approaches, omics data are first analyzed to identify scores (e.g., abundance or fold-change) for each variable, which are then mapped to nodes and/or edges of a network. Identification of the maximum scoring subnetwork then proceeds using either a heuristic (such as *simulated annealing* (SA) or *genetic algorithm*) or exact approach.

Simulated annealing

Simulated annealing is an algorithm for optimization of a function in a large search space. In numerical optimization, the simulated annealing algorithm proceeds by exploring the search space initially in a *hot* system, allowing it to search more exhaustively and make more erratic parameter changes at each step. At these higher temperatures, a worse solution may be temporarily accepted in order to escape a local optimum and subsequently identify a more globally optimal solution. As the system cools, however, the algorithm searches an increasingly narrower space and reduces acceptance of worse solutions. While simulated annealing cannot necessarily identify the single globally optimal solution, it can typically find a very good solution, even in large, noisy search spaces.

Genetic algorithm

Genetic algorithms are a heuristic approach to numerical optimization that take inspiration from microevolutionary processes. The general approach to optimizing a function with a genetic algorithm involves defining a fitness function (i.e., an objective function), and randomly generating a set of candidate solutions (i.e., *individuals*). With each iteration (*generation*) of the algorithm, the fitness of each individual is evaluated and the most fit are subjected to changes (*mutations*) and passed on to the next generation. The algorithm exits when a satisfactory solution has been obtained, or the maximum number of generations has been reached.

The main difference between heuristic and exact approaches is that exact approaches ensure that the optimal solution (i.e., maximally scoring subnetwork) is identified, at considerably higher computational expense. Heuristic methods do not guarantee the optimal solution, but typically return a satisfactorily good solution with substantially reduced computing time. Whether a heuristic or exact approach is used, a starting subnetwork is selected and then nodes are added/removed and the subnetwork is re-scored until an optimal solution is identified. Significance of the subnetwork is determined by, for instance, randomly sampling subnetworks of the same size and calculating an expected enrichment score. The jActiveModules algorithm is a well-known HCS approach to identification of active subnetworks. A related approach, GXNA, uses a modified significance calculation to account for non-independence of individual node p values. Variants of HCS approaches that make use of edge scores place importance on coordination of activity (possibly condition-specific, such as differential gene–gene co-expression between two conditions) in a subnetwork rather than the activity of the nodes alone, which is in line with the fundamental meaning of an interaction network.

NCP approaches seek to cluster or partition a network in conjunction with the node scores from a high-throughput dataset. One such algorithm, called JointCluster, identifies densely connected clusters both in a molecular interaction network and a co-expression network (i.e., a network with edges between strongly co-expressed genes, as calculated from a high-throughput dataset). Then, the maximally concordant clusters between the two networks are returned, thus identifying those that are densely interconnected and also highly co-expressed. Another NCP approach, explicitly involving the topological feature of modularity, involves first partitioning a global interaction network into its optimal modular structure (e.g., using simulated annealing). Then, each resulting module can be used directly as a gene set in GSA. During the partitioning step, it is also possible to add weights to the network edges (such as confidence scores or co-expression values) which will encourage highly weighted edges to be contained within rather than between modules. Naturally, the results of this method depend strongly on the modular structure of the starting network. For instance, the optimal modular structure of the network may contain one or more very large modules, which may be difficult to interpret in terms of biological function. In part for this reason, it is important to closely examine the edge content of the network being studied, as a network with a very high density of interactions will produce an optimal partition with low modularity. Therefore, if the starting network contains many low-confidence interactions, or interactions that are irrelevant for the specific system being studied (e.g., tissue-specific interactions) then strong modules may not easily be recovered.

NIF approaches to active subnetwork identification bear resemblance to dynamic simulation in that they explicitly consider information flow through a network, albeit at lower level of detail than dynamic simulation. Starting from a set of seed genes (such as a set of differentially expressed genes) NIF methods follow the path of flow through the network, identifying those subnetworks that experience the highest flow. NIF approaches

Example 4.13. [Active subnetwork identification]
We use our set of 898 skeletal muscle genes significantly correlated with fasting glucose levels (as we used in our example of GSA). To identify active subnetworks with a modularity-based TS approach, we start with a network of metabolic and protein–protein interactions from the Reactome database (www.reactome.org). This network includes 7024 genes/proteins (5069 of which are present in our complete microarray dataset) and 49,195 interactions among them. We first partition this network into topological modules using SA. This results in 25 distinct modules ranging in size from 58 to 927 genes (note that these numbers would change slightly if the SA algorithm were run again, since it is a heuristic approach). Using each module as a gene set, we can implement an OR GSA approach to identify those modules that are significantly enriched in our 898 *hits* (369 of which are present in the overall network). This analysis indicates that three of the 25 modules are significantly enriched in blood glucose correlated genes.

Module ID	Hits	Module size	*P* value	BH
1	41/369	302/5069	<1e-16	6e-04
2	19/369	124/5069	5e-04	0.0068
3	8/369	46/5069	0.0051	0.0428

Figure 4.15 illustrates the most significant subnetwork from our analysis. By applying a standard OR GSA approach to all genes in this subnetwork, we can obtain a clear understanding of the biological processes represented therein. Figure 4.16 illustrates the Reactome pathway models that are significantly enriched in this subnetwork, highlighting both the number of pathway genes occurring in the subnetwork, and the percent coverage of each pathway (e.g., a pathway with 50% coverage has 50% of its genes in the subnetwork). In general, these results appear to mirror the results obtained from direct GSA of our gene list (Sect. 4.5) in that the module pathways include a number of IRS-related pathways, as we saw in our GSA example. However, the network analysis results provide greater detail, as the resultant subnetworks contain explicit information on the overlap and intersection of related pathway models. This feature of pathway overlap is illustrated in Fig. 4.17, which shows a hierarchical clustering of each pathway model in our most significant subnetwork, based on similarity in gene content (considering all genes in the pathway models, rather than being restricted to those found in our subnetwork). It is clear from this diagram that some of these pathway models show very strong or even complete overlap. This is an important feature to consider when modeling from a systems perspective and supports the choice, or at least examination, of subnetwork models in analysis of biological systems.

may propagate the flow forward through the network, to identify possible response subnetworks, or may trace the flow backward, to identify possible regulator subnetworks. Naturally, for NIF methods it is important to use a network with directionality information for the edges. The HotSpot

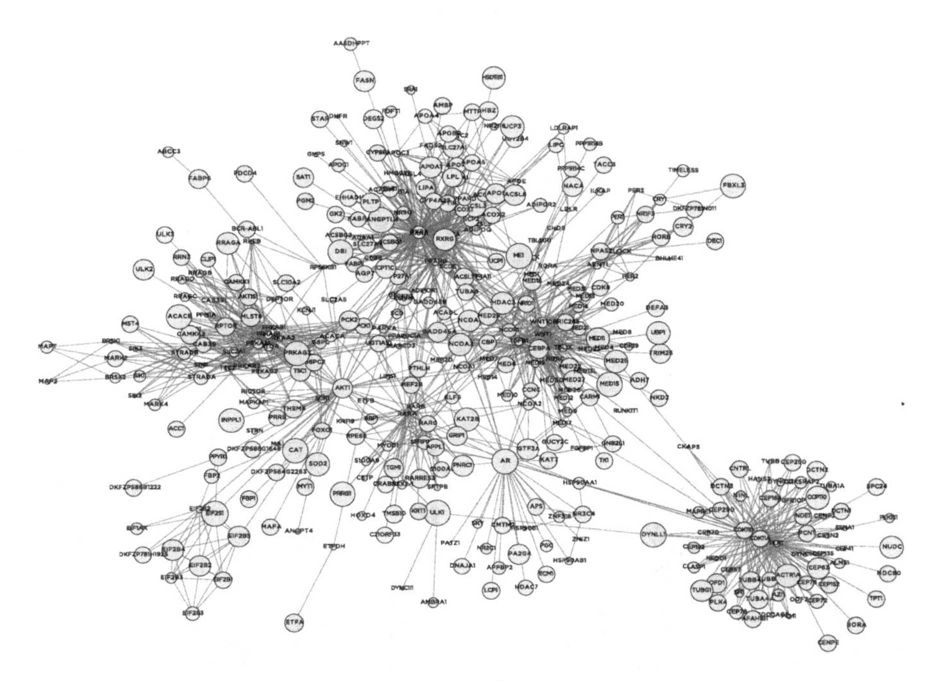

Figure 4.15 Most significantly enriched module from active subnetwork analysis of blood glucose correlated genes. Node size corresponds with significance level, with larger nodes being more significantly correlated with blood glucose.

algorithm is an NIF approach that uses the concept of heat diffusion when propagating a signal through a network. For instance, when a signal is propagated along a single chain of nodes with no side branches, very little heat in the signal is lost. However, if the signal passes through a very highly connected node, the signal diffuses across all subsequent paths, thus losing strength within each individual path. The TieDie algorithm uses a similar heat diffusion concept, but takes as input both a source and target set of genes and aims to forward-propagate and back-propagate the signal between them, in order to increase confidence in the result. In general, NIF approaches benefit from careful consideration of the "input" nodes in the analysis, as the signal from a very large set of input nodes can potentially spread across an entire network. However with precise input (possibly restricting input nodes to those within a limited number of degrees of separation from each other) NIF methods can produce highly informative subnetworks that directly represent the flow of information through the network.

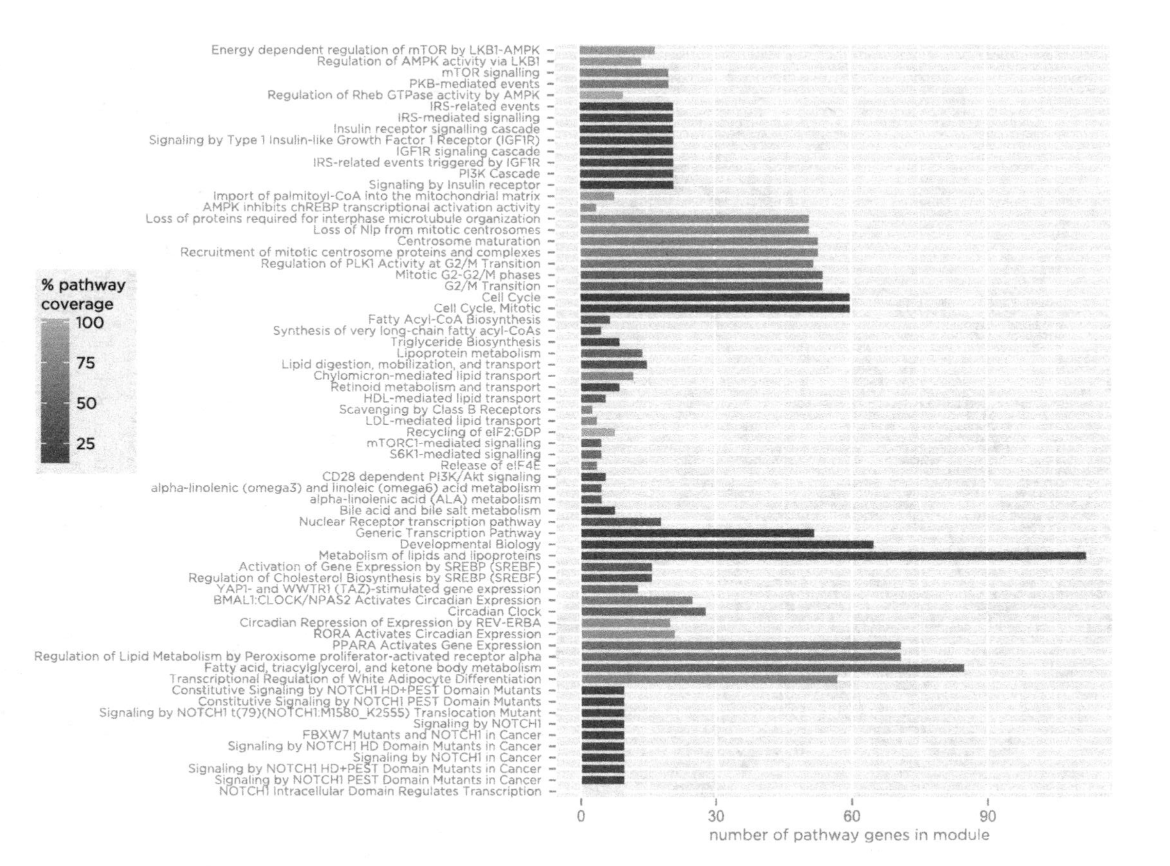

Figure 4.16 Significantly enriched Reactome pathway models within module from Fig. 4.15. The number of pathway genes found in the model are described on the x-axis, and the bar shading illustrates the % coverage of each pathway within the module.

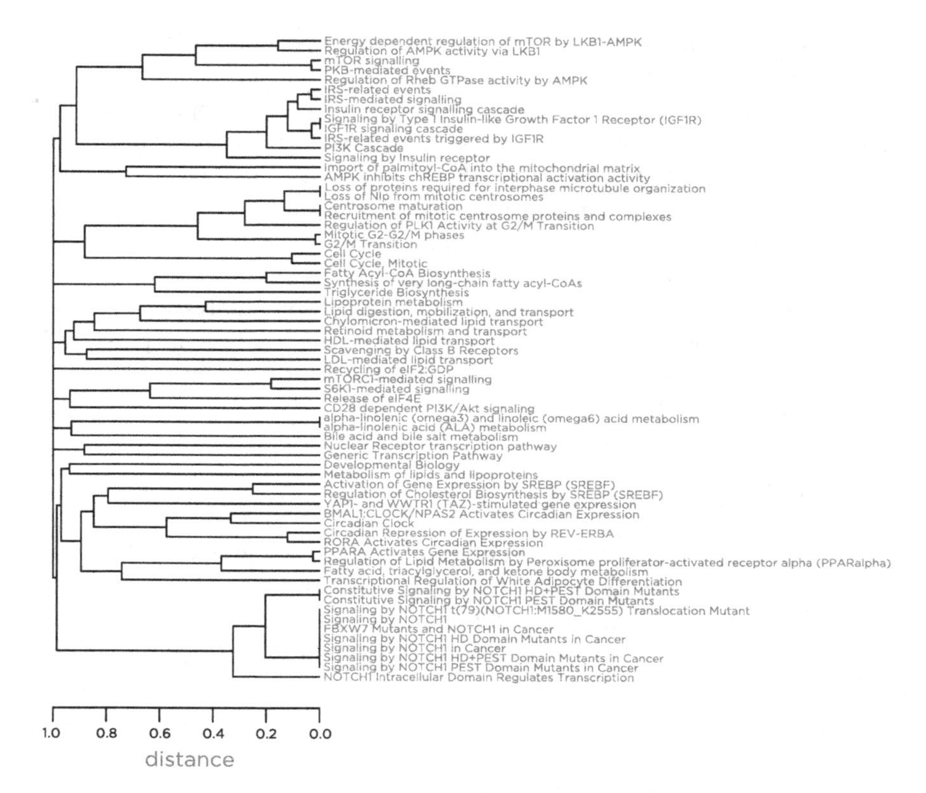

Figure 4.17 Hierarchical clustering of pathways occurring in the module from Figure 4.15, based on overlapping gene content between pathway models.

Regardless of the method used for active subnetwork identification, it is important to further examine any identified active subnetworks in order to determine their biological function. To do this, GSA methods can be used to comprehensively profile each module.

4.7 Summary

In this chapter we have introduced a range of statistical and bioinformatic approaches that can guide the analyst from omics data to biological insight and new hypotheses. Naturally, we have not discussed every method available, but we have aimed to present a range of methods to highlight the diversity of approaches that may be taken to high-throughput data analysis, and to provide the reader with a broad foundation for understanding more specialized approaches. Ultimately, it is important for the analyst to appreciate the magnitude of information in a high-throughput dataset, as well as the complexity of biological systems. As such, approaching the analysis from multiple angles may reveal important insight on the biological system being studied. For instance, while a regression analysis may identify variables that are significantly correlated with a phenotype of interest, clustering may reveal that the samples group together in unexpected ways, and principal component analysis may highlight groups of correlated variables that drive this clustering pattern. Pathway and network analysis are critical for biological interpretation of results, and network analysis in particular yields discrete subnetworks that may serve as the starting point for dynamic modeling efforts.

4.8 Further reading

We have provided here a brief introduction to static analysis of omics data; however, the reader may benefit from exploring statistical analysis more deeply in [Hastie *et al.* (2009)], with a more practical yet detailed introduction in [Faraway (2004)]. In particular, [Hastie *et al.* (2009)] provide an excellent discussion of a variety of methods (such as logistic regression, analysis of variance, regression trees, neural networks and support vector machines) that, in the interest of space, could not be discussed here. A detailed discussion of multivariate statistics, including vector and matrix operations, can be found in [Carroll and Green (1997)]. Comprehensive reviews of GSA approaches are found in [Hung *et al.* (2012); Khatri *et al.* (2012)], and for a deeper discussion of network analysis and algorithms for active subnetwork identification, the reader can consult [Raval and Ray (2013); Mitra *et al.* (2013); Liu *et al.* (2012)].

Chapter 5

Dynamic modeling technologies

We survey the main technologies used for modeling the dynamics of biological systems. We start with the more frequently used equation-based approaches that rely on classical mathematics. Then we move to network-based modeling that has a mixed background in mathematics and engineering. Rewriting systems and automata-based modeling are derived from formal language theory. Finally, we introduce graphic notations. Algorithmic modeling is dealt with in the next chapter. Hereafter, we will focus on molecular interaction modeling, if not otherwise specified. We will assume that the outcome of a simulation is the dynamic behavior of a model, i.e., to approximate reproduction of time behavior of a real system (see Chapter 8).

5.1 Equation-based approaches

Mathematical modeling of dynamic systems dates back to 1666–1672 with the works of Liebnitz and Newton, and has been largely used in physics and engineering to represent systems that evolve over time (e.g., Newton's laws of motion). The behavior of a system is represented by a set of equations whose variables represent the values of the attributes of interest. The main technology in this class is ordinary differential equations and their variants of partial differential equations (if space has to be incorporated in the description of the system) and stochastic differential equations (if the behavior of the system is affected by noise or randomness).

5.1.1 *Differential equations*

A *differential equation* (DE) relates an unknown function in one or more variables to its derivatives (the highest degree of the derivatives determines the order of the differential equation). If noise terms are not included in the equation, the solution of the equation is a *deterministic* relation over some continuously varying quantities (function) and their changing rate in space/time (derivatives).

Ordinary differential equation

Let $f : \mathbb{R} \to \mathbb{R}$ be an unknown function, $t \in \mathbb{R}$, and $F : \mathbb{R}^{n+1} \to \mathbb{R}$ be a given function, $n \in \mathbb{N}$. If $f^{(n)}$ stands for the nth derivative of f,

$$F(t, f(t), f^{(1)}(t), \ldots, f^{(n-1)}(t)) = f^{(n)}(t)$$

is an *explicit* ordinary differential equation (ODE) of *order* n. An *implicit* ordinary differential equation is

$$F(t, f(t), f^{(1)}(t), \ldots, f^{(n-1)}(t), f^{(n)}(t)) = 0$$

with $F : \mathbb{R}^{n+2} \to \mathbb{R}$. An ODE not depending on t is *autonomous*. If F can be expressed as a linear combination of the derivatives $f^{(i)}$ possibly depending on t, i.e.,

$$f^{(n)} = \sum_{i=0}^{n-1} a_i(t) f^{(i)} + r(t)$$

with $a(t)$ and $r(t)$ (the source term) continuous functions, then the ODE is *linear*. When the source term is null, the equation is *homogeneous*, otherwise it is *non-homogeneous*.
The first derivative is sometimes written df/dt and hence a first-order ODE is $df/dt = F(t, f)$ or $df/dt = F(t, f; p_1, \ldots, p_k)$ to highlight the k constant parameters of the equation.
A function $g : I \to \mathbb{R}$, $\emptyset \neq I \subset \mathbb{R}$, is a *solution* of the ODE if g is differentiable n times on I and $F(t, g(t), g^{(1)}(t), \ldots, g^{(n)}(t)) = 0$, with $t \in I$. A solution $h : J \to \mathbb{R}$, $I \subset J$ of the same ODE is an *extension* of g if $h(t) = g(t), \forall t \in I$. A solution with no extension is a *maximal* solution. A solution defined on $\mathbb{R}$ is a *global* solution.
A *general* solution contains n arbitrary variables corresponding to the constants of integration. A *particular* solution is obtained by assigning values to the arbitrary variables to fulfill the *boundary conditions*. A solution that cannot be derived by the general solution is *singular*.
If $f : \mathbb{R} \to \mathbb{R}^m$, we have a *system* of ODEs of *dimension* m and $F : \mathbb{R}^{mn+1} \to \mathbb{R}^m$.

The simplest form of differential equations are the *ordinary* ones (see the box on the previous page). They have been extensively used to model biological systems, mainly in their non-linear form.

The applicability of this modeling technology is determined by the validity of two assumptions:

(i) the system is a *well-stirred* collection of interacting molecules within a fixed reaction volume, and
(ii) the abundance of molecules can be expressed as a continuous function of time.

The well-stirred assumption considers the distribution of molecules to be homogeneous in space and the diffusion time to be significantly smaller than the reaction time. This is necessary to make the probability of collision (precondition for interaction) of any two molecules in the system uniform. Cells are crowded environments filled by molecules in discrete numbers, hence both assumptions do not hold in general. We need to determine conditions under which the approximation obtained through ODEs (due to invalidity of assumptions) is still acceptable. ODE assumptions are reasonable only if the observation time of the phenomenon considered is some order of magnitude larger than the overall traveling time of the molecules in the system volume.

Applicability of ODEs

Cell diameter is roughly 10^{-5} m and volume 5×10^{-13} l. Assuming that the diffusion of a protein in the cytoplasm is from 10 to 20 times slower than in aqueous solutions (where the diffusion constant is $D = 10^{-7}$ cm^2/s), the average time to traverse the cell varies from 1 min to 2 min. Small molecules like metabolites move faster. Therefore the well-stirred assumption is applicable on processes of time scales of hours or greater.
Assuming that the average concentration of a protein in a cell is 10 nM, we have $10^{-8} \times A \times \text{vcell} = 3000$ molecules per cell, where $A = 6.022 \times 10^{23}$ is the Avogadro number and here $\text{vcell} = 5 \times 10^{-13}$ l. We suggest 3000 molecules to be the minimal amount needed to safely approximate discrete numbers of molecules with continuous functions.

The assumption (ii) is imposed by the mathematical formulation of the ODEs and requires continuous functions for ensuring the existence of the derivatives.

Discrete numbers of molecules can be approximated by continuous functions when the cardinality of the species involved in the system is huge. A consequence of assumption (ii) is that the dynamic behavior described by ODEs provides only deterministic descriptions (the solution of the equations is always the same) and average values of the concentrations of species. Fluctuations in the number of molecules must therefore be irrelevant to the overall behavior of the system for ODEs to be a reasonable technology. Hints on the applicability of ODEs in molecular modeling are reported in the box on the previous page.

To model the dynamics of molecular systems with ODEs, we assume that each component A is associated with a continuous function $[A] : \mathbb{R} \to \mathbb{R}$ that maps time points — the domain of $[A]$ — to concentrations — the codomain of $[A]$.

The intuition behind an ODE of the form $d[A]/dt = h_1[B] - h_2[C]$ is that the concentration of A increases with the concentration of B according to a factor h_1 and decreases with the concentration of C according to a factor h_2. In other words, B can produce A with a rate h_1 and C can consume A with a rate h_2. The rate functions h_1 and h_2 are real valued and in the simplest cases can even be constant functions (e.g., for a constant increase factor we can set $h_1(t) = 1$ and for an exponential increase factor we can set $h_1(t) = 1 - e^{-kt}$). Example 5.1 is a simple application of ODE modeling.

Example 5.1. [ODE model of the Lotka–Volterra two-species interaction] Consider an ecosystem with two species: rabbits (R) and foxes (F). Rabbits are prey and foxes are predators. The assumptions in the Lotka–Volterra model are that the environment is not changing during the process (there is continuously available food for the prey), the growth rate of a species depends on the number of its individuals, predators eat prey only. The equations describing the variation of the number of rabbits and foxes due to their interactions over time are

$$d[R]/dt = h_1[R] - (h_2[F])[R]$$
$$d[F]/dt = -h_3[F] + (h_4[R])[F]$$

where h_1 is the growth rate of prey, $h_2[F]$ is the rate at which predators eat prey, h_3 is the death rate of predators and $h_4[R]$ is the growth rate of predators due to eating prey.

ODEs — Elementary templates

Synthesis of a molecule A that has a constant synthetizing rate is modeled as $d[A]/dt = k$. *Autocatalytic synthesis* is when A is needed to produce itself and it is modeled as $d[A]/dt = h_s[A]$ with h_s being the synthesis rate function.

Degradation of a molecule A usually depends on how many As are in the system and it is modeled as $d[A]/dt = -h_d[A]$.

Binding of two different molecules A and B to form a complex (dimerization) AB is modeled as $d[AB]/dt = h_b[A][B]$. Components A and B will decrease their amount when binding and it is modeled as $d[A]/dt = -h_b[A][B]$ and $d[B]/dt = -h_b[A][B]$.

Unbinding of two molecules AB into the components A and B is modeled by the two equations $d[A] = h_u[AB]$ and $d[B] = h_u[AB]$. The complex AB decreases its amount when unbinding into the original components and it is modeled as $d[AB]/dt = -h_u[AB]$. Reversibility of binding and unbinding is modeled by coupling the equations

$$d[A]/dt = h_u[AB] - h_b[A][B]$$
$$d[B]/dt = h_u[AB] - h_b[A][B]$$
$$d[AB]/dt = h_b[A][B] - h_u[AB].$$

Modification of an attribute of a component (e.g., phosphorylation of a site of a molecule A) is interpreted as the creation of a new species modeled by introducing a new variable A^m. Reversibility changes A^m into A again yielding $d[A]/dt = -h[A] + h_r[A^m]$ coupled with $d[A^m]/dt = -h_r[A^m] + h[A]$.

The main reactions that we may want to model with ODEs are synthesis and degradation, binding and unbinding and modifications (e.g., phosphorylation or dephosphorylation). The box above reports some tips on how to model these reactions with ODEs.

Once we know how to model the elementary reactions, we need to connect the dots for modeling complex networks of interactions. We need to define at least one equation for each component in the system. The resulting equations form a system of ODEs (see Examples 5.2 and 5.3).

Example 5.2. [ODE — Enzymatic reaction]
An enzyme-mediated catalysis reaction is represented by

$$E + S \underset{k_2}{\overset{k_1}{\rightleftharpoons}} ES \xrightarrow{k_3} EP \xrightarrow{k_4} P + E.$$

The enzyme E binds its substrate S to produce a temporary complex ES that undergoes some structural modification yielding a variant EP. The enzyme and the substrate can bind and unbind with rates k_1 and k_2, respectively. After the production of EP with rates k_3, the reverse reaction almost never happens because it is largely energetically unfavorable. The same is true for the production of the product P and release of the enzyme with rate k_4. The set of ODEs modeling the enzymatic reaction is

$$\begin{aligned}
d[S]/dt &= k_2[ES] - k_1[S][E] \\
d[E]/dt &= k_2[ES] + k_4[EP] - k_1[S][E] \\
d[ES]/dt &= k_1[S][E] - (k_2 + k_3)[ES] \\
d[EP]/dt &= k_3[ES] - k_4[EP] \\
d[P]/dt &= k_4[EP].
\end{aligned}$$

There is an equation for each reactant and product. The elements in the right side of the equations are positive for the arrows producing the component on the left side and negative for the arrows consuming the component. The constants are determined by the rates of the chemical reactions. For instance, in the first equation we have $k_2[ES] - k_1[S][E]$ because ES produces S with rate k_2, and S and E can be consumed with rate k_1 to produce ES.

Example 5.3. [ODE — Gene regulatory network]
Consider the cross-inhibition network depicted in Fig. 5.1. Gene a and gene b inhibit each other by the binding of the proteins B and A to part of their promoter regions, expressed by the other gene. The set of ODEs modeling this gene regulatory network is

$$\begin{aligned}
d[A]/dt &= k_{sA}h([B]) - k_{dA}[A] \\
d[B]/dt &= k_{sB}h([A]) - k_{dB}[B]
\end{aligned}$$

where k_{sX} and k_{dX} are the synthesis and degradation rates of protein X expressed by gene x. The dynamics are mediated by the sigmoidal function $h([X])$ (see Hill kinetics and Fig. 5.4) that makes the system *bistable*, i.e., there are two stable equilibrium points (either there is a high concentration of protein A and a low concentration of protein B or the opposite case).

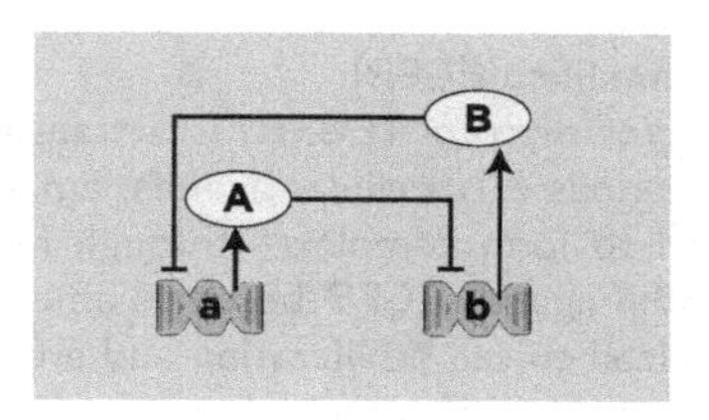

Figure 5.1 A simple gene regulatory network.

The way in which we model modifications with differential equations is a major limitation of this technology. In fact, we need as many variables for each component as the number of its states (configurations), i.e., as the number of all combinations of the values of the attributes of the component. The dynamic of a biological system is modeled by as many equations as the number of variables we need for the configurations of the components.

Combinatorial explosion of differential equations

A component with n attributes, each of which can assume k different values, has k^n different configurations. Therefore we need k^n variables (and equations). If $n = 9$ and $k = 2$, we need $2^9 = 512$ variables (and equations). A complex of m of this components may generate $(k^n)^m = k^{n \times m}$ configurations. Therefore we need $k^{n \times m}$ variables (and equations). If $m = 2$, $n = 9$ and $k = 2$, we need $2^{18} = 262144$ variables.

We can easily generate a number of equations that makes the model difficult to read and maintain, even considering simple biological systems (see Examples 5.4 and 5.5). This phenomenon is known as *combinatorial explosion* of the size of the model with respect to the number of components in the modeled system. The main cause of combinatorial explosion is that mathematical functions and variables are state-less representations of components that do not maintain any identity of biological elements. Therefore, each time we modify a component slightly, we need a new function/variable to represent the modification. Combinatorial explosion is exacerbated by binding/unbinding interactions. In fact, the number of configurations of complexes formed by binding reactions is the product of the configurations of the constituent components. Combinatorial explosion requires us to introduce a further assumption for the applicability of ODEs, which is purely practical and is not due to the theoretical limitations of ODEs. It is reported in the box on page 140 under item (iii).

Example 5.4. [ODE — simplified EGFR]
The epidermal growth factor receptor (EGFR) is a transmembrane protein, able to sense the presence of various extracellular ligands, among which the epidermal growth factor (EGF), and to form complexes through reversible homodimerization. Upon binding with the ligand, EGFR becomes able to activate intracellular pathways, some of which lead to cell proliferation and are thus relevant for tumor growth. When an EGFR receptor is in a homodimer, it can become active through a mechanism called auto-trans-phosphorylation (see Fig. 5.2).

EGFR has nine phosphorylation sites and hence $2^9 = 512$ different configurations that becomes $2^{18} = 262{,}144$ in the case of homodimers. We consider the extremely simplified case in which each arm has only two tyrosine residues.

Let A_{0000} denote the homodimers of EGFR ready to activate auto-trans-phosphorylation. The first two indexes represent the left arm status and the second two indexes represent the right arm status (0 is free and 1 is phosphorylated). Note that this is just a naming convention used to understand from the name of the state-less mathematical variables the biological entities represented. The 16 equations are:

$$\frac{d[A_{0000}]}{dt} = -h_p[A_{0000}] + h_d([A_{1000}] + [A_{0100}] + [A_{0010}] + [A_{0001}])$$
$$\frac{d[A_{1000}]}{dt} = -(h_p + h_d)[A_{1000}] + \frac{h_p}{4}[A_{0000}] + \frac{h_d}{2}([A_{1100}] + [A_{1010}] + [A_{1001}])$$
$$\vdots$$
$$\frac{d[A_{1100}]}{dt} = -(h_p + h_d)[A_{1100}] + \frac{h_p}{3}([A_{1000}] + [A_{0100}]) + \frac{h_d}{3}([A_{1110}] + [A_{1101}])$$
$$\vdots$$
$$\frac{d[A_{1110}]}{dt} = -(h_p + h_d)[A_{1110}] + \frac{h_p}{2}([A_{1100}] + [A_{0110}] + [A_{1010}]) + \frac{h_d}{4}[A_{1111}]$$
$$\vdots$$
$$\frac{d[A_{1111}]}{dt} = -h_d[A_{1111}] + h_p([A_{0111}] + [A_{1011}] + [A_{1101}] + [A_{1110}])$$

where we (wrongly) assume that the rate of phosphorylation h_p is the same for all the sites independently of the current configuration (similarly for the dephosphorylation h_d).

The exact phosphorylation dynamics could be modeled by indexed function rates that depend on the position of phosphorylation or dephosphorylation (e.g., h_p^{0100} for the phosphorylation of the second residue on the left arm). In this case, each term of the equations would have its specific rate and no fractional rate would be needed. Note that dots stand for all the equations obtained by permutations of positions of 1 and 0 in the case of 1, 2 and 3 phosphorylated sites.

Example 5.5. [ODE — further simplified EGFR]
A further simplification with respect to Example 5.4 is just to count the number of phosphorylated sites without considering the exact position of the residues. This simplification reduces the configurations from 2^{18} to 20, but it does not account for the fact that different ligands induce phosphorylation at different sites and that different configurations determine different fates. Let A_{00} denote the homodimers of EGFR ready to activate auto-trans-phosphorylation, with the indexes counting the number of phosphorylated sites, but it is blind to the exact location of the phosphorylation (e.g., a homodimer with six total phosphorylated residues will be further phosphorylated on the left or the right arm with the same rate and probability). We assume that the rate functions depend on the number of tyrosine residues already phosphorylated (e.g., h_d^{36} is the dephosphorylation rate of an homodimer with three residues phosphorylated on the left arm and six on the right arm). The 20 equations are:

$$
\begin{aligned}
\frac{d[A_{00}]}{dt} &= -h_p^{00}[A_{00}] + h_d^{01}[A_{01}] + h_d^{10}[A_{10}] \\
\frac{d[A_{10}]}{dt} &= -(h_p^{10} + h_d^{10})[A_{10}] + \frac{h_p^{00}}{2}[A_{00}] + \frac{h_d^{11}}{2}[A_{11}] + h_d^{20}[A_{20}] \\
\frac{d[A_{01}]}{dt} &= -(h_p^{01} + h_d^{01})[A_{01}] + \frac{h_p^{00}}{2}[A_{00}] + \frac{h_d^{11}}{2}[A_{11}] + h_d^{02}[A_{02}] \\
\frac{d[A_{ij}]}{dt} &= -(h_p^{ij} + h_d^{ij})[A_{ij}] + \frac{h_p^{(i-1)j}}{2}[A_{(i-1)j}] + \frac{h_p^{i(j-1)}}{2}[A_{i(j-1)}] \\
&\quad + \frac{h_d^{(i+1)j}}{2}[A_{(i+1)j}] + \frac{h_d^{i(j+1)}}{2}[A_{i(j+1)}], 1 \leq i, j \leq 8 \\
\frac{d[A_{89}]}{dt} &= -(h_p^{89} + h_d^{89})[A_{89}] + \frac{h_d^{99}}{2}[A_{99}] + h_p^{79}[A_{79}] \\
\frac{d[A_{98}]}{dt} &= -(h_p^{98} + h_d^{98})[A_{98}] + \frac{h_d^{99}}{2}[A_{99}] + h_p^{97}[A_{97}] \\
\frac{d[A_{99}]}{dt} &= -h_d^{99}[A_{99}] + h_p^{89}[A_{89}] + h_p^{98}[A_{98}].
\end{aligned}
$$

For the sake of completeness, we list below the equations describing the EGF:EGFR binding and the homodimer formation:

$$
\begin{aligned}
\frac{d[EGF:EGFR]}{dt} &= -h_u[EGF:EGFR] + h_b[EGF][EGFR] \\
\frac{d[A_{00}]}{dt} &= -(h_{ud} + h_p^{00})[A_{00}] + h_{bd}[EGFR]^2 + h_d^{01}[A_{01}] + h_d^{10}[A_{10}].
\end{aligned}
$$

Note that we added to $d[A_{00}]/dt$ the negative term with rate h_{ud} for the unbinding of the homodimer and the positive term representing the homodimer formation. The same would have been done in Example 5.4 to $d[A_{0000}]/dt$.

ODE assumptions

(i) the system is a well-stirred collection of interacting molecules within a fixed reaction volume,
(ii) molecule abundances can be expresses as continuous functions of time, and
(iii) molecules have a very limited number of attributes and the size of the dynamically formed complexes is limited.

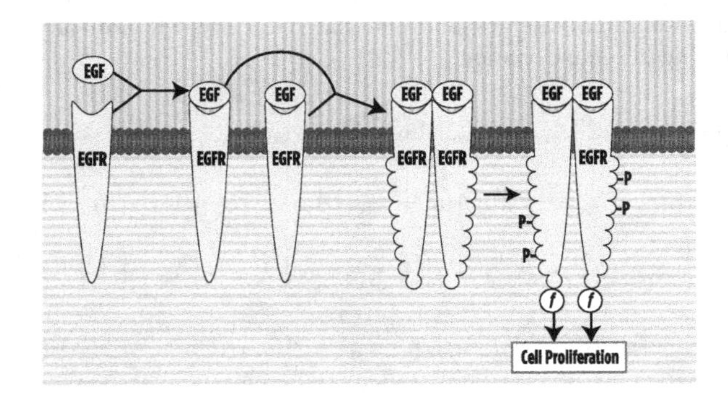

Figure 5.2 EGF-mediated EGFR phosphorylation. The bubbles on the EGFR transmembrane receptor represent tyrosine residues that can be phosphorylated. Homodimer formation is enabled by EGF binding EGFR. Homodimerization then enables auto-transphosphorylation. The circle with the f inside is a condition enabling the cell proliferation and in this case the condition is that at least two tyrosine residues are phosphorylated. Examples 5.4 and 5.5 simplify reality by assuming that both EGFR arms of a homodimer must be bound to EGF.

A main issue in applying ODE technology is the identification of the rate functions that account for the dynamics of the reactions by fitting the experimental observations well. We conclude the survey of ODEs with a brief description of the main rate functions used in describing the kinetics of the system.

Many biochemical reactions are mediated by catalytic molecules (*enzymes*) that allow the reaction to happen or increase the rate of the reaction (see also Example 5.2). The *substrate* (the molecule that is changed by the enzyme) is selectively converted by an enzyme into a different species called *product*. The enzymes are usually very specific and the transformation of the substrate into the product is usually assumed to be irreversible. The

Michaelis–Menten kinetics

The 1-substrate-1-product reaction with one active site mediated by an enzyme

$$E + S \underset{k_2}{\overset{k_1}{\rightleftharpoons}} ES \xrightarrow{k_3} P + E$$

assumes that the formation of ES is relatively fast, that the equilibrium is reached rapidly and that the production rate of P is the rate-limiting step of the overall system. Formally, $k_2 \gg k_3$ so that the variation over time of $[ES]$ is almost 0. Under the assumption $[S] \gg [E_{tot}] = [E] + [ES]$, the Michaelis–Menten equation is a single reaction abstraction of the catalytic reaction

$$E{+}S \xrightarrow{v} P{+}E,\ V = V_{max}\frac{[S]}{[S]+K_m},\ V_{max} = k_3[E_{tot}],\ K_m = \frac{k_3+k_2}{k_1}.$$

Note that K_m is $[S]$ that gives half the maximal rate (see Fig. 5.3). Assume that I the enzyme inhibitor, $K_I = \frac{[E][I]}{[EI]}$ and $\alpha = (1 + I/K_I)$. Michaelis–Menten kinetics with *competitive inhibition* is (adding the reactions below to the system)

$$E + I \rightleftharpoons EI, \quad V = V_{max}\frac{[S]}{[S]+\alpha K_m}.$$

Michaelis–Menten kinetics with *non-competitive inhibition* is (adding the reactions below to the system)

$$E + I \rightleftharpoons EI, \quad EI + S \underset{E}{\rightleftharpoons} IS, \quad EIS \rightleftharpoons ES$$

$$V = \frac{V_{max}}{\alpha}\frac{[S]}{[S]+K_m}.$$

Michaelis–Menten abstraction of the enzymatic reaction (see the box above and Fig. 5.3) has the advantage that its parameters (V_{max} and K_m) are easier to measure than the rarely available rate constants (k_1, k_2 and k_3). The Michaelis–Menten constant K_m is the concentration of the substrate S at half the maximal rate V_{max} of the reaction (i.e., it is the substrate concentration at which half the enzyme active sites are filled by S molecules). The maximal rate V_{max} is the rate at which the total concentration of the enzyme is in the ES complex.

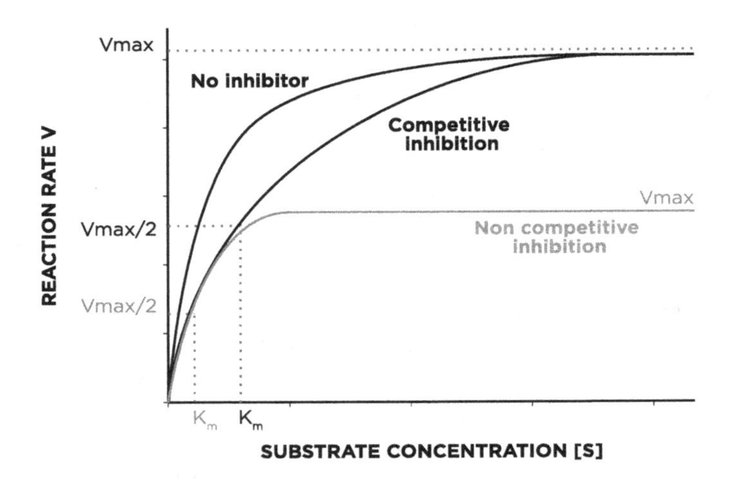

Figure 5.3 Michaelis–Menten function with and without inhibitors.

The concentration of the substrate needs to be higher (and therefore K_m is larger) to reach V_{max} with competitive inhibition (substrate and inhibitor compete for the same active site of the enzyme). Instead, with non-competitive inhibition (the inhibitor binds the enzyme on a different site with respect to the substrate, so that the activities of the inhibitor and of the substrate are independent and do not affect each other). K_m is unchanged because the corresponding sites are still active and free. Consequently, V_{max} is reduced for all the concentrations of the substrate $[S]$ because of the additional reactions the enzyme can undergo making it temporarily unavailable to bind the substrate. The effects of inhibition on Michaelis–Menten kinetics are depicted in Fig. 5.3.

Another important phenomenon occurring in biological systems is *cooperativity*. This phenomenon is exhibited by enzymes and receptors that have multiple binding sites and that change their affinity upon binding of a ligand (e.g., increased affinity of hemoglobin for oxygen on the remaining three binding sites after the first oxygen is bound; transcription factors composed of repeated subunits reach full activation after multiple binding to the target). The *Hill equation* (see the box on the next page) is a good approximation of cooperativity by using a sigmoidal response curve to replace a sequence of reactions with a single reaction. The intermediate complexes of the multistep reaction are hidden in the Hill function as the enzyme–substrate complex is hidden in the Michaelis–Menten equation.

Hill kinetics

The cooperative binding of n A molecules to a scaffold molecule B is often represented by

$$nA + B \underset{k_u}{\overset{k_b}{\rightleftharpoons}} A_nB.$$

At equilibrium, the ratio between bound scaffold molecules and total scaffold molecules is given by the Hill function (Fig. 5.4)

$$Y = \frac{[A_nB]}{[B] + [A_nB]} = \frac{[A]^n/K_D}{1 + ([A]^n/K_D)} = \frac{[A]^n}{K_D + [A]^n} = \frac{K_A[A]^n}{1 + K_A[A]^n}$$

where

$K_D = k_u/k_b = ([A]^n[B])/[A_nB]$ and $K_A = 1/K_D = [A_nB]/([A]^n[B])$.

An alternative formulation is

$$Y = \frac{[A]^n/K_D}{1 + ([A]/K_{0.5})^n}$$

where $K_{0.5} =$ is $[A]$ at which half of the scaffold molecules are bound.

The Hill equation assumes that n ligands are simultaneously binding a scaffold protein (e.g., a receptor), which is not physically possible for $n > 1$. The value of the Hill constant n describes the cooperativity of ligand binding as follows:

$n > 1$: *positive cooperativity* — binding of one ligand increases affinity for further ligands;
$n = 1$: *no cooperativity*;
$n < 1$: *negative cooperativity* — binding of one ligand decreases affinity for further ligands.

The steepness of the sigmoidal curve depends on the value n and has the inflection point at $K_D = 1/K_A$.

Additional details on biochemical kinetics are in Sect 5.2.1.

We already noted that many biological processes occurring in the cell expose an inhomogeneous distribution of molecules, violating assumption (i) of ODEs. As a consequence, the time for a protein to traverse the cell is long compared to the reaction time. A solution is to keep track of the spatial position of the molecules within the model. A possible technology

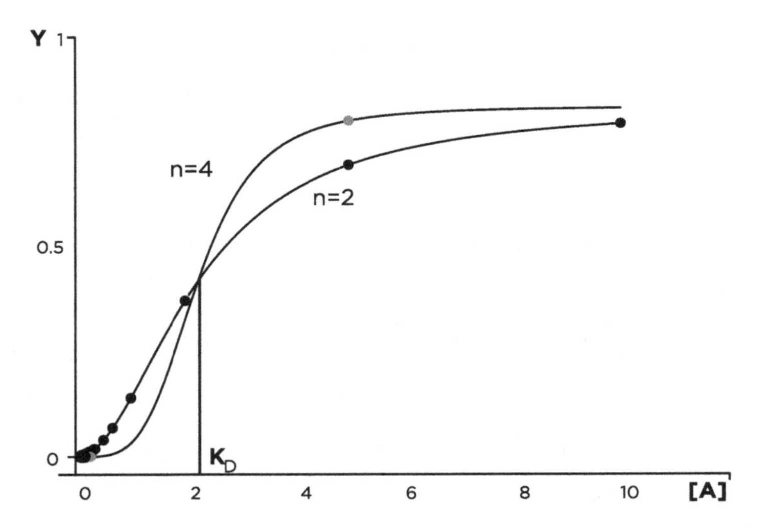

Figure 5.4 The Hill function Y is the ratio between the scaffold molecules B (e.g., a receptor) bound to molecules A (e.g., a ligand) and the total scaffold molecules. The parameter n determines the level of cooperatively and affects the steepness of the function. Note that $n = 1$ reduces to Michaelis–Menten kinetics.

to address spatiality of systems is *partial differential equations* (PDEs), i.e., differential equations that contain at least one partial derivative.

To consistently apply PDEs, we still need the assumptions (ii) and (iii) introduced for ODEs. In fact, the representation of molecules is still continuous (so we need large numbers of individual molecules to neglect the effect of approximation to the closest integer) and the number of variables and equations needed is still determined by the set of possible configurations of each molecule.

Partial differential equation

Let $f : \mathbb{R}^q \to \mathbb{R}$ be an unknown function, and $x_1, .., x_q \in \mathbb{R}$ independent variables. If $\partial^n f / \partial X_n$, with $X_n \subseteq \{x_1, .., x_q\}$ and $|X_n| = n \leq q$, stands for the nth partial derivative of f with respect to the variables in X_n,

$$F(x_1, .., x_q, f, \frac{\partial f}{\partial x_1}, .., \frac{\partial f}{\partial x_q}, \frac{\partial^2 f}{\partial x_1^2}, .., \frac{\partial^2 f}{\partial x_1 \partial x_q}, ..) = 0$$

is a partial differential equation (PDE). If f is a continuous function of space and time, it is sometimes called *field*.

Reaction-diffusion general template

Let $\mathbf{u} = (x, y, z)$ denote a point in space, $f_i(\mathbf{u}, t), i \in \{A, B\}$, denote the density of species i at point $\mathbf{u}$ in space and at time t, D_i be the diffusion constant of species i, $r_{AB}(f_A, f_B)$ and $r_{BA}(f_B, f_A)$ denote the reactions between A and B. Recall that the gradient of a scalar field is a vector field that points in the direction of the greatest rate of increase of the scalar field, and whose magnitude is the greatest rate of change. The gradient operator

$$\nabla = \frac{\partial}{\partial x}\mathbf{x} + \frac{\partial}{\partial y}\mathbf{y} + \frac{\partial}{\partial z}\mathbf{z}$$

where $\mathbf{x}, \mathbf{y}, \mathbf{z}$ are the unit vectors in their respective directions. The gradient of a scalar field f is

$$\nabla f = \frac{\partial f}{\partial x}\mathbf{x} + \frac{\partial f}{\partial y}\mathbf{y} + \frac{\partial f}{\partial z}\mathbf{z}.$$

The Laplace operator in a three-dimensional space $\nabla^2 = \partial^2/\partial x^2 + \partial^2/\partial y^2 + \partial^2/\partial z^2$ represents the flux density of the gradient of a function. We can interpret the Laplace operator of a function f at a point $\mathbf{u}$ as the rate at which the average value of f over spheres centered at $\mathbf{u}$, deviates from $f(\mathbf{u})$ as the radius of the sphere grows.

The general PDE template of the two interacting species, in space and time, is

$$\frac{\partial f_A(\mathbf{u}, t)}{\partial t} = D_A \nabla^2 f_A(\mathbf{u}, t) + r_{AB}(f_A, f_B)$$

$$\frac{\partial f_B(\mathbf{u}, t)}{\partial t} = D_B \nabla^2 f_B(\mathbf{u}, t) + r_{BA}(f_B, f_A)$$

where the first term on the right side of the equations describes species diffusion and the second term describes the reactions.

A PDE is a relation involving a set of unknown functions of several independent variables and their partial derivatives with respect to those variables (see the box on the previous page). Independent variables are used in practice to represent spatial coordinates of objects and time.

One of the most commonly used approaches to model the dynamics of biological systems over space are the so-called *reaction-diffusion* systems that date back to Fick and Turing. The state of the system is represented by continuous functions of space and time, called *fields*. Each field represents the density or concentration of each species involved in the system. As already observed for ODEs, we need a different field for each possible configuration a species can transit. We also need to add to each field for each molecule a reaction term that describes how the molecule can interact with the other components. The reaction term will depend on the density of the reactant species and on their position in the spatial description.

Diffusion is modeled over volumes of the system that are small enough

Currents

Currents denote the numbers of particles traversing a surface per unit of time. Formally, a current is a three-dimensional vector $\mathbf{j}$. The net change of particle numbers in a small volume due to transport is expressed by

$$\nabla \cdot \mathbf{j} = \frac{\partial}{\partial x} j_x + \frac{\partial}{\partial y} j_y + \frac{\partial}{\partial z} j_z$$

where ∇ is the gradient operator in three dimensions and j_i, $i \in \{x, y, z\}$, is the current through a surface perpendicular to direction i.
The creation and destruction of particles is captured by source and sink terms, usually given by the kinetics equation of the reactions occurring in the system. The continuity equation for the evolution of a particle density is

$$\frac{\partial \rho}{\partial t} + \nabla \cdot \mathbf{j} = s$$

where $\rho = dc/dV$ is the amount of a scalar variable c per unit volume (density) and s is a source ($s > 0$) or a sink ($s < 0$) term. If we use currents to represent conserved quantities such as energy, it will be $s = 0$.

to capture the flow of the molecules from one volume to the neighboring ones and that are large enough to be insensitive to spatial fluctuations.

The generalization of reaction-diffusion systems is given by *continuity equations*. These equations encode that the number of molecules in a small volume can change due to their transport into or out of the volume element or due to their synthesis or degradation within the volume element (e.g., a reaction between A and B in volume v can be interpreted as synthesis in v of the complex $A : B$ and the degradation in v of an A molecule and a B molecule). Continuity equations can be used to describe the transport of all conserved quantities (here number of molecules or mass). A main

Example 5.6. [Applying partial differential equations: Lotka–Volterra]
We enrich Example 5.1 by adding a two-dimensional grid space over which predators and prey can move. Each cell of the grid we create in this way accounts for the predators and prey in the cell. We can define a reaction-diffusion process through the Laplace operator to model interaction and movement of the individuals. Assume that a vector $\mathbf{r} = (x, y)$ denotes the position of a prey or predator. Then, the resulting equations are

$$\partial[R(\mathbf{r})]/\partial t = D_R \nabla^2 [R(\mathbf{r})] + h_1[R(\mathbf{r})] - (h_2[F(\mathbf{r})])[R(\mathbf{r})]$$
$$\partial[F(\mathbf{r})]/\partial t = D_F \nabla^2 [F(\mathbf{r})] - h_3[F(\mathbf{r})] + (h_4[R(\mathbf{r})])[F(\mathbf{r})]$$

where D_R and D_F are the diffusion constants for rabbits and foxes and ∇^2 is the Laplace operator in two dimensions.

concept involved in continuity equations is that of current (see the box on the previous page).

The differential equations explored so far are all deterministic. Therefore, if we need to model a system that has inherent noise in its behavior (e.g., population dynamics or gene regulation), we must resort to different

Stochastic differential equations

A *stochastic differential equation* (SDE) is an ordinary differential equation in which some of the terms are stochastic processes, so that the solution is still a stochastic process. The general form of an SDE is

$$dx = f(x, t)dt + g(x, t)dw(t)$$

where f and g are arbitrary functions of x depending on time t and w is a random function of time, usually white noise represented by a standard brownian motion. In the strongest sense, $w(t)$ represents a white noise if the value $w(t)$ is a random variable independent of $w(t'), \forall t' < t$. Brownian motion is described by $w(t) = \int_0^t dw(s)$.

technologies. An extension of differential equation to deal with stochasticity is obtained by adding a term to each equation that describes the noise of the system. The resulting equations are called *stochastic differential equations* (SDEs).

Example 5.7. [Applying stochastic differential equations: Lotka–Volterra]
The assumptions in Example 5.1 are not true in practice, when the environment is affected by noise fluctuations. An option for coping with environmental noise is to consider a perturbation of the growth rate with white noise by replacing h_1 in the equation of the rabbits with $h_1(t) + \sigma(t)w(t)$, where $w(t)$ is a brownian motion white noise. The resulting equations becomes

$$d[R] = (h_1[R] - (h_2[F])[R])dt + \sigma(t)dw(t)$$
$$d[F]/dt = -h_3[F] + (h_4[R])[F].$$

5.1.2 *Difference equations*

Differential equations consider continuous variables. Some systems have discrete valued attributes that are only approximated by continuous variables. If we need a precise account of the value of a discrete attribute, we can adopt the discrete variant of differential equations called *difference equations.* Difference equations are recurrence relations that given the initial state of a system and some parameter values, uniquely determine all the future states of the system. We usually call the measurement points steps

Difference equations

Assume that x_t is a value associated with an attribute of a system at step t and that F is a function. Then, $x_{t+1} = F(x_t)$ is a difference equation. The function F may refer to many previous states of the system as in $x_{t+1} = F(x_t, x_{t-1}, .., x_{t-k})$. The number of states in the history needed to determine the next value of x is the *order* of the difference equation ($k + 1$ above). The difference equation is *non-linear* if the function F depends on non-linear combinations of its arguments.

to distinguish them from the time used in differential equations.

There is a straighforward method for mapping from ordinary differential equations to difference equations implementing Euler's method (see Example 5.8).

Classical applications of difference equation are population studies in which we want to count the elements of a population with a given characteristic (see Example 5.9).

Example 5.8. [Discretization of differential equations]
The ODE $f^{(1)} = F(t, f)$ can be discretized into a difference equation by applying Euler's method with step h. The step says that we assume that $f_i = f(t_0 + ih)$ with t_0 the initial time of the system. The discretized version of the ODE is then $f_{n+1} = f_n + hF(t_n, f_n)$.

Example 5.9. [Applying difference equations: Lotka–Volterra]
Consider the Lotka–Volterra model of a two-species ecosystem as described in Example 5.1. The corresponding difference equations are

$$R_{n+1} = R_n + h_1 R_n - h_2 F_n R_n \qquad F_{n+1} = F_n - h_3 F_n + h_4 R_n F_n$$

where R_n (F_n) denotes the number of rabbits (foxes) at step n and the parameter h_i has the same meaning described in Example 5.1.

5.2 Rewriting systems

Rewriting systems (see the box on the next page) or reduction systems are a set of objects plus a set of relations called *rewriting rules* that define how the objects can be transformed. Rewriting rules can be non-deterministic because they can be applied in different ways, or more than one rule can be applied to the same configuration of the system. The rewriting rules define the dynamics of the modeled system. Rewriting systems can be classified into two main categories: *term rewriting* and *graph rewriting*. A term rewriting system is a rewriting system where the set of objects are terms, i.e., objects that can be represented as trees of symbols, usually specified through a grammar (see Sect. 6.1). A special case of term rewriting is string rewriting where the objects have a flat linear structure. A main class of term rewriting systems used in modeling of biological systems is that of chemical reactions, introduced in the next section.

A generalization of term rewriting is graph rewriting, where the objects of the rewriting systems are graphs. The graph rewriting rules are based on pattern-matching applied to an host graph. The left part of a rewriting rule is the *graph pattern* and the right part is the *replacement graph*. The host graph is searched for all the subgraphs matching the pattern in the left side of a rule and one, some or all of them (depending on the rewriting strategy) are replaced by the right part of the rule. From an application perspective, finding the subgraphs of a host graph that match the pattern of a rule corresponds to solving the subgraph isomorphism problem.

Rewriting systems

A rewriting system is a pair $\langle O, \rightarrow \rangle$ with $\rightarrow \subseteq O \times O$. O is the set of objects or elements of the rewriting system describing all possible configurations of the system modeled and the relation $\rightarrow$ defines the rewriting rules to map (rewrite) one configuration into another. An object $x \in O$ is *reducible* if there exists $y \in O$ such that $(x, y) \in \rightarrow$, usually written $x \rightarrow y$. The objects that are not reducible (irreducible) are in *normal form*. An object y is the normal form of x, if $x \rightarrow^* y$ with y in normal form and $\rightarrow^*$ the reflexive and transitive closure of $\rightarrow$. If each object has at least a normal form, the rewriting system is *normalizing*. Two objects x and y are *joinable* $(x \downarrow y)$ if $\exists z \in O.(x \rightarrow^* z \wedge y \rightarrow^* z)$. Quantitative rewriting systems $\langle O, \rightarrow, \mathbb{R} \rangle$ with $\rightarrow \subseteq O \times \mathbb{R} \times O$ are obtained by labeling the rewriting rules with quantities. Symmetric, quantitative rewriting rules are usually written

$$x \underset{k_2}{\overset{k_1}{\rightleftharpoons}} y \text{ as a shorthand for } x \xrightarrow{k_1} y \text{ and } x \xleftarrow{k_2} y$$

with $k_1, k_2 \in \mathbb{R}$. Terminology and definitions of rewriting systems apply to quantitative rewriting systems as well.

Example 5.10. [Term rewriting: DNA as a string]
A strand of DNA can be represented as a string over the alphabet $\{A, G, T, C\}$ and can be generated by the following term rewriting system starting from a non-terminal start symbol DNA

$$DNA \rightarrow A \,|\, G \,|\, T \,|\, C \,|\, A\,DNA \,|\, G\,DNA \,|\, T\,DNA \,|\, C\,DNA$$

The vertical bar denotes alternative rules or productions and DNA can be rewritten as each one of the alternatives. A, G, T, C are called terminal symbols because they do not allow further reductions. The sequences of rewritings

$$DNA \rightarrow A\,DNA \rightarrow A\,A\,DNA \rightarrow A\,A\,T\,DNA \rightarrow A\,A\,T\,A$$

generates the strand $AATA$. At each step, the rightmost DNA is replaced by one of the alternatives in the definition of the rewriting rules. Note that this rewriting system corresponds to a regular grammar in Chomsky's hierarchy.

Before examining some examples of rewriting systems, we introduce some relevant properties related to the existence and computation of normal forms (see the box on the next page). *Confluence* is an important property

Example 5.11. [Graph rewriting]
Consider the rule depicted in Fig. 5.5(A). It states that each pair of edges connecting two nodes can be replaced by a single edge provided that the rate of the replacing edge is the sum of the rates of the replaced edges. The rule is context-free, i.e., it can be applied independently of how the source and target nodes of the replaced edges are connected to the rest of the graph to which they belong. If the initial graph is the one depicted in Fig. 5.5(B), two rewriting steps (Fig. 5.5(C–D)) eliminate all the double edges, generating the rates $k_1 = h_2 + h_3$ and $k_2 = h_6 + h_7$.

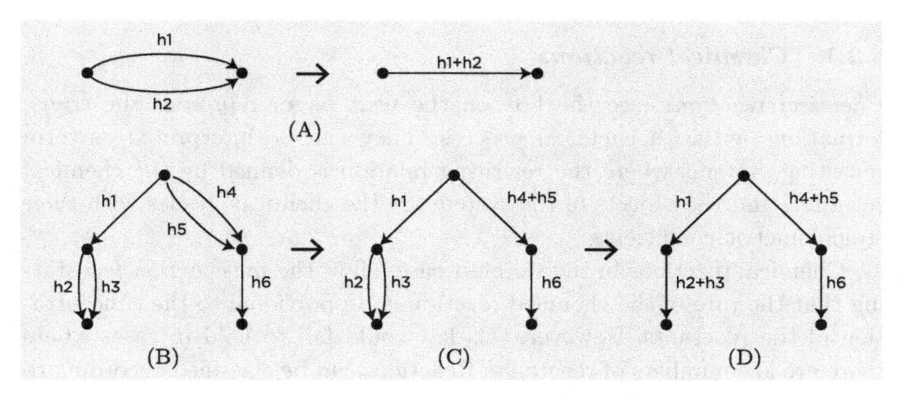

Figure 5.5 A simple graph rewriting rule (upper part) and its application to an initial graph (lower part).

because it ensures that the normal form of an object is unique if it exists. Furthermore, terminating rewriting systems are always normalizing. Finally, convergent rewriting systems exhibit a unique normal form for each object in O.

Rewriting systems: properties

A rewriting system is *confluent* if for all $w, x, y \in O$, $w \to^* x$ and $w \to^* y$ implies $x \downarrow y$. It is *locally* or *weakly* confluent if for all $w, x, y \in O$, $w \to x$ and $w \to y$ implies $x \downarrow y$.
A rewriting system is *terminating* or *noetherian* if there is no infinite chain of rewritings $x_0 \to x_1 \to x_2 \to \ldots$. A terminating and locally confluent rewriting system is confluent.
A confluent and terminating rewriting system is *convergent*.

Example 5.12. [Properties of DNA as a string]
Let $dna = \{A, C, T, G\}^*$ and $O = dna \cup \{wDNA \mid w \in dna\}$. Then $\langle O, \rightarrow \rangle$, where $\rightarrow$ is defined by the alternative productions in Example 5.10, is neither confluent nor terminating because $A\,DNA$ and $C\,DNA$ are not joinable and $DNA \rightarrow A\,DNA \rightarrow A\,A\,DNA \ldots$ is not finite if at each step DNA is replaced by $A\,DNA$.

In the next two subsections we introduce two special cases of rewriting systems: chemical reactions and P-systems.

5.2.1 *Chemical reactions*

Chemical reactions (see the box on the next page) represent the transformations within a chemical process. They can be interpreted as term rewriting systems where the rewriting relation is defined by the chemical reactions and the objects of the system are the chemical species with their stoichiometric coefficients.

Chemical reactions in the simplest case follow the *mass action law* stating that the rate of the chemical reaction is proportional to the concentration of the reactants. However, this law could fail to hold in cases where there are low numbers of reactants. Reactions can be classified according to their order. The order of a reaction with respect to a reactant is the exponent of the concentration of that reactant in the rate equation. The order of the reaction is the sum of the concentration exponents of the reactants. For instance, the order of the reaction in the box on the mass action law is $x + y$.

The unit of measure of the rate constant depends on the order of the reaction. In the general case of a reaction of order n, the unit of measure is $mol^{1-n} l^{n-1} s^{-1}$ or $(l/mol)^{n-1} s^{-1}$ equivalently. In fact, $rate = k[X]^n$ with unit of measure $mol\, l^{-1} s^{-1}$ and unit of measure of $[X]^n$ being $(mol\, l^{-1})^n$. Therefore the unit of measure of $k = rate/[X]^n$ is $(mol\, l\, s^{-1})/(mol\, l^{-1})^n = mol^{1-n} l^{n-1} s^{-1}$. Sometimes *molarities* $M = mol\, l^{-1}$ is used as a unit of measure.

Figure 5.6 describes the templates of the reactions classified according to their order and specifies the unit of measure of the rate constants of the reactions. The table also shows the corresponding ODE that models the reactions. Therefore the table shows the correspondence between chemical reactions and ODEs and provides a mapping algorithm between the two continuous, deterministic formalisms. Note that the table also shows the

Chemical reactions

A *chemical reaction* is a transformation of a finite set of chemical species into a different set of chemical species, e.g.,

$$2A + 3B \xrightarrow{k} C + 2D.$$

The species are usually separated by + symbols. The cardinality of the species on both sides of the reaction may be greater or equal to zero. The species on the left of the reaction are the *reactants* (or *reagents*) and the species on the right of the reaction are the *products* of the reaction. Reactions with no reactant are *synthesis* reactions and reactants with no product are *degradation* reactions.
Reactants and products in chemical reactions are equipped with coefficients denoting their abundance in the reaction to enforce laws of conservation of mass and charge. The coefficients are called *stoichiometric* because *reaction stoichiometry* studies the relationship between the quantities of reactants and products in a chemical reaction. A set of reactions can be represented by a *stoichiometric matrix* that has a row for each species in the system and a column for each reaction. The entry (X, j) corresponding to the species X in the reaction j is the algebraic sum of the stoichiometric coefficients of X in reaction j with a positive sign for products and a negative sign for reactants. Columns can be labeled with reaction rates.
The reaction rate for a reactant or product is the amount of the chemical that is formed or removed (in moles) per unit time per unit volume. The rate constant expresses the proportionality between the reaction rate and the reactant concentrations. Therefore, rate = k[X].

conversion of the deterministic rate in the chemical reactions into the corresponding stochastic rate.

Zero-order reactions model the synthesis of substances from null species or add new species to a system. *First-order* reactions (sometime called *unimolecular*) model the elementary conversion of an unstable molecule A into a different state B (radioactive decay is a classical example). *Second-order* reactions describe interactions between two components of the system.

Before approaching more complex dynamics, we explain the correspondence between the chemical reactions and the ODEs reported in Fig. 5.6,

Mass action law

The law of mass action for a chemical reaction $mA + nB \xrightarrow{k} C$ associates the rate $r = k[A]^x[B]^y$ with the reaction with x, y, experimentally determined. The constant k is the rate constant. The equation defining r is the *rate equation.*

Example 5.13. [Chemical reactions: Lotka–Volterra]
The rewriting rules below describe the dynamics of the Lotka–Volterra system by stating that a rabbit reproduces if it eats grass (first rule), that a fox reproduces if it eats rabbits (second rule) and that a fox can die (rule three).

$$R + G \xrightarrow{h_1} 2R + G \qquad F + R \xrightarrow{h_2} 2F \qquad F \xrightarrow{h_3} 0.$$

The constants in the chemical reactions are the rates driving the dynamics. Assuming that the reactions are numbered one to three from left to right, the stoichiometric matrix of the set of reactions is:

	$(\mathbf{1}, \mathbf{h_1})$	$(\mathbf{2}, \mathbf{h_2})$	$(\mathbf{3}, \mathbf{h_3})$
R	$+1$	-1	0
G	0	0	0
F	0	$+1$	$-1.$

Note that when the same species is both a reactant and a product in the same reaction, the stoichiometric coefficients are algebraically summed. The coefficient of R in the first reaction is $1 = 2 - 1$ because two copies of R are generated and one copy is consumed. Similarly the coefficient of G is zero because a copy of G is produced and one is consumed. The third reaction is a degradation of F and hence its coefficient is -1.

starting from first-order reactions. The conversion rate of molecules of type A into molecules of type B is a Markov process represented by

$$num(A) \longrightarrow num(A) - 1, \qquad num(B) \longrightarrow num(B) + 1$$

where $num(A)$ and $num(B)$ represent the numbers of individual molecules A and B in the system. The left-hand side of the reactions denote the number of molecules before the chemical reaction and the numbers on the right-hand side denote the cardinality of the individual molecules after the chemical reaction. Since the probability of converting a given molecule A

Order	Template	Unit of measure	ODE	Stochastic rate
0	$\varnothing \xrightarrow{k} A$	mol $l^{-1}s^{-1}$	d[A]/dt = k	kN_AV
1	$A \xrightarrow{k} B$	s^{-1}	d[A]/dt = −k[A] d[B]/dt = k[A]	k
2	$2A \xrightarrow{k} B$	l $mol^{-1}s^{-1}$	d[A]/dt = $-k[A]^2$ d[B]/dt = $k[A]^2$	$2k/(N_AV)$
2	$A+B \xrightarrow{k} C$	l $mol^{-1}s^{-1}$	d[A]/dt = −k[A][B] d[B]/dt = −k[A][B] d[C]/dt = k[A][B]	$k/(N_AV)$

Figure 5.6 Order of chemical reactions, unit of measures of rates, mapping with ODEs and transformation of deterministic rate into stochastic rate. N_A is the Avogadro number and V is the reaction volume.

does not depend on the other molecules in the system, we have that the probability of the state transition (with states represented by vectors of numbers of individual molecules)

$$[num(A), num(B)] \longrightarrow [num(A) - 1, num(B) + 1]$$

is $k * num(A)$, where k is a constant, positive real value. In an infinitesimal time dt, the variation in the number of individual molecules is

$$d\,num(A) = -k * num(A)dt, \qquad d\,num(B) = k * num(A)dt$$

and to derive equations that are valid independently of the volume of the reaction chamber, we divide both sides of the equations above by the volume V of the system resorting to concentrations $[A] = num(A)/V$ and $[B] = num(B)/V$. The resulting equations are

$$d[A]/dt = -k[A], \qquad d[B]/dt = k[A].$$

Similar reasoning applies to the other chemical reactions. We can therefore state a general algorithm to translate a set of chemical reactions into a set of equivalent ODEs (Algorithm 5.1). The first `for each` in the algorithm deals with the reactants that do not compare as product as well in the same reaction; the second iteration is for the products that are also not reactant and the last `for each` handles reactants that are also products in the same reaction. An application of the algorithm to a set of chemical reactions is in Examples 5.14 and 5.15.

All chemical reactions are theoretically reversible and the preferred direction is determined by thermodynamic conditions. The template for

Algorithm 5.1: From chemical reactions to ODEs

Input: A set of p chemical reactions

$$\begin{aligned} n_{11}X_{11} + \cdots + n_{q_1 1}X_{q_1 1} &\xrightarrow{k_1} m_{11}Y_{11} + \cdots + m_{r_1 1}Y_{r_1 1} \\ &\vdots \\ n_{1i}X_{1i} + \cdots + n_{q_i i}X_{q_i i} &\xrightarrow{k_i} m_{1i}Y_{1i} + \cdots + m_{r_i i}Y_{r_i i} \\ &\vdots \\ n_{1p}X_{1p} + \cdots + n_{q_p p}X_{q_p p} &\xrightarrow{k_p} m_{1p}Y_{1p} + \cdots + m_{r_p p}Y_{r_p p}. \end{aligned}$$

Output: A set of as many ODEs as the number of different species appearing as reactant or product in the set of chemical reactions.

begin
  for $i \in [1..p]$ **do**
    for *each reactant* $n_{ji}X_{ji}, X_{ji} \neq Y_{li}, l \in [1..r_i]$ **do**
      generate $d[X_{ji}]/dt = -k_i \prod_{z=1}^{q_i} [X_{zi}]^{n_{zi}}$;
    end
    for *each product* $m_{ji}Y_{ji}, Y_{ji} \neq X_{li}, l \in [1..q_i]$ **do**
      generate $d[Y_{ji}]/dt = m_{ji}k_i \prod_{z=1}^{q_i} [X_{zi}]^{n_{zi}}$;
    end
    for *each reactant* $n_{ji}X_{ji}, X_{ji} = Y_{li}, l \in [1..r_i]$ **do**
      generate $d[X_{ji}]/dt = k_i(m_{li} - n_{ji}) \prod_{z=1}^{q_i} [X_{zi}]^{n_{zi}}$;
    end
  end
  Rearrange the equations to have a single equation for each species in the system;
end

these reactions is

$$A + B \underset{k_r}{\overset{k_f}{\rightleftharpoons}} C$$

which originates the three ODEs

$$d[A]/dt = -k_f[A][B] + k_r[C] = d[B]/dt, \quad d[C]/dt = k_f[A][B] - k_r[C].$$

Reversibility of reactions strongly correlates with the notion of *equilibrium* or *steady state* that is a configuration in which there is no change in the amount of individual molecules A, B and C. As a consequence, the derivative with respect to t of the concentrations of the three molecules will be 0, i.e., $d[A]/dt = d[B]/dt = d[C]/dt = 0$. This condition allows us to define

Example 5.14. [Chemical reactions to ODEs]
Consider the set of chemical reactions in which we arranged species in a matrix-like structure so that the same species is always in the same position.

$$
\begin{array}{lllll}
1: 2A + 3B + C & \xrightarrow{k_1} & A & + 2C & \\
2: A + B + 2C & \xrightarrow{k_2} & & 2B + 2C & \\
3: & \xrightarrow{k_3} & & B & \\
4: A \qquad + 3C & \xrightarrow{k_4} & & & D \\
5: 2A + 3B & \xrightarrow{k_5} & & &
\end{array}
$$

In the empty positions we can assume to have a 0 coefficient. For instance, reaction 4 can be written as

$$A + 0B + 3C \xrightarrow{k_4} 0A + 0B + 0C + D.$$

The stoichiometric coefficient of the reactants and products used in Algorithm 5.1 are collected in the matrixes

$$\mathbf{n} = \begin{pmatrix} 2 & 3 & 1 \\ 1 & 1 & 2 \\ 0 & 0 & 0 \\ 1 & 0 & 3 \\ 2 & 3 & 0 \end{pmatrix} \quad \text{and} \quad \mathbf{m} = \begin{pmatrix} 1 & 0 & 2 & 0 \\ 0 & 2 & 2 & 0 \\ 0 & 1 & 0 & 0 \\ 0 & 0 & 0 & 1 \\ 0 & 0 & 0 & 0 \end{pmatrix}.$$

For each $w \in [1..5]$, the species are

$$X_{1w} = A, X_{2w} = B, X_{3w} = C \text{ and } Y_{1w} = A, Y_{2w} = B, Y_{3w} = C, Y_{4w} = D.$$

We start with the first reaction and we identify the reactants that are not appearing as products as well (i.e., $X_{21} = B$) and we generate the equation

$$d[B]/dt = -k_1 \prod_{z=1}^{3} [X_{z1}]^{n_{z1}} = -k_1 [A]^2 [B]^3 [C].$$

The second for each in the algorithm looks for products that are not reactants of the reaction, and no products of reaction 1 satisfy this condition. Hence we move to third iteration and handle the intersection of reactants and products. For instance, $X_{11} = A$ generates the equation

$$d[A]/dt = k_1 (m_{11} - n_{11}) \prod_{z=1}^{3} [X_{z1}]^{n_{z1}} = k_1 (1-2) [A]^2 [B]^3 [C] = -k_1 [A]^2 [B]^3 [C].$$

Example 5.15. [Chemical reactions to ODEs — continued]
Iterating the Algorithm 5.1 for each reaction, we generate the following equations:

$$d[A]/dt = k_1(m_{11} - n_{11}) \prod_{z=1}^{3} [X_{z1}]^{n_{z1}} = -k_1[A]^2[B]^3[C]$$
$$d[B]/dt = -k_1 \prod_{z=1}^{3} [X_{z1}]^{n_{z1}} = -k_1[A]^2[B]^3[C]$$
$$d[C]/dt = k_1(m_{31} - n_{31}) \prod_{z=1}^{3} [X_{z1}]^{n_{z1}} = k_1[A]^2[B]^3[C]$$

$$d[A]/dt = -k_2 \prod_{z=1}^{3} [X_{z2}]^{n_{z2}} = -k_2[A][B][C]^2$$
$$d[B]/dt = k_2(m_{22} - n_{22}) \prod_{z=1}^{3} [X_{z2}]^{n_{z2}} = k_2[A][B][C]^2$$
$$d[C]/dt = k_2(m_{32} - n_{32}) \prod_{z=1}^{3} [X_{z2}]^{n_{z2}} = 0[A][B][C]^2$$

$$d[B]/dt = k_3 \prod_{z=1}^{3} [X_{z3}]^{n_{z3}} = k_3[A]^0[B]^0[C]^0 = k_3$$

$$d[A]/dt = -k_4 \prod_{z=1}^{3} [X_{z4}]^{n_{z4}} = -k_4[A][B]^0[C]^3 = -k_4[A][C]^3$$
$$d[C]/dt = -k_4 \prod_{z=1}^{3} [X_{z4}]^{n_{z4}} = -k_4[A][B]^0[C]^3 = -k_4[A][C]^3$$
$$d[D]/dt = k_4 \prod_{z=1}^{3} [X_{z4}]^{n_{z4}} = k_4[A][B]^0[C]^3 = k_4[A][C]^3$$

$$d[A]/dt = -k_5 \prod_{z=1}^{3} [X_{z5}]^{n_{z5}} = -k_5[A]^2[B]^3[C]^0 = -k_5[A]^2[B]^3$$
$$d[B]/dt = -k_5 \prod_{z=1}^{3} [X_{z5}]^{n_{z5}} = -k_5[A]^2[B]^3[C]^0 = -k_5[A]^2[B]^3.$$

Next step is rearranging the equations so that we have just one equation for each species in the system:

$$d[A]/dt = -k_1[A]^2[B]^3[C] - k_2[A][B][C]^2 - k_4[A][C]^3 - k_5[A]^2[B]^3$$
$$d[B]/dt = -k_1[A]^2[B]^3[C] + k_2[A][B][C]^2 + k_3 - k_5[A]^2[B]^3$$
$$d[C]/dt = k_1[A]^2[B]^3[C] - k_4[A][C]^3$$
$$d[D]/dt = k_4[A][C]^3.$$

the *equilibrium constant* from the equations above as

$$K_{eq} = k_f/k_r = \lim_{t \to +\infty} \frac{[A]_t[B]_t}{[C]_t}$$

and to compute the equilibrium concentration of C in terms of the initial concentrations of A and B and the value of the equilibrium constant

$$[C] = \frac{K_{eq} + [A]_{t_0} + [B]_{t_0}}{2} - \sqrt{(\frac{K_{eq} + [A]_{t_0} + [B]_{t_0}}{2})^2 - [A]_{t_0}[B]_{t_0}}.$$

The concentration of C is derived by applying the mass conservation law from which it is $[A] + [C] = [A]_{t_0}$ and $[B] + [C] = [B]_{t_0}$.

Steady states are the fixed points (solutions) of the algebraic equations $\vec{f}(\vec{x}) = 0$ where $\vec{x}$ is a vector of the concentrations of all the species in the system (e.g., $\vec{x} = ([A], [B], [C])$) and $\vec{f}$ is a rate vector collecting the right-hand side of the ODEs. Consider the ODEs for the equilibrium reactions above and assume that the indexes of f and x identify the corresponding vector position, it is

$$\vec{f}(\vec{x}) = (\; f_1(\vec{x}) = -k_f x_1 x_2 + k_r x_3,$$
$$f_2(\vec{x}) = -k_f x_1 x_2 + k_r x_3,$$
$$f_3(\vec{x}) = k_f x_1 x_2 - k_r x_3).$$

5.2.1.1 *Stochastic mass action kinetics*

We conclude the section on chemical reactions by addressing the conversion of deterministic rate constants into stochastic rate constants (see Fig. 5.6). The main difference between deterministic and stochastic approaches is the way in which the number of molecules in a system is measured. Deterministic representations rely on concentrations (a real number) measured in moles per liter, while stochastic approaches are based on the number of molecules (an integer). If a species X has concentration $[X]$ in a volume of V litres there are $[X]V$ moles and hence $N_A[X]V$ molecules, where N_A is the Avogadro number.

An issue that must be addressed in the conversion is the change in the units of measure of the model parameters. Since stochastic kinetics relies on numbers of molecules, the unit of measure of the stochastic constant c corresponding to the deterministic on k must replace concentrations in the unit of measure of k with *molecules* in the unit of measure of c. The outcome of this conversion will vary depending on the order of the reaction because the order determines the unit of measure of k. We explain the conversion for the reactions in Fig. 5.6.

Mass action stochastic kinetics

The rate equation of mass action is replaced by *reaction propensity* (sometimes called stochastic rate equation). A *redundancy function* h defines the number of distinct reactant combinations from the state of the system and a reaction R. The *propensity function* for the reaction R is $a = hc$, where c is the stochastic constant rate of R. The probability that an R reaction will occur in the volume V in the infinitesimal interval $[t, t + dt)$ is adt and depends on the state of the system at time t.

The unit of measure of the mass action constant k of *zero order* reactions is $mol \cdot l^{-1} \cdot s^{-1}$ and the concentration of the species A is produced at a rate of $k\ mol \cdot l^{-1}$. If we fix a reaction volume V and we convert concentrations into molecule numbers, the production rate of A molecules is $N_A V k = c$ measured in $molecules{\cdot}s^{-1}$.

The unit of measure of the mass action constant k of *first-order* reactions is s^{-1} and the concentration of the species B is produced at a rate $k[A]$ that, transformed into molecules per second, is $N_A V k[A] = k|A|$ (recall that $[A] = |A|/(N_A V)$). The propensity function for first-order reactions is $a = c|A|$ because the redundancy function of a single species corresponds to the cardinality of that species (each molecule could react). The propensity function is the rate of growth of B measured in molecules per second, hence we impose $k|A| = c|A|$ and we conclude $k = c$.

From deterministic rates to stochastic rates

Consider a generic chemical reaction of order n

$$a_1 A_1 + a_2 A_2 + \ldots + a_n A_n \xrightarrow{k} C.$$

The stochastic rate corresponding to k is

$$c = \left(\frac{k}{(N_A V)^{n-1}}\right) \cdot \prod_{j=1}^{n} a_j!$$

The stochastic rate constant of second-order reactions changes from dimerization to homodimerization. Consider dimerization (two molecules from two different species interact to form a product) first. The unit of measure of the mass action constant k of *dimerization* reactions is $mol^{-1} \cdot l \cdot s^{-1}$ and the concentration of the species C is produced at a rate of $k[A][B]\ mol \cdot l^{-1}$ that, converted into molecules per second, is $N_A V k[A][B] = (k|A||B|)/(N_A V)$. The propensity function for dimerization is $a = c|A||B|$, because $h = |A||B|$ computes all the combinations of molecules from A and B. The propensity function is the rate of the growth of C measured in molecules per second, hence we impose $(k|A||B|)/(N_A V) = c|A||B|$ and we conclude $k/(N_A V) = c$. Consider now *homodimerization* (two molecules from the same species interact to form a product). The unit of measure of the mass action constant k of *homodimerization* reactions is still $mol^{-1} \cdot l \cdot s^{-1}$ and the concentration of the species A is consumed at a rate of $k[A]^2$ $mol \cdot l^{-1}$ that, converted into molecules per second, is $N_A V k[A]^2 = (k|A|^2)/(N_A V)$. The propensity function for homodimerization is $a = c|A|(|A|-1)/2$, because $h = |A|(|A|-1)/2$ computes all the combinations of two molecules from A. The propensity function is the rate of growth of C which, measured in molecules per second, is $c|A|(|A|-1)/2$, hence we impose $(k|A|^2)/(N_A V) = c|A|(|A|-1)/2$ and we conclude that $2k/(N_A V) = c$ under the approximation $|A|(|A|-1) \sim |A|^2$ that holds for large numbers of molecules A.

5.2.2 *P-systems and membrane computing*

P-systems are multiset rewriting systems usually based on a membrane structure that is a hierarchically arranged set of labeled membranes (which is a tree structure). Each membrane determines a *region* and there is a membrane defining the boundaries of system and enclosing all the others. A membrane system can be represented by a string of labeled matching parentheses.

Each region contains a multiset of symbols corresponding to the chemicals in a solution. In the basic case, the system dynamics is specified by multiset rewriting rules called *evolution rules*, written $u \to v$, with u, v multiset of objects. The rules may have more than one object per side (for instance $uxy \to vx$ is a rule).

Rules are applied by relying on a maximally parallel selection strategy, i.e., the selected rule is applied in as many instances as the multiplicity of objects in the system allows.

Evolution rules describe the reactions that can happen inside a compartment. Additional *communication* rules are used to manage transport of objects between membranes. There are *symport* and *anti-port* rules. Symport rules describe the active transport of two molecules together across a membrane, while anti-port rules describe the situation in which two molecules pass simultaneously through a membrane channel in opposite directions. A generalization to multisets of arbitrary size for these rules leads to:

- (x, in) and (x, out) model the flow of the objects in x within a membrane or outside a membrane and these are the the templates for symport rules;
- $(x, in; w, out)$ states that x objects flow into a membrane while simultaneously the w objects flow outside the same membrane and this is the template for anti-port rules.

P-systems with active membranes

We assume a set of polarization labels for membranes $\{+, -, 0\}$ with metavariable e. Then, a *P-system with active membranes* is

$$\Pi = (O, H, \mu, w_1, \ldots, w_m, R)$$

where O is the *alphabet* of the objects, H is a finite set of *labels* for membranes, μ is the *membrane structure* with all the membranes initially having neutral polarization (0), $\forall i \in \{1, \ldots, m\}, w_i \in O^*$ are strings of multiset of objects placed in the m regions of μ, $m \geq 1$ is the initial *degree* of the system and R is a finite set of rules of the forms:

(1) $[_h u \rightarrow v]_h^e, h \in H, e \in \{+, -, 0\}, u, v \in O^*$. It is a membrane and charge-dependent rule that does not modify membranes.
(2) $a[_h]_h^{e_1} \rightarrow [_h b]_h^{e_2}, h \in H, e_1, e_2 \in \{+, -, 0\}, a, b \in O$. It is an *in* communication rule; the object a is introduced in the membrane h and changed; also the charge can be changed by the rule.
(3) $[_h a]_h^{e_1} \rightarrow [_h]_h^{e_2} b, h \in H, e_1, e_2 \in \{+, -, 0\}, a, b \in O$. It is an *out* communication rule symmetric to the previous one.
(4) $a \rightarrow [_h v]_h^e, h \in H, e \in \{+, -, 0\}, a \in O, v \in O^*$. It is a *membrane creation* rule.
(5) $[_h a]_h^e \rightarrow b, h \in H, e \in \{+, -, 0\}, a, b \in O$. It is a *membrane dissolving* rule.
(6) $[_h a]_h^{e_1} \rightarrow [_h b]_h^{e_2} [_h c]_h^{e_3}, h \in H, e_1, e_2, e_3 \in \{+, -, 0\}, a, b, c \in O$. It is a *membrane division* rule.
(7) $[_{h_1} a]_{h_1}^{e_1} [_{h_2} b]_{h_2}^{e_2} \rightarrow [_h c]_h^{e_3}, h_1, h_2, h \in H, e_1, e_2, e_3 \in \{+, -, 0\}$, $a, b, c \in O$. It is a *membrane merging* rule.
(8) $[_{h_1} a]_{h_1}^{e_1} [_{h_2}]_{h_2}^{e_2} \rightarrow [_{h_2} [_{h_1} b]_{h_1}^{e_1}]_{h_2}^{e_2}, h_1, h_2 \in H, e_1, e_2 \in \{+, -, 0\}$, $a, b \in O$. It is an *endocytosis*.
(9) $[_{h_2} [_{h_1} a]_{h_1}^{e_1}]_{h_2}^{e_2} \rightarrow [_{h_1} b]_{h_1}^{e_1} [_{h_2}]_{h_2}^{e_2}, h_1, h_2 \in H, e_1, e_2 \in \{+, -, 0\}$, $a, b \in O$. It is an *exocytosis*.

The last set of rules needed to complete the definition of the dynamics of P-systems are the *membrane handling* rules. These rules define the dynamics of membranes themselves that can change dynamically by forming new membranes, dividing, dissolving, etc. These rules also include, on the left and right side of the reactions, membranes represented as a string of labeled matching parentheses enriched with polarization information (see the box on the previous page and examples).

Example 5.16. [P-systems: Lotka–Volterra]
The dynamics of a Lotka–Volterra system in a single location l_0 can be expressed in P-systems by relying on the following rules:

$$[_{l_0} RG \rightarrow RRG]^0_{l_0}, \quad [_{l_0} FR \rightarrow FF]^0_{l_0}, \quad [_{l_0} F \rightarrow \emptyset]^0_{l_0}.$$

Adding a compartment l_1 with the same reactions as above, and assuming that only foxes are moving from one to another, we have the additional rules

$$[_{l_0} F]^0_{l_0} \rightarrow [_{l_0}]^0_{l_0} F, \quad F \rightarrow [_{l_1} F]^0_{l_1}$$

and the symmetric pair to move from l_1 to l_0.

The quantitative stochastic version of P-systems is simply obtained by associating rates with rules.

5.3 Network-based approaches

Network modeling has been largely used in biology. *Boolean networks* are the most adopted technology, especially in the field of gene regulatory networks. Another approach that aims at describing the dynamics of systems both qualitatively and quantitatively is *Petri nets.*

5.3.1 *Boolean networks*

Boolean networks emerge as a *qualitative* modeling technology aiming at overcoming the lack of precise quantitative information of reaction mechanisms and kinetic parameters that makes it difficult to develop models with other technologies like differential equations.

The goal of Boolean networks is to describe *qualitative dynamical properties* that are invariant with respect to system behavior.

A Boolean network is made up of a set of entities (represented by the nodes of a graph) that can be in two alternative states: active (represented by 1) or inactive (represented by 0). The arcs connecting the nodes are oriented and usually classified into activating (arrow-ended) and inhibitory arcs (bar-ended). The dynamics of the network are expressed by a Boolean

Qualitative dynamical properties

A property is *invariant* with respect to a set of operations if it is not affected by the execution of the operations.
A *qualitative dynamical property* is a property that is invariant over time to a set of reaction mechanisms and kinetic parameters.

next-state function for each entity that depends on the status of the entities that are the source of the arcs pointing to the considered entity. The update of the state of the network can be either synchronous (the entities update their states together and simultaneously) or asynchronous (the entities update their states independently). The dynamics of Boolean networks can be completely specified by associating a truth table to each node of the network. An example of Boolean network is given in Fig. 5.7.

Boolean networks

A *Boolean network* is $B = \langle N, E, \mathcal{F} \rangle$, where N is a set of nodes that can assume values in $\{0,1\}$, $E = A \cup I$ is a set of activating (A) and inhibitory (I) arcs such that $A \cap I = \emptyset$. $\mathcal{F} = \bigcup_{n \in N} f_n : \{n_i \,|\, (n_i, n) \in E\} \rightarrow \{0,1\}$ is the family of *Boolean next-state* functions defined for each node and defining the dynamics of the network. Considering all the possible values that n_i can assume, f_n is a truth table.
Synchronous Boolean networks update the status of all nodes simultaneously and exhibit a deterministic behavior, while *asynchronous Boolean networks* update the status of the nodes independently and exhibit a nondeterministic behavior.

A compact representation of the behavior of a Boolean network can be obtained by representing the truth tables through Boolean expressions in disjunctive normal form (a disjunction of one or more conjunctions of literals) and then applying logic minimization. We represent each state by a Boolean term where a variable x represents the fact that the corresponding entity is in state 1 and $\overline{x}$ means that the corresponding entity is in state 0.

The disjunctive normal forms for each node of the Boolean network are obtained by applying Algorithm 5.2 where $[n]$ denotes the next state of n, $\overline{x}$, $x + y$ and xy represent the Boolean operators *not*, *or* and *and*, respectively. Note that line (1.1) in Algorithm 5.2 initializes both $[n]$ and $[\overline{n}]$ to 0 being the neutral element of $+$.

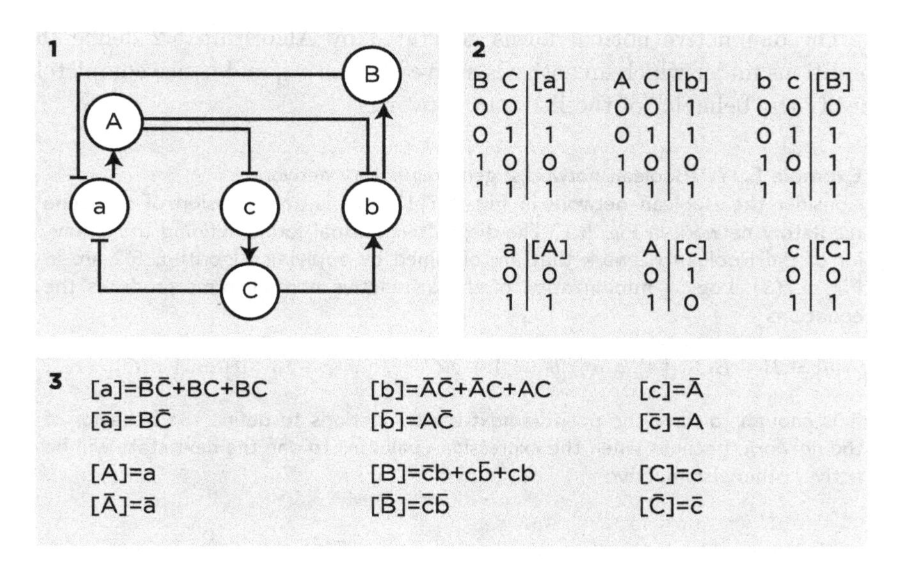

Figure 5.7 A Boolean network (1) and its truth tables (2). The notation $[n]$ means the next state of n. The logical equations corresponding to the truth tables are in part (3).

Algorithm 5.2: From truth tables to disjunctive normal forms

Input: A set of truth tables f_n associated with the N nodes of a Boolean network.
Output: A set of Boolean expressions in disjunctive normal form.

begin
  for $n \in N$ **do**
    $[n] = [\overline{n}] = 0$;
    for *each row* i *in* f_n **do**
      let $n_1, .., n_k$ be all the columns in f_n different from n;
      if $f_n(i, n) = 1$ **then**
        $[n] = [n] + x_1 \cdot .. \cdot x_k$
      else
        $[\overline{n}] = [\overline{n}] + x_1 \cdot .. \cdot x_k$
      end
    end
  end
end
with $x_i = n_i$ if $f_n(i, n_i) = 1$ or $x_i = \overline{n}_i$ otherwise.

The disjunctive normal forms generated by Algorithm 5.2 define the conditions under which an entity is active or inactive, and hence completely specify the behavior of the Boolean network.

Example 5.17. [Boolean networks: gene regulatory network]
Consider the Boolean network in Fig. 5.7(1) that is an extension of the gene regulatory network in Fig. 5.1. The disjunctive normal forms defining the behavior of the Boolean network that are obtained by applying Algorithm 5.2 are in Fig. 5.7(3). Logical minimization of the disjunctive normal forms produces the equations

$$[a] = \overline{B} + BC \quad [A] = a \quad [b] = \overline{A} + AC \quad [B] = c + \overline{c}b \quad [c] = \overline{A} \quad [C] = c.$$

It is enough to have the positive next-state functions to define the behavior of the network, because when the expression evaluates to one the next state will be active, otherwise inactive.

The idea of logical minimization is to merge iteratively the terms of the disjunctive normal forms that differ for the status of just a single variable (see Example 5.18).

Example 5.18. [Logic minimization of disjunctive normal forms]
Consider the disjunctive normal form $\overline{a}bc + abc + \overline{b}c$. By merging the terms that differ for the status of a single variable iteratively, we obtain

$$\overline{a}bc + abc + \overline{b}c \;\rightarrow\; bc + \overline{b}c \;\rightarrow\; c.$$

5.3.2 *Petri nets*

Petri nets (PNs) have been used extensively in modeling distributed, concurrent systems and have been also adopted to model biological systems. They represent in a simple graphical form the evolution of systems by highlighting both the state configurations and the transitions between them. A Petri net coincides with a directed, bipartite graph.

A Petri net is made up of *places*, *transitions* and *arcs*. Transitions specify the events that may occur and the places are the enabling conditions for the events. Arcs connect places and transitions (see Fig. 5.8).

A subset Q of places of N is called a *trap* if each transition that removes a token from a place in Q also adds a token to some place in Q. A subset Q of places of N is called a *syphon* if each transition that adds a token to

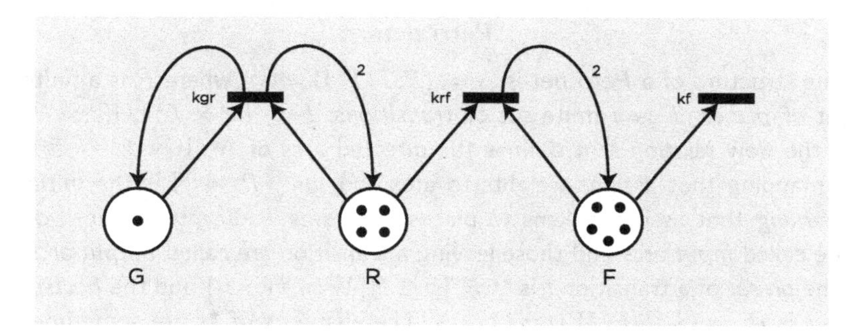

Figure 5.8 A Petri net where places contain tokens and transitions are labeled by stochastic rates. The numbers on the arcs denote the weight. Arcs without label are assumed to have weight 1.

Q also removes a token from Q. When a syphon is empty, it will remain empty in each reachable marking.

Biological entities in a process are usually represented by the places of a Petri net, while the reactions that modify the states of the systems are modeled by the transitions (see Example 5.19). The tokens in the places represent occurrences of molecules. States usually represent passive components of the system, while transitions denote active components. Arcs model abstract relationships between places and transitions.

Dynamic modeling of biological systems cannot abstract time. A *timed Petri net* (TPN) is a Petri net that allows delay in the firing of transitions and uses a clock to count time. Each transition is associated with an integer $i \geq 0$ called *delay*. Although timed transitions are the most common timed extensions to Petri nets, time may also be associated with

- *places*: tokens generated in an output place become available after a delay has elapsed;
- *tokens*: tokens are decorated with a time-stamp that determines when a transition can fire (time-stamps can be incremented after firing);
- *arcs*: flow delay of tokens is associated with arcs.

Timed Petri nets manage discrete time, but sometimes biological systems are approximated by ODEs assuming continuous time. The corresponding tool in the network framework is a *continuous timed Petri net* (CTPN).

Stochasticity is a main feature of biological systems. To take variability into account, *stochastic Petri nets* (SPNs) have been introduced. SPNs

Petri nets

The structure of a *Petri net* is $N = \langle P, T, F, W, m_0 \rangle$, where P is a finite set of *places*, T is a finite set of *transitions*, $F \subseteq (P \times T) \cup (T \times P)$ is the *flow relation* that defines the directed arcs of N, $W : F \to \mathbb{N}$ is a mapping that assigns weights to arcs and $m_0 : P \to \mathbb{N}$ is the *initial marking* that assigns tokens to places. The arcs leading to a transition are called *input arcs* and those leaving a transition are called *output arcs*. The *preset* of a transition t is ${}^\bullet t = \{p \in P \mid W(p,t) > 0\}$ and the *postset* of t is $t^\bullet = \{p \in P \mid W(t,p) > 0\}$. The elements of ${}^\bullet t$ are sometimes called *input places* and the elements of $t^\bullet$ *output places*.
The state of a Petri net is a distribution of tokens over the places, i.e., a marking. The *dynamic behavior* of the Petri net is then represented by laws determining the changes of markings and reflected by the *enabled transitions*. A marking enables a transition if the weight of the transition is smaller than the number of tokens in the source place. Formally, (p,t) is enabled if $W(p,t) \leq m(p)$ where m is the current marking. The behavior of a Petri net is nondeterministic if a marking enables more than one transition.
A *step* in the Petri net is the change of a marking into another by the firing of an enabled transition, written $m \xrightarrow{t} m'$. The new marking is defined as $\forall p \in N \,.\, m'(p) = m(p) - W(p,t) + W(t,p)$.
A marking m_n is *reachable* from the initial marking if there is a sequence of steps from m_0 to m_n and $RM(m_i) = \{m_j \mid m_i \to^* m_j\}$ is the set of the *reachable markings* from m_i.
In the graphical notation, places are represented by circles, transitions by bars or boxes, and arcs by arrows connecting circles and boxes. The weight is depicted by labeling the arrows (the default weight, not written on the arcs, is 1) and tokens are represented by dots inscribed into circles. The dynamics can be also defined as graph rewriting.

extend Petri nets by associating each transition with a stochastic rate representing the unique parameter of a probabilistic exponential distribution (see Fig. 5.8). Some basic notions on stochastic processes are recalled in Chapter 8.

Stochastic Petri nets can be related directly to chemical reactions. We show the relationship on the Lotka–Volterra example and we sketch a translation algorithm between chemical reactions and stochastic Petri nets.

Timed Petri nets

The structure of a *timed Petri net* is $TPN = \langle P, T, F, W, m_0, d \rangle$, where $\langle P, T, F, W, m_0 \rangle$ is a Petri net and $d : T \to \mathbb{Z}^+$, the delay function. Markings are parametrized with respect to time x such that $m(p, x)$ means the marking of place p at time x. Enabling of transitions is also dependent on time, so that (p, t) is enabled at time x if $W(p, t) \leq m(p, x)$. If (p, t) is enabled at time x and $d(t) = i$, the new marking after the firing of t is $\forall p \in N, m(p, x+i) = m(p, x) - W(p, t) + W(t, p)$. Sometimes timed Petri nets admit *immediate transitions* that do not depend on time and fire as soon as they become available.

Continuous timed Petri nets

The structure of a *continuous timed Petri net* is $CTPN = \langle P, T, I, O, W, S_i, S_o, m_0 \rangle$, where P and T have the usual meaning as in PN, $I \subseteq (P \times T)$ are the input arcs, $O \subseteq (T \times P)$ are the output arcs, $W : (P \times T) \to \mathbb{R}^+$ is the weight function of the input arcs, $S_i : (P \times T) \to \mathbb{R}^+$ and $S_o : (T \times P) \to \mathbb{R}^+$ are the input and output speed, respectively. A transition fires continuously producing continuous tokens as long as its condition on weight and marking is satisfied, and speeds are used to update time.

The informal description of the relationship between the chemical reaction of the Lotka–Volterra model in the Example 5.13 and the Petri net in Fig. 5.8 can be pushed further ahead to devise a mapping between the two formalisms (see Algorithm 5.3).

There is also an algebraic interpretation of Petri nets. Each Petri net can be represented by a matrix N. We assume an ordering $p_1, .., p_n$ on the places and $t_1, .., t_m$ on the transitions according to their indexes. The matrix is indexed by places on the rows and transitions on the columns. The entries of the matrix are $n_{ij} = W(t_j, p_i) - W(p_i, t_j)$. Each marking m can be represented as a column vector M whose entries are $a_i = m(p_i)$. We can now express the dynamic updating of markings in linear algebra notation as $M'(p_i) = M(p_i) + n_{ij}$ for each place p_i and each step $M \xrightarrow{t_j} M'$.

The linear system $N' \cdot x = 0$, where N' is the transpose of N, has a solution $\mathbf{n}$ that associates a constant n_i with each place p_i. The vector $\mathbf{n}$

Stochastic Petri nets

The structure of a *stochastic Petri net* is $SPN = \langle P, T, F, m_0, r\rangle$, where P, T and F have the usual meaning as in PN and $r : T\times(P \to \mathbb{N}) \to \mathbb{R}^+$ is the *rate function*. If the rate function is marking independent, it is $r : T \to \mathbb{R}^+$. Under the assumption that r defines exponential distributions, it is possible to associate a *continuous time Markov chain* (CTMC) with an SPN by defining its state space $RM(m_0)$ and by associating with each arc connecting m_i to m_j in a CTMC the sum of the rates of the transitions leading from m_i to m_j in an SPN. The generator matrix of a CTMC is

$$q_{ij} = \begin{cases} \sum_{k\in E_j(m_i)} r_k & \text{if } i \neq j \\ -q_i & \text{if } i = j \end{cases}$$

where $q_i = \sum_{k\in E(m_i)} r_k$, $E_j(m_i) = \{h \mid h \in E(m_i) \wedge m_i \xrightarrow{t_h} m_j\}$ and $E(m_i) = \{h \mid \exists j.\, m_i \xrightarrow{t_h} m_j\}$.

Example 5.19. [Petri nets: Lotka–Volterra]
The Petri net describing the behavior of the Lotka–Volterra system modeled in Example 5.13 as a system of chemical reactions is depicted in Fig. 5.8. We associate a place with each species we are interested in: foxes, rabbits and grass. We then model their interactions by three transitions corresponding to the chemical reactions in Example 5.13. The stoichiometry of the reaction is represented in the Petri net by the weights on the arcs. For instance, the integer 2 labeling the arc from the transition labeled kgr states that a rabbit (a token in the place for rabbits) can be consumed to produce two new tokens in the same place. The interpretation is that a rabbit can reproduce.

defines the *place invariants* of the Petri net. In fact, for each reachable marking m, it holds that $\sum_i n_i m_0(p_i) = \sum_i n_i m(p_i)$.

Similarly, the linear system $N \cdot x = 0$ has a solution $\mathbf{m}$ that associates a constant m_j with each transition t_j. The vector $\mathbf{m}$ defines the *transition invariants* of the Petri net. Transition invariants are related to sequences of steps that reproduce the initial marking, i.e., the final marking of the sequence of steps is again m_0. We define the *counting vector* of a sequence of steps as the $\mathbf{m}$-dimensional vector $\mathbf{c}$ such that c_j is the number of transitions t_j occurring in the sequence. Then, if $\mathbf{m}$ is the counting vector of a sequence of steps σ, σ reproduces the initial marking. The matrix N that represent a Petri net corresponds to the stoichiometric matrix of the corresponding system of chemical reactions, while the linear equations $N' \cdot x = 0$ and $N \cdot x = 0$ correspond to the flux and species balance at steady state.

There is also an algorithm mapping ODEs into stochastic Petri nets. Figure 5.9 shows an application of Algorithm 5.3 to an enzymatic reaction (see also Example 5.2). The very same result is obtained by starting from the ODE representation of the enzymatic reaction and applying Algorithm 5.4.

There are many extensions of PNs that are relevant to model biological systems. We mention here *colored Petri nets* and *fluid Petri nets.*

The primary goal of colored Petri nets is to enhance the ability of representing data types of Petri nets by allowing more token types (distinguished by different colors) and by defining guard conditions on such types to decide which are the enabled transitions. They are essentially mixing the ability of the graphical formalism of Petri nets to represent synchronization issues between processes and the ability of programming languages to define and manipulate data types as well as conditionally control the flow of instruction execution.

Fluid Petri nets enhance Petri nets by allowing places with a fluid amount of tokens (continuous value). The outcome of the extension is to have a mixed model that manages systems with both discrete and

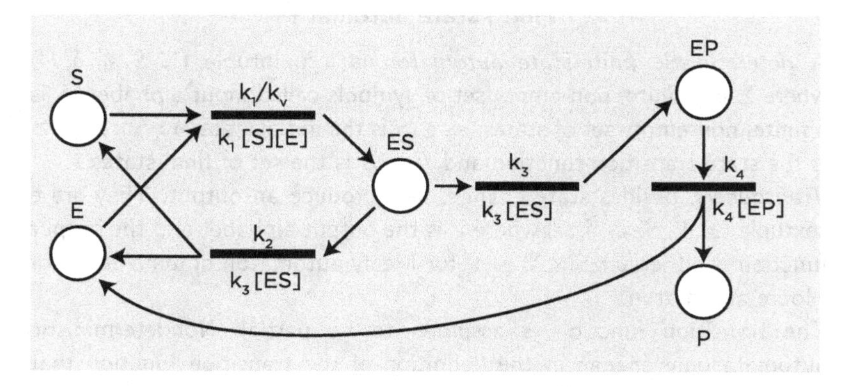

Figure 5.9 Application of Algorithm 5.3 to the enzymatic reaction in the Example 5.2. The very same result is obtained by Algorithm 5.4 starting from the ODE representation of the enzymatic reaction. For simplicity places (circles) and transitions (square boxes) contain the ODE variables and terms that generate them.

Algorithm 5.3: From chemical reactions to stochastic Petri nets

Input: A set of chemical reactions CE.
Output: A stochastic Petri net with behavior equivalent to CE.

begin
generate a place in the Petri net for each species in CE;
generate a transition in the Petri net for each reaction in CE;
for *each transition t in the Petri net generated by a reaction r in CE* **do**
add an arc between each place representing the reactants in r and the transition t;
add an arc between the transition t and the places representing the products of r
end
associate with each arc the stoichiometric coefficient of the corresponding reaction;
for *each* $p \in P$ **do**
$m_0(p) := n$ with n abundance of the species that generated p in the initial configuration of CE
end
label each transition t with $k * (m_0(p_1))^{w_1} * \ldots * (m_0(p_n))^{w_n}$, where k is the stochastic kinetic constant obtained by transforming the constant associated with the reaction in CE that generated t, ${}^{\bullet}t = \{p_1, .., p_n\}$, and w_i is the weight associated with the arc (p_i, t)
end

Finite-state automata

A *deterministic finite-state automaton* is a quintuple $(\Sigma, S, s_0, \delta, F)$ where Σ is a finite, non-empty set of symbols called input alphabet, S is a finite, non-empty set of states, $s_0 \in S$ is the initial state, $\delta : S \times \Sigma \rightarrow S$ is the state transition function and $F \subseteq S$ is the set of final states.
Transducers, besides state change, also produce an output. They are e sextuple $(\Sigma, \Gamma, S, s_0, \delta, \omega)$ where Γ is the output alphabet and the output function is either $\omega : S \times \Sigma \rightarrow \Gamma$ for Mealy automaton or $\omega : S \rightarrow \Gamma$ for Moore automaton.
The transition function is assumed to be partial. Nondeterministic automata only change in the definition of the transition function that becomes $\delta : S \times \Sigma \rightarrow 2^S$. Stochastic automata are obtained by defining a stochastic transition rate $\delta : S \times \Sigma \times \mathbb{R} \rightarrow S$, where $\mathbb{R}$ is the unique parameter defining an exponential distribution.

Algorithm 5.4: From ODEs to stochastic Petri nets

Input: A system of ODEs.
Output: A stochastic Petri net with behavior equivalent to the ODEs.

begin
  generate a place in the Petri net for each variable in the left sides of each equation;
  generate a transition in the Petri net for each unique term on the right sides of the ODEs without duplications and label the transition with the constant k' obtained by transforming the constant k in the corresponding term into a stochastic rate (see Fig. 5.6);
  for *each occurrence of the term x with constant k in the ODEs that generates transition t* **do**
    if *the sign of x is $+$* **then**
      add an arc from t to the place corresponding to the variables occurring in the left side of the equation containing the occurrence of x considered
    end
    if *the sign of x is $-$* **then**
      add an arc from each place corresponding to the variables in the occurrence of x considered to the transition t
    end
  end
end

continuous states. We do not study colored Petri nets and fluid Petri nets in details, but pointers to the relevant literature are given in Sect. 5.8.

5.4 Automata-based approaches

Finite-state automata (FSA) are a formal model of computation used to study and design algorithms and electronic devices. The automaton can be in only one of a finite number of states at a time and can change its current state when triggered by an event or a condition with a transition. FSA are defined by the list of states they can transit and by the event and conditions triggering state transitions. The computational power is limited by the memory of the automaton, corresponding to the number of states. If the automaton produces an output at each state change, it is usually called a transducer and can be either of Mealy type (the output depends on the current state and an input) or of Moore type (the state change depends only on the current state).

Stochastic finite-state automata (SFSA) are obtained by FSA by replacing the transition function with a stochastic one where the additional parameter defines a probabilistic distribution.

FSA have inspired many formalisms reviewed so far such as Boolean networks and Petri nets. They find a major representative in cellular automata and are the theoretical basis of language-based formalisms (see Chapter 6).

Cellular automata

A *cellular automaton* is a quadruple $M = (\Gamma, Q, N, \delta)$ where Γ is the *interconnection graph* (typically an n-dimensional grid), Q is a set of *states*, $N : \Gamma_n \times \mathbb{N} \to 2^{\Gamma_n}$ (Γ_n being the nodes of Γ) is the *neighborhood* (a mapping that identifies the set of cells neighboring within a distance equal or less to the second parameter of a given cell) and $\delta : \Gamma_s \times 2^{\Gamma_s} \to \Gamma_s$ (Γ_s being the state of the nodes of Γ) is the *local dynamics* (or transition function). The pair (Γ, Q) is the *cell space* of M. A *configuration* x of a CA is a mapping $x : \Gamma_n \to Q$ that assigns a state to each node of Γ.

The transition function computes the next state of each cell $\delta(x(i), x(N(i,r))) = s', i \in \Gamma_n$, where $r \in \mathbb{N}$ is the parameter defining the distance for the neighborhood.

The *global dynamics* of CA is a mapping $\gamma : C \to C$ from the set of the configurations to itself. The global state is the union of all the local states.

5.4.1 *Cellular automata*

Cellular automata (CA), originally invented by von Neumann, have a *space* represented by a uniform n-dimensional grid of identically programmed cells containing some data and interacting with one another in a neighborhood. Time advances in discrete steps and rules (usually a finite-state machine) define how each cell computes its new state given its current state and the states of its closest neighbors based on a distance notion (see the box above). To manage the cells on the edge of the grid, we assume a toroidal arrangement of cells so that the extreme on the right is connected with the extreme on the left and the extreme on the top is connected with the extreme on the bottom of the grid.

Example 5.20. [Cellular automata: game of life]
The most famous example of cellular automata is the *game of life*. The CA is a two-dimensional grid in which each cell has two states: dead or alive. An alive cell stays alive if it has two or three alive neighbors, otherwise it dies. A dead cell becomes alive when it has exactly three alive neighbors.

Example 5.21. [Cellular automata: neighborhood]
Given a cell i represented in a bidimensional grid by its cartesian coordinates (x_i, y_i) and a distance measure r we can define the *von Neumann neighborhood* of radius r as $N(i,r) = \{j \in \Gamma_n \mid |x_j - x_i| + |y_j - y_i| \leq r\}$. The *Moore neighborhood* of radius r $N(i,r) = \{j \in \Gamma_n \mid |x_j - x_i| \leq r,\ |y_j - y_i| \leq r\}$. Figure 5.10 reports graphically the neighborhood of a cell in the case $r = 1$.

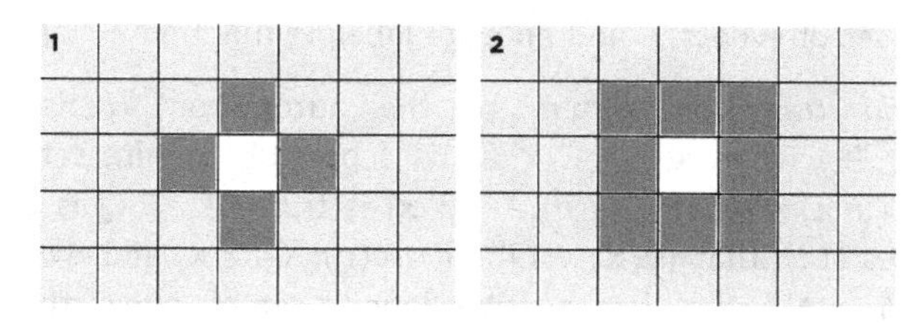

Figure 5.10 The gray cells are the neighborhood of the black cells. The von Neuman neighborhood is on the left and the Moore neighborhood is on the right with $r = 1$.

All the cells compute their next state together according to the same algorithm (the same FSA) and update their state. It is possible to implement stochastic cellular automata by replacing the deterministic transition function with a stochastic transition function as done for state machines.

5.4.2 *Hybrid automata*

Hybrid automata is a formalism introduced to model mixed discrete-continuous systems. The dynamics of hybrid automata consists of continuous (flows) and discrete (jumps) changes, but it can be abstracted by discrete *timed transition systems* that model flows by transitions recording the source, target and duration of changes. Hybrid automata can be composed relying on a parallel composition operator to build models iteratively.

Hybrid automata

A *hybrid automaton* is a made up of:

— A finite set of real-valued variables $X = \{x_1, .., x_n\}$ where n is the *dimension* of X. $\dot{X} = \{\dot{x}_1, .., \dot{x}_n\}$ denotes first derivatives of variables during continuous change and $X' = \{x'_1, .., x'_n\}$ denotes the values of the variables at the end of discrete changes.
— A finite directed multigraph (V, E). The nodes V are cooled *control modes* and the edges E *control switches*.
— Three node labeling functions i, inv and fl. i and inv assign to each node a predicate with free variables in X. fl assigns to each node a predicate whose free variables are in $X \cup \dot{X}$.
— An edge labeling function j that assigns to each edge a predicate with free variables in $X \cup X'$.
— A finite set of events Σ and an edge labeling function $ev : E \to \Sigma$.

The *hybrid transition system* of the automaton H is $S^t_H = \langle Q, Q^0, A, \xrightarrow{a} \rangle$ where $Q \subseteq V \times \mathbb{R}^n$ is a possibly infinite set of *states* such that $(v, \mathbf{x}) \in Q$ iff $inv(v)[X := \mathbf{x}]$ is true; $Q^0 \subseteq Q$ is the set of *initial states* such that $(v, \mathbf{x}) \in Q^0$ iff $inv(v)[X := \mathbf{x}]$ and $i(v)[X := \mathbf{x}]$ are true; $A = \Sigma \cup \mathbb{R}_{\geq 0}$ is a possibly infinite set of *labels*; the labeling relation $\xrightarrow{a} \subseteq Q \times A \times Q$ is defined as follows:

— $\forall \sigma \in \Sigma, (v, \mathbf{x}) \xrightarrow{\sigma} (v', \mathbf{x}')$ iff $\exists e = (v, v') \in E$, $ev(e) = \sigma$ and $j(e)[X, X' := \mathbf{x}, \mathbf{x}']$ is true;
— $\forall \delta \in \mathbb{R}_{\geq 0}, (v, \mathbf{x}) \xrightarrow{\delta} (v, \mathbf{x}')$ iff $\exists f : [0, \delta] \to \mathbb{R}^n$ differentiable with first derivative f' and such that $f(0) = \mathbf{x}$, $f(\delta) = \mathbf{x}'$ and $\forall \epsilon \in (0, \delta)$ both $inv(v)[X := f(\epsilon)]$ and $fl(v)[X, \dot{X} := f(\epsilon), f'(\epsilon)]$ are true.

The *abstract hybrid transition system* of H is $S^a_H = \langle Q, Q^0, B, \xrightarrow{b} \rangle$ where Q and Q^0 are as before; $B = \Sigma \cup \{\tau\}, \tau \notin \Sigma$; the labeling relation $\xrightarrow{\sigma}$ is as before and $(v, \mathbf{x}) \xrightarrow{\tau} (v', \mathbf{x}')$ iff $\exists \delta \in \mathbb{R}_{\geq 0}$ such that $(v, \mathbf{x}) \xrightarrow{\delta} (v, \mathbf{x}')$.

5.5 Relationship between continuous and stochastic models

Much of this chapter is about the interactions between stochastic and continuous formalisms. An approximation in between the two extremes of the

Composition of hybrid automata

A *consistency check* for two labeled transition systems S_1 and S_2 is an associative, partial function $\otimes : \xrightarrow{a_1}_{S_1} \times \xrightarrow{a_2}_{S_2} \rightarrow A_1 \cup A_2$ where A_1 and A_2 are the labeling alphabet of the two transition systems.

The product of the two transition systems with respect to $\otimes$ is $S_1 \times S_2 = \{Q_1 \times Q_2, Q_1^0 \times Q_2^0, range(\otimes), \xrightarrow{a}\}$ such that $\forall a \in range(\otimes), (q_1, q_2) \xrightarrow{a} (q_1', q_2')$ iff $\exists a_1 \in A_1, a_2 \in A_2. \otimes (a_1, a_2) = a, q_1 \xrightarrow{a_1} q_1', q2 \xrightarrow{a_2} q_2'$.

The *composition* of two hybrid automata H_1 and H_2 produces the hybrid transition system $S^t_{H_1||H_2} = S^t_{H_2} \times S^t_{H_2}$ with respect to $\otimes$ defined as

$$\otimes(a_1, a_2) = \begin{cases} a_1 & \text{if } a_1 = a_2 \text{ or } (a_1 \in \Sigma_1/\Sigma_2 \text{ and } a_2 = 0) \\ a_2 & (a_2 \in \Sigma_2/\Sigma_1 \text{ and } a_1 = 0). \end{cases}$$

spectrum is the chemical master equation (CME — see the box on page 314 in Chapter 8). It is used to show that the exact description of systems by stochastic formalisms averages to the CME solution over many runs of the system.

5.6 Diagrammatic modeling

A graphical notation focuses on the intuitive description of the system. Modeling with a graphical language is an activity comparable to producing flow charts, video animations or illustrations in print media.

Most graphical languages use symbols to represent different types/classes of nodes in the model (e.g., enzymes, cytokines, metabolites, etc.). This is useful up to a point: the more symbol choices a user has, the harder the model becomes to decode. Therefore, a minimal alphabet of symbols generating a language still expressive enough to model most of the relevant biological systems is a desirable characteristic of a graphical formalism. Relevant examples are SBML and SBGN. The Systems Biology Mark-up Language (SBML) is an XML-based file format, developed by a consortium including experts from computer science, mathematics and biology. The Systems Biology Graphical Notation (SBGN) is closely related to SBML and is a standardized graphical representation for signaling pathways, metabolic networks and gene regulatory networks.

In its basic form, graphical languages consist of edges and nodes, where nodes represent entities within the cell and edges stand for an interaction between two or more entities. A cell is formed of several well-known compartments, such as the cytoplasm, the nucleus, the mitochondria and organelles. Graphical languages should consider support for these structures. As models become larger, biological compartments provide a natural way to cluster and subdivide nodes. Graphical models describing biological processes will almost always have a set of edges crossing. The eye can easily follow the course of curvy edges, which is not the case for straight lines due to 90° cross angles.

Illustrations of biological concepts are widely used in textbooks and journals. These graphical representations are rich in information and have shaped the expectation for graphical modeling languages. Indeed, BioCarta is perceived as having the best pathway diagrams, since its illustrations are very close to textbook style. The labeling of edges and nodes has an important impact on the aesthetic appearance of a graphical model. For nodes, the challenging aspects are long protein names, and how they are graphically resolved. Edges, on the other hand, are an ideal space to provide more details about kinetic reactions.

There has been a considerable effort to produce graphical languages with a well-defined semantic, allowing researchers to write mathematical models of biological systems without the need for acquiring programming skills. There has been little research in the actual choice of symbols, neither for the standard SBGN nor for the proprietary languages. We argue that current approaches to graphical languages could be improved in two ways: first by making the actual symbols less cluttered, and second, by integrating symbols that are already familiar in the biological field.

Following an analysis of existing graphical formalisms, we propose a new graphical language Style. It has been designed by surveying biology textbooks and journal articles combined with interviewing scientists in the field. A minimal set of symbols was obtained via an iterative process. Style consists of six space types, five element types and three arrow types.

One of the arguments that emerged for most existing graphical languages was their low information density. Figure 1 in [Le Novère *et al.* (2009)] shows how the same graphical symbol can be used with different meanings and how a meaning can be associated with many different symbols.

In our design proposal for Style we aimed for a small number of biologically meaningful symbols which are able to express a wide variety of biological models and for a graphical solution that is compact and similar to textbook illustrations.

5.6.1 *Elements*

Figure 5.11 presents the five elements of Style, namely small molecules, proteins, nucleic acids, generic elements and undefined precursors. Each element has a distinctive shape and color, allowing the eye to immediately spot differences.

Informed by textbooks and interviews, we reserved the round shape for proteins (see ellipse in Fig. 5.11), and select a contrasting shape for small molecules (rectangle). Nucleic acids represent molecules carrying genetic information (genes or transcripts). These elements are depicted via a simplified double helix structure, outlining their close relationship with DNA. The undefined precursors are small elements from which cells synthesize, or peptides into which proteins degrade. In addition, Style offers a generic element, allowing the language to integrate custom components.

Each element comes in two versions: normal (first row in Fig. 5.11) and spiky (second row). This graphical difference highlights a state change for a single element: for example Fig. 5.12 (left) shows an activation of the protein RAF and (right) a chain of phosphorylation, where the element phosphorylates twice until it reaches the final configuration P". Naming the final protein P" in a round shape would work, but by changing only the shape and maintaining the same name, a user can express graphically that one is looking at the same protein but in a different state. Further, spiky shapes have the potential to guide the eye immediately to interesting points within a pathway.

Style permits duplicates. A single element can influence more than one element: for example in Fig. 5.12 (right), the activated enzyme E influences

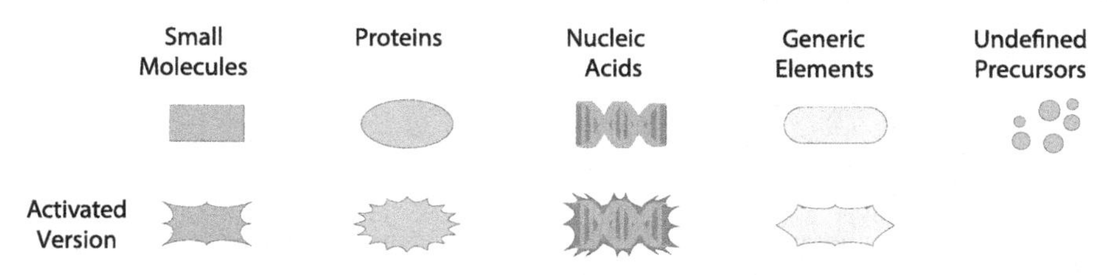

Figure 5.11 The five species elements in Style; top row shows inactive elements, bottom row presents the same elements in their activated form. Reproduced with permission from [Gostner *et al.* (2014)].

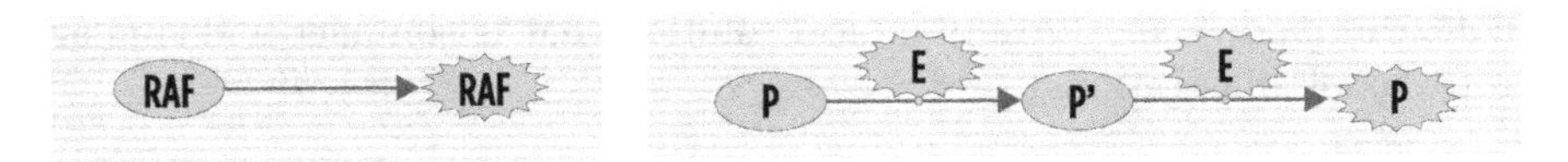

Figure 5.12 Left: a reaction that makes an inactive element active. Right: example of enzymatic reactions where one substrate catalyzes two reactions. Reproduced with permission from [Gostner *et al.* (2014)].

two reactions. A solution alternative to the one depicted (admitting duplicates) is representing each element only once in the pathway by using lines (similar to wiring lines in engineering) to indicate their related reactions (as implemented in CellDesigner). We exposed users to the wiring alternative and they reported difficulties in understanding the meaning: their first guess was that the lines represent reactions rather than the influence of the enzyme on the reaction.

Style proposes a simple cell template, with pre-defined elements such as extracellular space, cell membrane, cytosol, nucleus, mitochondria, organelles and a generic element (Fig. 5.13). All space elements form background information.

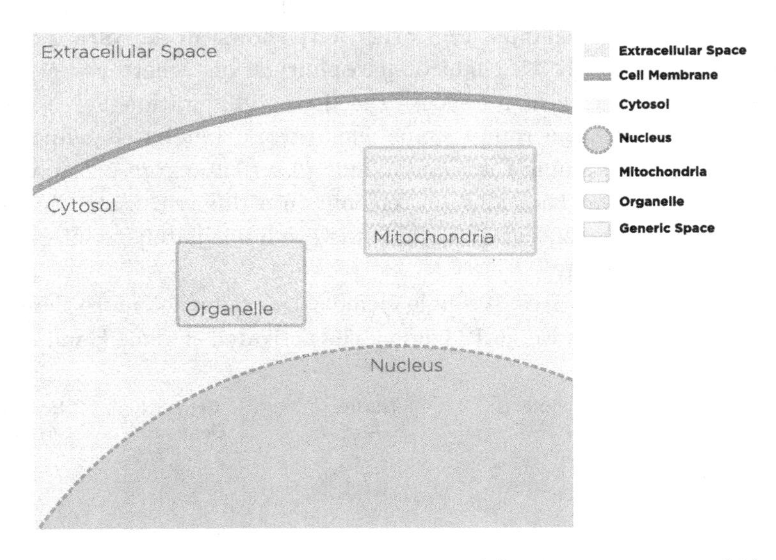

Figure 5.13 Textures and patterns for illustrating different compartments within a cell. Each compartment is characterized by a volume and will be populated by the elements representing proteins, small molecules, DNA-related matter, generic elements and raw material. Each element will be characterized by its concentration or number of molecules. Reproduced with permission from [Gostner *et al.* (2014)].

Figure 5.14 Style reactions. From left to right: modify, inhibit and translocate. Reproduced with permission from [Gostner *et al.* (2014)].

5.6.2 *Reactions*

Style consists of three distinct reaction types (Fig. 5.14), namely modification (left), inhibition (middle) and translocation (right) between different cell compartments. Style uses the context of elements to break down the meaning into finer-grained reactions, rather than introducing new types of arrows.

5.6.2.1 *Modification*

The simplest reaction, such as a substrate RAF becoming activated ($RAF \rightarrow RAF\ Activated$), can be expressed by connecting two shapes with a modify arrow as in Fig. 5.12 (left). If the reaction is performed with the aid of an enzyme, as for example in $P+E \rightarrow P'+E \rightarrow P\ Activated+E$, then the supporting enzyme is placed directly on the reaction arrow as in Fig. 5.12 (right). Style also permits a group of enzymes to influence a reaction, by stacking elements one behind the other, and relating them with a logical operator (Fig. 5.15).

Modification reactions often require stoichiometry information, such as $A + A \rightarrow B$. For such cases, Style proposes an effective use of arrow labels indicating the required number of elements (Fig. 5.16 left). Novel complexes are formed by association of two or more elements, for example $A + B \rightarrow C$. There, the graphical representation consists of two modify arrows merging into one, expressing an association (Fig. 5.16 middle). When inverting the direction, the modify arrow expresses a dissociation (Fig. 5.16 right). It is common practice to have unnamed complexes, meaning that an association does not have an explicit name. In such cases, Style duplicates the starting elements and surrounds them

Figure 5.15 Approaches on how to handle graphically multiple catalyzers for a single reaction. Besides the classical logical operators (the three cases on the left), the user can also define his own composing function (last on the right). Reproduced with permission from [Gostner *et al.* (2014)].

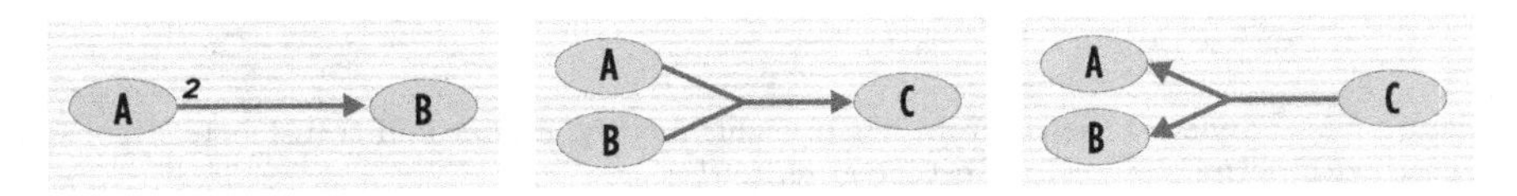

Figure 5.16 Association and dissociation between two or more species. In the leftmost case, the number on the arrow close to protein A is the stoichiometry of the reaction (2 As are needed to form 1 B). Reproduced with permission from [Gostner *et al.* (2014)].

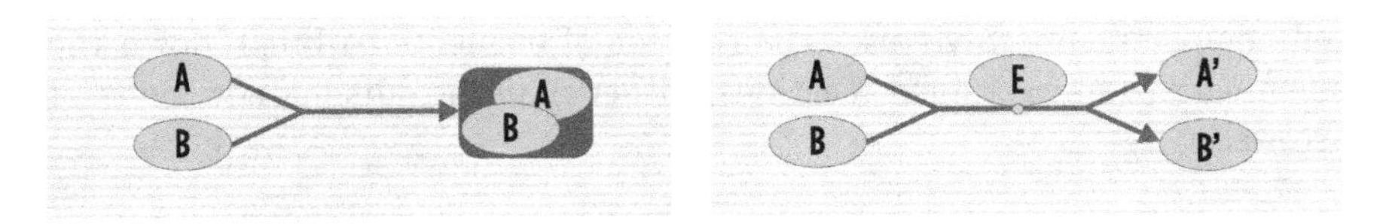

Figure 5.17 Left illustrates how complexes are visualized. Right shows a reaction with cofactors. Reproduced with permission from [Gostner *et al.* (2014)].

with a rectangle (Fig. 5.17 left). Finally, reactions with cofactors, such as $A + B(\textit{cofactor}) \rightarrow A'(\textit{product}) + B'(\textit{cofactor modified})$, facilitated by the enzyme E, is still expressed via a modify arrow (Fig. 5.17 right). As these examples illustrate, Style proposes context information to define the semantics of the language, thereby keeping the number of different symbols at a minimum.

5.6.2.2 *Inhibition*

Inhibition is attached to enzymatic reactions via a T-bar arrow (Fig. 5.18). Style allows the inhibition arrow pointing directly to the enzyme of a reaction (Fig. 5.18 right) because this is often advantageous from a graphical point of view (though both pictures have semantically the same meaning). There are a range of different inhibitions, such as Michaelis–Menten (in its

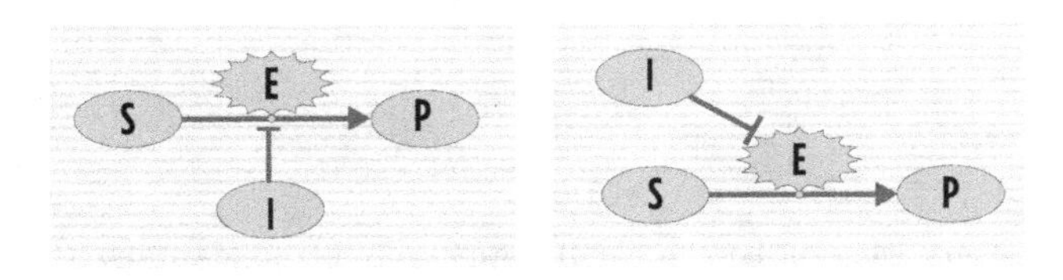

Figure 5.18 Inhibition reactions: here the element I inhibits the reaction between S and P. Reproduced with permission from [Gostner *et al.* (2014)].

several forms, such as cooperative, competitive and non-competitive inhibition), mass action law or custom defined functions. Style aims to depict a general inhibition, while the exact details are omitted for clarity.

5.6.2.3 *Translocation*

Style proposes two types of translocation: when an element moves unaided from one space to another (passive transport, often gradient-driven — Fig. 5.19 left), and when a protein/complex in a membrane transfers an

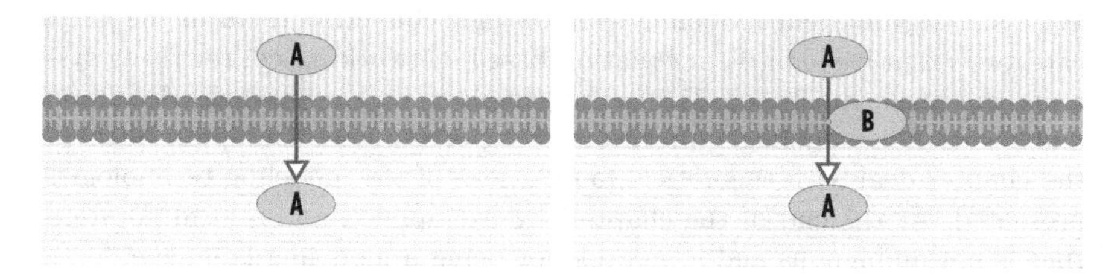

Figure 5.19 Translocation of A through a membrane to a different compartment. Left: passive translocation; right: active translocation aided by the element B. Translocations can move objects from one compartment to another and in both directions. Note that the presence of the protein B within the membrane makes membranes first-class compartments as well. Reproduced with permission from [Gostner *et al.* (2014)].

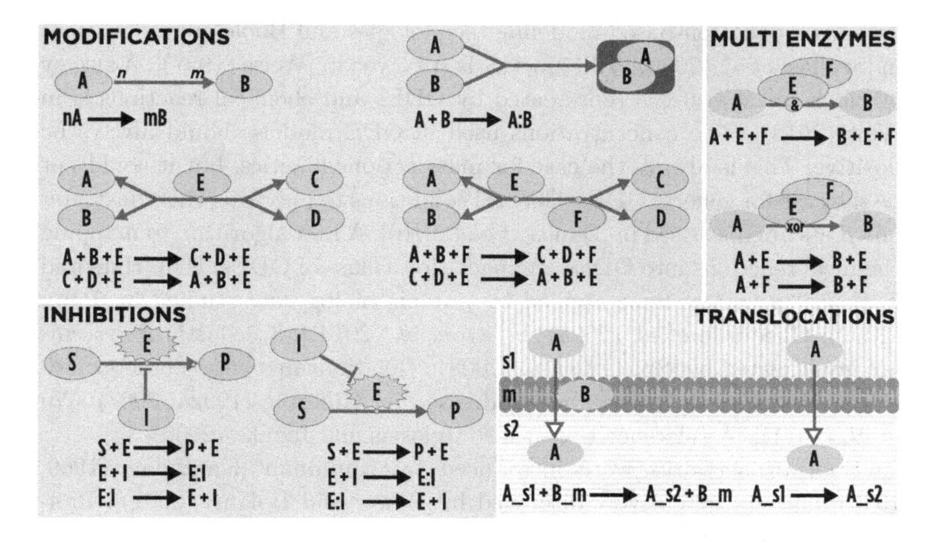

Figure 5.20 From Style to chemical reactions and back.

element from one space to another (active transport — Fig. 5.19 right). It is important to note that the name of the translocating protein or complex remains the same.

It is possible to map Style diagrams into chemical reactions as illustrated in Fig. 5.20. An example of the application of this translation is in Example 6.36.

5.7 Summary

We discussed different modeling technologies, clustering them in the main categories that we identified: equation-based, network-based, rewriting systems and automata-based. For each technology we surveyed the definition and intuition in modeling applications by reporting some examples. Boxes in the chapter deal with theoretical foundations of the modeling technologies presented. At the end of the chapter, we exemplified a graphical notation that is not ambiguous and is suitable for being a modeling language.

5.8 Further reading

Early adoption of ODEs in biological modeling dates back to the early 20th century [Volterra (1926)] and a seminal paper is also [Noble (1960)]. Further details on equation-based modeling technologies and Boolean networks are in [Szallasi *et al.* (2006)]. Hill kinetics is surveyed in [Weiss (1997)]. A survey of biochemical systems represented by ODEs and chemical reactions is in [Voit (2013)]. The concentrations used in ODE models should always be positive. This is always the case for mass actions kinetics, but it could not be the case for general ODEs. General conditions to imposing this constraint on ODEs are discussed in [Danos *et al.* (2010)]. A first algorithm to mapping chemical reactions into ODEs and back (for a class of ODEs) is in [Hárs and Tóth (1979)]. P-systems applied to systems biology are studied in [Păun (2002); Gheorghe *et al.* (2008); Frisco *et al.* (2014)]. Stochastic P-systems are investigated by [Spicher *et al.* (2008)]. Relationships between deterministic and stochastic kinetics are addressed in [Gillespie (1992a, 2007); Wu *et al.* (2011)]. A reference text is also [Iglesias and Ingalls (2009)].

Boolean networks were introduced by Kauffman [Kauffman (1969, 1993)] and are extensively discussed in [Bower and Bolouri (2001)]. Random and probabilistic extensions are dealt with in [Gershenson (2004)].

Relationships between ODE and Boolean models are analyzed in [Krumsiek *et al.* (2010)].

Petri nets have been used extensively to model biological systems [Goss and Peccoud (1998)] and there are many variants. A general introduction to Petri nets is [Reisig (1985)]. Timed extensions of Petri nets, both discrete and continuous, can be found in [Reisig and Rozenberg (1998); David and Alla (2005)]. Stochastic Petri nets and their variants are reported in [Aimone Marsan *et al.* (1995); Haas (2002)]. A reference collecting various approaches based on Petri nets is [Koch *et al.* (2011)]. In particular, Chapter 3 [Reisig (2011)] provides a general overview of Petri nets, Chapter 6 introduces timed Petri nets, continuous timed Petri nets and hybrid functional Petri nets [Saito *et al.* (2011)], Chapter 7 [Mura (2011)] introduces stochastic petri nets and a mapping from chemical reactions, Chapter 11 [Hardy and Iyengar (2011)] for a mapping of ODEs into Petri nets. Other extensions of Petri nets exist: colored Petri nets are introduced in [Jensen (1997)] and fluid Petri nets are considered in [Trivedi and Kulkarni (1993)].

A good introduction to the algorithmic nature of biological processes and to the use of automata theory to model them is [Bauer and Martinez (1974)]. Relevant work on the usage of interacting automata to model biological systems and to compare them with classical ODE or chemical reaction representations is in [Cardelli (2008a,b,c)]. Additional material on the relations between stochastic formalisms (including language-based formalisms) and continuous models is in [Lecca *et al.* (2013)]. Cellular automata are studied in [Weimar (1998); Schiff (2008)].

Hierarchical modeling and coherent mapping between different representations is addressed in [Mangold *et al.* (2005); Faeder (2011)].

An interesting requirement analysis for graphical representation of biological systems is [Saraiya *et al.* (2005)]. Style is introduced in [Gostner *et al.* (2014)].

Details on the biological content of this chapter can be found in [Alberts *et al.* (2002)] and [Lodish *et al.* (2004)]. The biology of EGFR is extensively described in [Sigismund *et al.* (2013)].

Chapter 6

Language-based modeling

We introduce modeling technologies based on programming languages under the assumption that biological elements are represented by pieces of code (processes, programs) and their interaction is modeled by message passing/data sharing or scope redesign of the corresponding programs. Dynamic evolution of biological systems is then represented by the state changes of the execution of the programs. The very first approach dates back to Fontana and Buss, who used λ-calculus to model biological systems. Subsequent works based on process calculi had more dissemination. First we introduce process calculi and their derived languages, then we introduce domain-specific languages (their formal semantics is given in Appendix C).

6.1 Process calculi

Process calculi describe systems by using a very limited number of primitives to build basic bricks called *processes*. Processes are composed by sequentializing action chains, declaring names or modeling alternatives through nondeterministic choices.

The interplay between name definition and scope of the definitions within a system gives rise to subtle issues. In particular, the meaning of a process is independent of the choice of its bound names. It is like changing the variable names in mathematical equations. Systems are represented by processes that are composed by algebraic operators for sequential, nondeterministic and parallel composition (indeed, process calculi are also called process algebras). Systems evolve over time through interaction represented via message-passing between processes (see the boxes).

Process calculi: prefix, definitions, substitution, α-conversion

Processes are built from *elementary actions* that are $snd(a, v)$ (send/write/output a value v on the channel a) and $rcv(a, x)$ (receive/input/read a value/name v from a channel a and instantiate all the future occurrences of the *target variable* x with v). The first argument of the action is sometimes called *subject*, while the second argument is called *object* of the action. Elementary actions are *complementary* if they are different, but act on the same channel.
Elementary actions can be chained into sequences or *sequentialized* through the infix (.) *prefix operator*. The process nil cannot perform any action and is used to terminate prefix chains $snd(a, v).rcv(a, x).nil$. Names can be defined through the *new name* operator $new\, a$ that is also used to clarify the *scope* in which a is known/visible/available/private in a system. The argument of a definition *bounds* the occurrences of the same name in the scope of the definition (e.g., P is the scope of the definition of a in $(new\, a)P$). The rcv action also binds the occurrences of its target variable x in the scope of the action and hence rcv acts as a definition of x. The same name defined in different parts of the system represents different things. The names that are not bound, are said to be *free* or public (see Example 6.1).
A *substitution* $Q\{v/x\}$ is a function that replaces all the free occurrences of x in Q with v. The substitution is *valid* if $bn(Q) \cap \{v\} = \emptyset$, where bn is a function that computes the bound names of the argument process. Substitutions that are not valid *capture names*, i.e., names that were free before the substitution become bound after. α-*conversion* changes bound names avoiding captures, i.e., $(new\, a)P$ is α-equivalent to $(new\, b)P\{b/a\}$, if $b \notin fn(P)$ with $fn(P)$ the function that computes the free names of P (see Example 6.2).

Example 6.1. [Free and bound names]
In $(new\, a)(snd(a, v).(new\, a)(rcv(a, x).snd(b, x).nil))$, the leftmost definition binds the a in the leftmost snd, the second definition of a binds the a in the rcv (the two as are different) and the rcv binds the x in the rightmost snd. The names v and b are free.

Example 6.2. [Substitutions and captures]
If $Q = rcv(a, v).snd(x, w).nil$, the substitution $Q\{v/x\}$ is not valid because the subject of the send that is free in Q would become bound by the object of the rcv after the substitution. Changing a into b in $(new\, a)snd(b, a).nil$ is not an α-conversion because captures b in the snd.

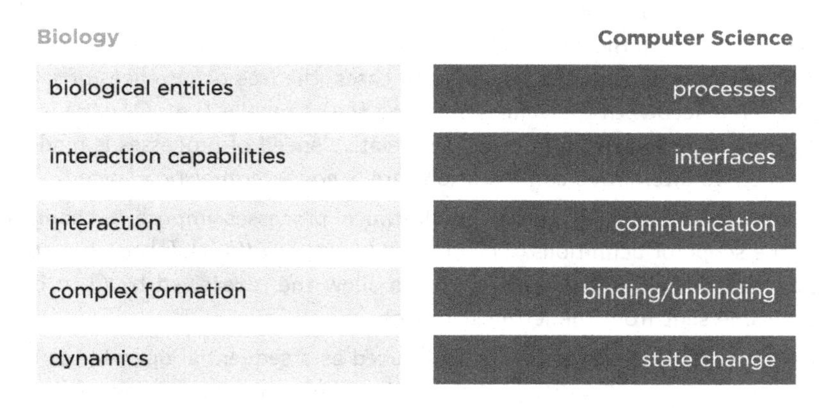

Figure 6.1 Metaphor for modeling biological systems with programming languages.

The behavior of biological systems is then represented by the configurations that the programs originate. The metaphor for modeling biological systems with process calculi is reported in Fig. 6.1. Process calculi are algorithmic and quantitative, interaction-driven, compositional, scalable and modular, address parallelism and complexity, and express causality.

Throughout the rest of the book we will use the *Backus–Naur form* (BNF) to describe the syntax of calculi and languages (see the box on page 191).

The meaning and the behavior of a system cannot be dependent on the way in which processes are listed in parallel compositions or whether they are on the left or right side of a choice. In other words, we need to define equivalence classes of processes that are syntactically different but semantically identical and define the behavior on these equivalence classes. Furthermore, the equivalence relation must be a congruence (a context-independent equivalence relation). We call the congruence *structural congruence* to highlight that the syntactical structure of a system does not affect its semantics.

Process calculi: parallel composition, choice, recursion

Processes may run independently *in parallel* and possibly *synchronize* on complementary actions. If P, Q and R are processes,

$$snd(a,v).P \,|\, rcv(a.x).Q \,|\, snd(a,w).R$$

may evolve into either $P \,|\, Q\{v/x\} \,|\, snd(a,w).R$ when the leftmost snd synchronize with the rcv or $snd(a,v).P \,|\, Q\{w/x\} \,|\, R$ when the rightmost snd synchronize with the rcv. In both cases, the free occurrences of x in Q are replaced with the value sent along the channel a (i.e., $Q\{v/x\}$ is a *substitution*). No assumption on the relative speed of processes is made, so the two alternative configurations are a *nondeterministic choice*.

Communication of bound names between processes impose a change in the scope of definitions. The process $(new\, a)snd(b,a).P \,|\, rcv(b,x).Q$ must become $(new\, a)(P \,|\, Q\{a/x\})$ to allow the a received by Q to be the same sent from the leftmost process.

Nondeterministic choice ($+$) is introduced as a sequential operator as in $snd(a,v).P + snd(a,w).R \,|\, rcv(a.x).Q$. The intuitive meaning of $+$ is that either the left or the right operand is selected and executed and the non-selected operand is discarded forever. Therefore the process above can evolve into either $P \,|\, Q\{v/x\}$ when the leftmost snd synchronizes with the rcv or $R \,|\, Q\{w/x\}$ when the rightmost snd synchronizes with the rcv. A main difference with | is that | does not affect the non-selected processes.

There are three alternatives for recursion. *Recursive processes* have the form $rec\, X.P$ where X is the recursive variable occurring in P and being replaced by the recursive process at each recursive call (e.g., $rec\, X.snd(a,b).X \to snd(a.b).rec\, X.snd(a,b).X$). The *bang operator* is defined by $!P = P \,|\, !P$ so that as many copies of P as needed are available. *Unique defining equations* $A(x_1,..,x_n) = P$ with $fn(p) \subseteq \{x_1,..,x_n\}$ and $A(x_1,..,x_n)$ occurring in P acts as recursive definitions by using A instantiated with actual parameters into system specifications.

We recall the brief history of this field by grouping calculi into three different generations. Here we consider stochastic π-calculus, β-binders and `BlenX` as representatives of the generations of calculi. Many other interesting formalisms exists and references to them are in Sect. 6.7.

Backus–Naur form

The syntax of programming languages is described by context-free grammars. A *grammar* is a quadruple $G = (N, \Sigma, P, S)$ where N is a finite set of *non-terminal* symbols, Σ is an alphabet of *terminal* symbols, $P \subseteq (N \cup \Sigma)^* \times (N \cup \Sigma)^*$ is the finite set of productions and $S \in N$ is the start symbol. We adopt the following conventions: $A, B, \ldots \in N$, $a, b, \ldots \in \Sigma$, $X, Y, \ldots \in (N \cup \Sigma)$, $x, y, \ldots \in \Sigma^*$ and $\alpha, \beta, \ldots \in (N \cup \Sigma)^*$. We often write $\alpha \longrightarrow \beta$ for a production $\langle \alpha, \beta \rangle \in P$. If $|\alpha| = 1$ for all the productions and $|\beta| > 0$, the grammar is context-free. By abuse of notation $\longrightarrow$ also denotes one rewriting step with reflexive and transitive closure $\longrightarrow^*$. The language defined by G is $L(G) = \{w \mid w \in \Sigma^*,\ S \longrightarrow^* w\}$.

To define a grammar, we list only its productions, starting with the productions having the start symbol on the left and collecting all the alternative productions to the right of the symbol $::=$ separated by vertical bars. This approach is called *Backus–Naur form* (BNF).

The syntax of the process calculus introduced in the previous boxes in BNF is (assuming an infinite countable set of names with metavariables a, b)

$$\begin{aligned} P &::= nil \mid A.P \mid (new\, a)P \mid P|P \mid P + P \mid !P \\ A &::= snd(a, b) \mid rcv(a, b). \end{aligned}$$

The start symbol is P that has six alternative productions and uses inside them the non-terminal symbol A to represent actions. The productions of A are listed second and have two alternatives.

Example 6.3. [Structural congruence]
The equivalences below hold according to the definition of structural congruence (see the box on the next page):

$$\begin{aligned} &[(new\, a)(snd(a, v).rcv(a, x).nil \mid rcv(c, w).nil)] + (new\, c)snd(c, z).nil \equiv \\ &[(new\, a)(snd(a, v).rcv(a, x).nil) \mid rcv(c, w).nil] + (new\, c)snd(c, z).nil \equiv \\ &[(new\, a)(snd(a, v).rcv(a, x).nil) \mid rcv(c, w).nil] + (new\, k)snd(k, z).nil \equiv \\ &(new\, k)[((new\, a)(snd(a, v).rcv(a, x).nil) \mid rcv(c, w).nil) + snd(k, z).nil] \equiv \\ &(new\, k)(new\, a)[(snd(a, v).rcv(a, x).nil) \mid rcv(c, w).nil) + snd(k, z).nil] \end{aligned}$$

where the second step is an α-conversion of the subject of the rightmost snd.

Structural congruence

The set P of processes is partitioned in equivalence classes with respect to the *structural congruence* defined by the minimal relation satisfying the following axioms

$$P \mid Q \equiv Q \mid P \qquad P + Q \equiv Q + P$$
$$P \mid nil \equiv P \qquad P + nil \equiv P$$
$$(P \mid Q) \mid R \equiv P \mid (Q \mid R) \qquad (P + Q) + R \equiv P + (Q + R)$$
$$!P \equiv P \mid !P \qquad P \equiv Q \text{ if } P, Q \text{ } \alpha\text{-equivalent}$$
$$(new\, a)P \mid Q \equiv (new\, a)(P \mid Q), a \notin fn(Q)$$
$$(new\, a)(new\, b)P \equiv (new\, b)(new\, a)P$$
$$(new\, a)P \equiv P \text{ if } a \notin fn(P).$$

6.1.1 *First generation*

For a long time the only calculus equipped with a quantitative implementation to model and simulate biological systems has been the *stochastic π-calculus.* Although successful in proving the potential of process calculi, the stochastic π-calculus suffers from some limitations in modeling biology. In fact, it was designed to model and analyze the performance of computer networks and hence it is strictly tuned to computer science. As a matter of fact, many tricks are needed to encode biology, and some natural primitives to describe common biological phenomena are missing (see Examples 6.4, 6.5 and 6.6).

Note that both interaction and complexation are based on a key-lock mechanism of affinity. In fact, the partners of a reaction must share the same channel and must perform synchronous complementary operations on it. Real biology does not work in the same manner. Interaction can happen even if the shape of the active domains that are going to react are not exactly complementary to one another (it is not always a key-lock mechanism of interaction), although with different strengths and probabilities depending on the concentration of the entities in the reaction volume.

Biological processes are driven by quantities that determine the rate of the reactions. Therefore, we must rule out nondeterminism so that the rates of the real systems are reflected in the selection of the corresponding instructions in our representations. The idea developed in the implementation of the stochastic π-calculus is to adapt the Gillespie algorithm for stochastic simulation of chemical reactions to action selection in the runtime of the language (see Sect. 8.3).

Stochastic π-calculus

The syntax of the *stochastic π-calculus* (Sπ) coincides with that presented in the box on Backus–Naur form on page 191, with an additional function $rate : \mathcal{N} \rightarrow \mathbb{R}^+$ that associates the unique parameter of an exponential distribution determining the rate of interactions/reactions with channel names ($\mathcal{N}$ denotes the countable infinite set of possible names). Furthermore, the *snd* and *rcv* actions are represented by ! and ? in infix notation, respectively (i.e., $a!b$ and $a?b$ corresponds to $snd(a, b)$ and $rcv(a, b)$; note that also bang is represented by !, but in prefix notation and with a process as argument). Finally, the *new* operator is sometimes written ν.

The structural congruence imposed on the syntax of the stochastic π-calculus is the minimal relation defined by the rules in the box on the previous page.

The intuitive semantics of the calculus is the same of that described in the two previous boxes on process calculi, except that nondeterminism is ruled out by implementing a race condition among the enabled actions using the rate information. The runtime of the calculus implements the Gillespie algorithm to select the next step to be performed (see Algorithm 8.7). The formal definition of Sπ is in Appendix C.3.

The π-calculus family of formalisms has been designed to model computer systems. Therefore, they exhibit some limitations when applied to biological systems.

Processes. The same syntactical concept is used to model different biological entities at different levels of abstraction like domains, proteins and whole systems.

Restriction. A primitive intended to declare names and identify their scope is used to represent compartments, membranes and complexes.

Complex/Decomplex. Creation and destruction of complexes of biological entities is not a primitive operation, but shows up from the interplay of private names and the variation of their scopes. It must be programmed through particular classes of communications. Furthermore, channel names are used both to define the boundaries of complexes as well as to manipulate the structure of programs.

Interaction. The interaction of biological entities can occur only if the entities share the very same channel name. This implies a

Example 6.4. [Modeling PPI networks in π-calculus]
Proteins are characterized by functional units called domains. For instance, a protein P with three domains (D_1, D_2 and D_3) can be represented by three processes with the same names as the corresponding domains that are composed in parallel to form a process representing the protein. The π-calculus uses the scope of names to identify boundaries and the binding is represented by sharing of private names (Fig. 6.2:1). No attachment or detachment of proteins is modeled directly through *ad hoc* primitives. Thus, our protein can be modeled as $A = (\nu ch)(D_1 \mid D_2 \mid D_3)$, where (νch) is the declaration of the new name ch with scope $(D_1 \mid D_2 \mid D_3)$.

Biological systems contain multiple copies of the same protein and they simply differ in the name of the private channel they share (Fig. 6.2:2). Note that even if the same name is used, due to the scope of definitions, each protein has a different shared private name (consider α-conversion). Indeed, $(\nu ch)(D_1 \mid D_2 \mid D_3) \mid (\nu ch)(D_1 \mid D_2 \mid D_3)$ is α-equivalent to $(\nu ch)(D_1 \mid D_2 \mid D_3) \mid (\nu ch_1)(D_1\{ch_1/ch\} \mid D_2\{ch_1/ch\} \mid D_3\{ch_1/ch\})$, assuming that $ch_1 \notin fn$ $(D_1 \mid D_2 \mid D_3)$. Multiplicity is modeled through parallel composition yielding $MultiA = A \mid .. \mid A$. Finally, a system contains multiple species so that it ends up in $Sys = MultiA_1 \mid .. \mid MultiA_n$.

The processes encoding the domains implicitly describe the interaction capabilities of the protein through the names used in their syntactic definition. Recall that π-calculus communications are determined by complementary actions (send "!" and receive "?") performed on the very same channel name that must be shared between the partners of the interaction. For instance, if $D_3 = a!y.D_3'$, the protein A can interact with all the other entities that are willing to perform a receive on the shared channel a, e.g., the protein B made up of the single domain $D_4 = a?w.Q_4'$ (Fig. 6.2:3, 4).

Example 6.5. [Complex formation in π-calculus]
The scope of private names can be manipulated to create and destroy different complexes during the execution. If we consider the system $((\nu ch)(\nu y)A) \mid B)$, where A and B are defined in Example 6.4, after the interaction of D_3 and D_4, the new system will be $(\nu y)((\nu ch)(D_1 \mid D_2 \mid D_3') \mid D_4')$. The residual of protein A and the residual of protein B after the interaction share the private name y that A sent and made available to B and therefore they are interpreted as being physically attached (Fig. 6.2:5).

The detachment of the proteins must be programmed as well by opening the scope of the name y. For instance, if $D_2 = a!y.D_2'$ and it performs the output of the private name with no other process willing to perform the input on the channel a, the resulting system no longer has a restriction on the name y which becomes available to the whole system and hence the two proteins detach because they do not share any private name anymore (Fig. 6.2:6).

Example 6.6. [Sensitivity of interactions in π-calculus]
To model different sensitivity of interactions between biological elements in π-calculus-like formalisms, we must explicitly write the whole set of alternatives relying on the choice operator in the processes. If the domain D_3 of protein A in the Example 6.4 can interact with n different domains, we have to specify it within the protein as $A = (\nu x)(D_1 \mid D_2 \mid a_1!y.D_3' + ... + a_n!y.D_3')$. Furthermore, any complementary domain must contain a receive operation on the very same channel a_i representing the compatibility between the two entities (Fig. 6.2:7).

perfect key-lock mechanism of interaction that is not realistic in the biological domain. In fact, the interaction can happen with different strengths or probabilities.

Low-level programming. The reversibility of reactions as well as the complementarity of complexation and decomplexation must be programmed through communication and name passing. All the arcs of the biological interaction network must be specified through a complementary pair of send and receive on the same channel rather than inferring it from the sensitivity or affinity of interaction of entities, as occurs in biology.

Incremental model building. Most of the information related to the interconnection structure of the entities and on the sensitivity/ affinity of interaction is hard-coded into the syntax of the processes through the send/receive pairs. Whenever new knowledge is discovered (e.g., on the capability of interaction of entities), a new piece of code must be produced and distributed in not parts of the specification that are not well identified with a negative impact on the scalability of the approach. In fact, the compositional modeling style of π-calculus-like calculi is simply a way of structuring a description of a system by identifying subsystems that run in parallel rather than a true scalable and compositional methodology.

Identity of entities. Biological entities can pass trough different states during their lifetime depending on the interactions in which they participate. In the π-calculus approach, prefixes are consumed and the identity between processes and biological entities is lost during the execution of the programs.

Implementation. Naming of channels could make two syntactically different programs represent the same semantic object, and hence the same entity. Indeed, the entities change their structure after

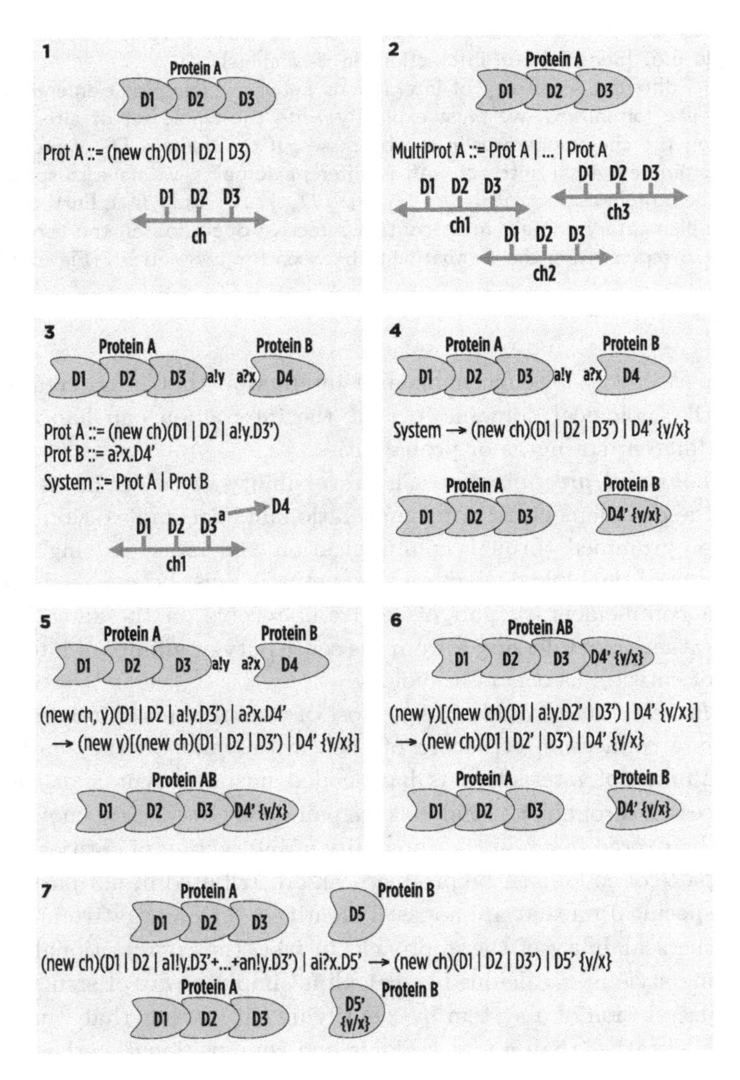

Figure 6.2 Protein interactions in π-calculus. 1. A protein. 2. Multiple copies of the same protein. 3–4. Protein–protein interaction. 5–6 Binding and unbinding. 7. Different sensitivity in interactions.

any execution step and their number is then continuously changing because processes can be created and updated dynamically. The computation of the structural congruence of subprograms is most of the time not efficient enough to ensure scalability.

Qualitative vs quantitative descriptions. Stochastic process calculi handle the kinetics of systems by associating channel names with rates. Frequently, the kinetic parameters for real biological interactions are not known and must be estimated to fit some phenotypical behavior that is measurable experimentally. Therefore, mixing quantitative and qualitative information forces the modeler to heavily rework the model to carry out parameter estimation and sensitivity analysis.
Space. Many biological phenomena are highly sensitive to the localization within the reference system volume of the reactants. Space must be hard-coded in the language.

We conclude this subsection with an example of a model of a feedback loop in stochastic π-calculus.

6.1.2 *Second generation of calculi for biology*

The second generation of process calculi directly defined to model biological systems improves the ease of expressing basic biological principles. The most representative languages are the κ-calculus and β-binders. The main step ahead is the 1:1 correspondence between biological entities and objects of these calculi that enables an easier tracing of biological components during simulations. The states through which an entity can pass are implicitly encoded in the program describing the internal behavior of objects, rather than being listed explicitly and represented with different placeholders, e.g., variables in ordinary differential equations (ODEs).

The κ-calculus is a formal, rule-based calculus of protein interactions. It was conceived to represent complexation and decomplexation of proteins, using the concept of shared names to represent bonds. Once the initial system has been specified and the basic reductions have been fixed, the behavior of the system is obtained by rewriting it. This kind of reduction resembles pathway activation.

β-binders (Fig. 6.5(1)) associates biological entities with boxes. Interaction is communication-based and can occur either within a box if the same channel name is used or between different boxes if they have compatible interfaces (even if they have different names and hence the interaction is over different channel names). In fact, typed interfaces of boxes enable promiscuity of interaction (sensitivity-/affinity-based rather than exact complementarity of channel names; see Fig. 6.5(2)). Note that types

Example 6.7. [A feedback loop in stochastic π-calculus]
We consider a molecular system, regulating gene expression by positive feedback (see Fig. 6.3). The system Sys includes two genes ($Gene\_A$ and $Gene\_TF$), their transcribed mRNAs (RNA_A and RNA_TF), the corresponding translated proteins (A and TF) and the degradation of both RNA and protein molecules. The events are mediated by interaction with cellular machineries for DNA transcription ($Transcr$), RNA translation ($Transl$) and RNA and protein degradation (RNA_D and Pr_D). Each of these interactions involves different molecular motifs and occurs at a different rate. In addition, A activates TF in a two-step mechanism. A binds TF, through A's BS domain, to form a complex and then A's K domain modifies the bound TF protein. Following modification and unbinding, TF can rapidly bind the transcription machinery using its newly acquired capabilities, causing faster promotion of transcription and closing a positive feedback loop.
Assume for simplicity's same that there is just one copy of each biological element. We start with the two genes plus the transcription, translation and degradation machineries (specialized on RNA and proteins)

$$Sys = Gene\_A \,|\, Gene\_TF \,|\, Transcr \,|\, Transl \,|\, RNA\_D \,|\, Pr\_D.$$

Transcription activates the production of RNA by interacting with the genes on two different channels: $basal$ and pA. The two channels represent normal and enhanced transcription by active TF (aTF)

$$\begin{aligned} Gene\_A &= basal?().(Gene\_A \,|\, RNA\_A) + pA?().(Gene\_A \,|\, RNA\_A) \\ Gene\_TF &= basal?().(Gene\_TF \,|\, RNA\_TF) + pA?().(Gene\_TF \,|\, RNA\_TF) \\ Transcr &= basal!().Transcr + pail?().pA!().Transr. \end{aligned}$$

RNA is generated by adding a copy of RNA in parallel with the gene that interacts with $Transcr$ in the residual of the interaction (i.e., in the process remaining after consuming the send and receive operations). The shift from normal to enhanced transcription is caused by a signal received on a channel $ptail$ by the transcription process (the signal is sent by aTF, see below). RNA can be either generated by the corresponding protein or be degraded by translation and degradation processes

$$\begin{aligned} RNA\_A &= utr?().(RNA\_A \,|\, A) + degm?() \\ RNA\_TF &= utr?().(RNA\_TF \,|\, TF) + degm?() \\ Transl &= utr!().Transl \\ RNA\_D &= degm!().RNA\_D. \end{aligned}$$

here do not play the same role played by names in π-calculus. In fact, the same type can be compatible with many different types, thus allowing interaction over different channels and between different boxes.

Example 6.8. [A feedback loop in stochastic π-calculus — continued]
Proteins are generated by adding a copy of the protein in parallel with the RNA that interacts with $Transl$ on channel utr in the residual of the interaction. Alternatively, RNA can be destroyed by interacting on channel $degm$ with RNA_D. Protein A has two domains (BS and K) that are used to bind TF and form the complex $A : TF$ and to activate TF, respectively. Accordingly to Example 6.4 and Fig. 6.2, we model A as a parallel composition of the two domains sharing three private channels (we need three channels because each one of them will deliver a different signal). We use a polyadic version of the calculus that allows for multiple values sent and received along channels (e.g., $bind!(bb1, bb2)$ and $bind?(x, y)$).

$$\begin{aligned} A &= (new\ bb1, bb2, bb3)(BS \mid K) \\ BS &= bind!(bb1, bb2).BSB + degp?().bb3!() \\ K &= bb2!(ptail).K + bb3?() \\ Pr\_D &= degp!().Pr\_D \end{aligned}$$

The binding domain BS can interact with TF on the channel $bind$ by sharing its private names ($bb1$ and $bb2$) thus forming a complex (see Example 6.5 and Fig. 6.2). The result is that the new process BSB (binding site bound) is activated. Alternatively, BS can degrade by interacting on channel $degp$ with the degradation process Pr_D. After this interaction, BS sends a message on $bb3$ to signal K to degrade as well so that the degradation of A is completed. This message must be immediate, i.e., must be delivered immediately after the interaction of BS and Pr_D. We model immediate actions by associating them with ∞ rate (see below). The main activity of K is to signal TF on their private channel $bb2$. Protein TF

$$TF = bind?(x, y).BTF + degp?()$$

can bind A by receiving $bb1$ and $bb2$ on the channel bind to instantiate within BTF the variables x and y in the receive action. Alternatively, TF can degrade by interacting with the protein degradation machine. The complex $A : TF$ is represented by the configuration

$$(new\ bb1, bb2)((new\ bb3)(BSB \mid K) \mid BTF\{bb1, bb2/x, y\})$$

where

$$\begin{aligned} BSB &= bb1!.BS \\ BTF &= x?().TF + y?(tail).x?().aTF(tail). \end{aligned}$$

The complex can break by releasing A and TF through the interaction on the private channel $bb1$ (recall that x in BTF will be instantiated with $bb1$).

Example 6.9. [A feedback loop in stochastic π-calculus — continued]
K activates TF by signaling on $bb2$ (that instantiates y in BTF). The interaction on $bb2$ transfers the channel $ptail$ to TF by instantiating the variable $tail$ in the residual of the receive action (i.e., $(x?().aTF)\{ptail/tail\}$). After this interaction, the complex can break and release an active version of TF that share owns the channel $ptail$

$$aTF(tail) = tail!().aTF(tail) + degp?().$$

This process can either activate enhanced transcription through its channel $ptail$ (that instantiates $tail$) or degrade. To complete the specification of the system, we need to define the function $rate$ that assign stochastic rates to channel names. We set

$$rate = \{(basal, 4), (utr, 1), (bind, 0.1), (bb1, 10), (bb2, 10), (bb3, \infty), (degm, 1), (degp, 0.1), (ptail, 100), (pA, 40)\}.$$

Assuming that

$$S = Transcr \,|\, Transl \,|\, RNA\_D \,|\, Pr\_D \text{ and } bbi = \{bb1, bb2, bb3\},$$

a possible sequence of transitions describing the behavior of the feedback loop is

$$Sys \longrightarrow S_0 \longrightarrow S_1 \longrightarrow S_2 \longrightarrow S_3 \longrightarrow S_4 \longrightarrow S_5 \longrightarrow S_6 \longrightarrow S_7$$

where

$$\begin{aligned}
S_0 &= Gene_A \,|\, RNA_A \,|\, Gene_TF \,|\, S \\
S_1 &= Gene_A \,|\, RNA_A \,|\, A \,|\, Gene_TF \,|\, S \\
S_2 &= Gene_A \,|\, RNA_A \,|\, A \,|\, Gene_TF \,|\, RNA_TF \,|\, S \\
S_3 &= Gene_A \,|\, RNA_A \,|\, A \,|\, Gene_TF | RNA_TF \,|\, TF \,|\, S \\
S_4 &= Gene_A | RNA_A | (new\ bbi)((BSB \,|\, K) \,|\, BTF\{bb1, bb2/x, y\}) \,| \\
&\quad Gene_TF \,|\, RNA_TF \,|\, S) \\
S_5 &= (new\ bbi)(Gene_A \,|\, RNA_A \,|\, BSB \,|\, K \,|\, Gene_TF \,|\, RNA_TF \,| \\
&\quad (bb1?().aTF(ptail) + bb3?())) \,|\, S \\
S_6 &= (new\ bbi)(Gene_A \,|\, RNA_A \,|\, A \,|\, Gene_TF \,|\, RNA_TF \,|\, aTF(ptail)) \,|\, S \\
S_7 &= (new\ bbi)(Gene_A \,|\, RNA_A \,|\, A \,|\, Gene_TF \,|\, RNA_TF \,|\, aTF(ptail)) \,| \\
&\quad pA!().Transcr \,|\, Transl \,|\, RNA_D \,|\, Pr_D.
\end{aligned}$$

The formal derivation of the transitions and their labeling with stochastic rates to match stochastic kinetics of reactions is described in Appendix C.

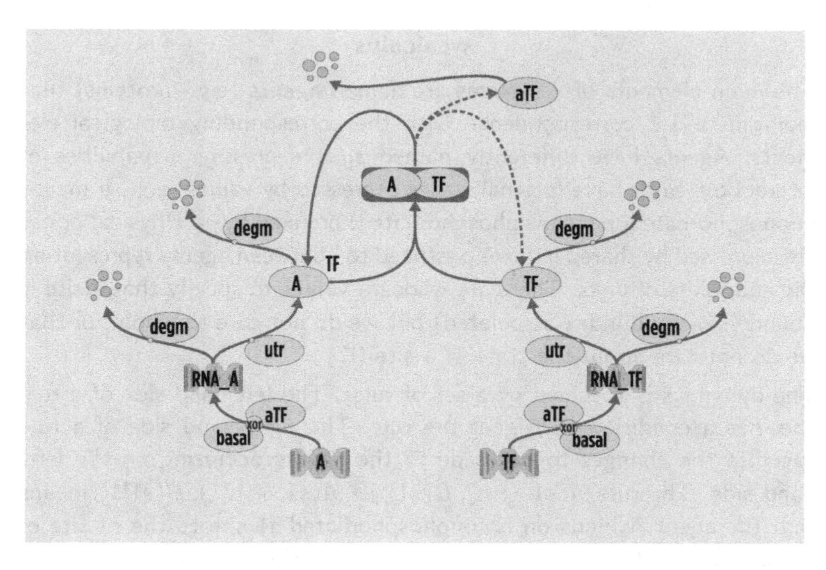

Figure 6.3 A feedback loop that enhances transcription of the genes coding for protein A and TF through an active form of TF. Activation of TF is caused by a kinase domain of A when A and TF are complexed.

The enclosing surfaces (boxes) of entities cannot be nested and maintain a strict correspondence between processes and biological entities. Primitives to manipulate binders can be included in the description of the internal behavior of boxes. It is possible to create new binders dynamically through the expose action and to change the status of a binder through the hide and unhide primitives (Fig. 6.5(3)).

Physical rearrangement of boxes is implemented by general split and join functions that operate on the structure of boxes and that are not inserted in the flow of control. This choice enables a higher level of nondeterminism and accommodates the specification of systems for which there is a considerable lack of knowledge. There is no need to specify merging or division of boxes as occurs in other approaches, because we only specify general conditions on the structure and on the status of the binders of boxes managed by the run-time of the language (Fig. 6.5(4)). The formal definition of β-binders is given in Appendix C.4.

Although the κ-calculus and β-binders solve many of the problems identified for the first generation of calculi, they still suffer from some drawbacks listed below (here we use the same categories used in the previous subsection, and those not listed are no longer considered an issue; those introduced

κ-calculus

The main elements of κ-systems are named *agents* (e.g., proteins) that maintain a 1:1 correspondence with the corresponding biological elements. Agents have differently named *sites* expressing capabilities of interaction. Sites have internal states expressed by values (e.g., u means unphosphorilated, p means phosphorilated) prefixed by $\sim$. Physical *bonds* are expressed by shared indexes postfixed to ! between agents representing the endpoints of links. There are wildcard values to specify that a site is bound (! with no index associated) but we do not care to whom, or that we do not care about the state of a site (!?).

The dynamics is specified by a set of rules. The left-hand side of a rule specifies a condition to trigger the rule. The right-hand side of a rule specifies the changes to be made to the agents occurring on the left-hand side. The rule $A(s1 \sim u), B(s1) \rightarrow A(s1 \sim u!1), B(s1!1)$ means that the agent A binds on its unphosphorilated s1 site to the s1 site of the agent B and the binding is represented by the index 1.

A κ-system is an initial set of rules with rates and a mixture of agents. The Gillespie algorithm stochastically selects the rule to apply in each configuration of the system. The κ-calculus has the philosophy *write what you need*: irrelevant context for the rules is always omitted in their definition.

Example 6.10. [Abstract activation of the *lymphocyte T helper*]
We consider the activation of the *lymphocyte T helper* (see Fig. 6.4). Lymphocyte T helpers (or helper T cells) are eukaryote cells belonging to our immune system. They play a central role by activating and controlling many specific defense strategies. Lymphocytes are normally inactive, and they start their activity only after being triggered by special events. Here, we will focus on the sequence of *phagocytosis–digestion–presentation* phases, and the *activation* of lymphocyte T helpers performed by macrophages.

Macrophages are cells that engulf a virus (phagocytosis). When this happens, the virus is degraded into fragments (digestion or lysis), and a molecule, the so-called antigen, is displayed on the surface of the macrophage (presentation or mating). The antigen may be recognized by a specific lymphocyte T helper, and this in turn activates the mechanisms of immune reply, a response specific to the recognized virus.

Examples 6.11 and 6.12 show the κ- and β-binder models.

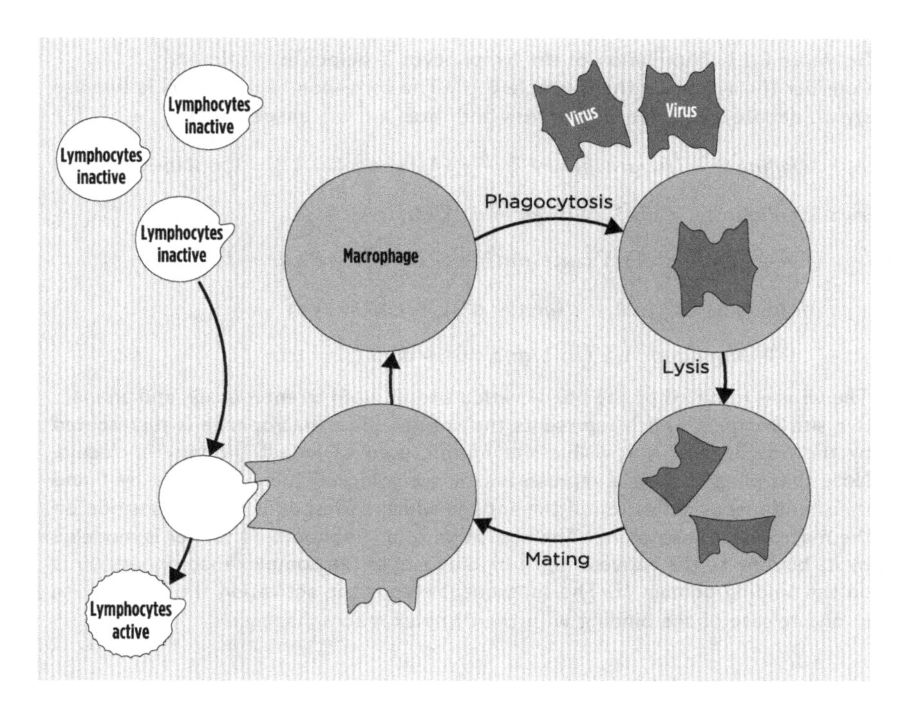

Figure 6.4 Abstract representation of the lymphocyte T helper activation. Viruses have inactive sites which represent the viral antigens. The process starts with the phagocytosis of the virus by the macrophage. The virus is then decomposed, and eventually viral antigens are moved to the surface of the macrophage. A lymphocyte T helper that binds to the macrophage becomes active and starts an immune reply.

here refer to issues raised by the second generation of calculi that were not present in the previous generation).

Complex/Decomplex. κ-calculus provides the best support for manipulating complexes. The main limitation is that the rules for creating complexes and manipulating them must all be specified by the modeler on the basis of the structure of the boxes. No support is provided by the implementation of the language to perform this fundamental mechanism of biological interaction automatically.

Low-level programming. Almost nothing is changed on this item with respect to the previous generation of calculi. The only improvement is that the interaction network can be implicitly modeled in β-binders by adequately defining the compatibility function between types.

Example 6.11. [Activation of the lymphocyte T helper in κ-calculus]
Consider the system in the Example 6.10. The corresponding κ-model is (we use i/a to denote a inactive/active sites and rates are left unspecified):

$$r_1 : M(phago), V(in, antigen \sim i) \xrightarrow{k_1} (M(phago!1), V(in!1, antigen \sim a))$$

$$r_2 : M(phago!1), V(in!1, antigen \sim a), T(ti, ir \sim i)$$
$$\xrightarrow{k_2} M(phago!1), V(in!1, antigen \sim a!2), T(ti!2, ir \sim i)$$

$$r_3 : M(phago!1), V(in!1, antigen \sim a!2), T(ti!2, ir \sim i)$$
$$\xrightarrow{k_3} M(phago!1), V(in!1, antigen \sim a), T(ti \sim a).$$

The virus is rendered by the box V with a site in, used to enter a cell, and an inactive site $antigen$, which represents the antigen. The macrophage is represented by the box M with the visible site $phago$, used to engulf the virus molecule. Note that phagocytosis is represented by a binding uniquely identified by 1 and the simultaneous activation of the antigen which corresponds to its exposition on the macrophage membrane. The activation of a lymphocyte T helper is modeled by its binding to the active antigen binding site of the complex macrophage:virus on the binding domain ti. The subsequent step is the activation the binding site ir and release of the binding with the complex macrophage:virus.

Interaction. The key-lock mechanism of interaction is relaxed in β-binders, while it is hidden in the complexation and decomplexation rules of κ-calculus.

Incremental model building. The identification of entities with boxes in both calculi makes the extension of models easier because the parts of the specifications to be changed are clearly identified.

Implementation. The association between boxes and biological entities simplifies the implementation of the stochastic simulation algorithms. However, the efficiency is still not acceptable, i.e., the generality of split and join functions of β-binders poses further difficulties to implementing the calculus efficiently.

Qualitative vs quantitative descriptions. A step forward is that rates in β-binders are no longer merged with the qualitative descriptions, but are externally specified through the compatibility function between the types of the interfaces (see also Fig. 6.5(2)). However, complex kinetics function cannot yet be incorporated within the models.

Example 6.12. [Activation of the lymphocyte T helper in β-binders]
Consider the system described in Example 6.10. The corresponding β-binders model is (assuming that each binder type is compatible with itself only, e.g., $comp(v_i, v_i) = r_i$ or $comp(Ant_j, Ant_j) = k_j$):

$B_M = \#(x, v_1)..\#(x, v_n)[x?w.\ \mathsf{expose}(a, \{w\}).(!a!s \mid P_{DIG})]$

$B_V = \#(x, v_1)[z!Ant_1.P_{INF}]$

$B_T = \#(x : Ant_1)\#(y : Ant_4)[a?y.P_{ACT}]$

and $f_{join}(\#(x, v_1)..\#(x, v_n), \#(x, v_1), P_1, P_2) = (\#(x, v_1)..\#(x, v_n), z/x, Id)$. Note that the lymphocyte T helper has compatibility with just two antigens (Ant_1 and Ant_4). The system will have many different binders for T helper cells and the macrophage will interact with the ones compatible with the presented antigen. The first step is the engulfment of the virus by the macrophage as defined by the $join$ function. The result is the removal of the virus box from the system and the new configuration of the macrophage (note the substitution of x with z imposed by the definition of the $join$).

$$B_{M'} = \#(x, v_1)..\#(x, v_n)[\ z?w.\ \mathsf{expose}(a, \{w\}).(!a!s \mid P_{DIG})\{z/x\} \mid z!Ant_1.P_{INF}\ ]$$

The new box $B_{M'}$ performs an intracommunication over z to transfer the antigen to the macrophage that exposes it (w instantiated by Ant_1) becoming:

$B_{M''} = \#(x, v_1)..\#(x, v_n)\#(a, Ant_1)[\ !a!s \mid P_{DIG})\{z/x\} \mid z!Ant_1.P_{INF}]$.

These steps enable the macrophage to communicate over the newly created binder with some lymphocyte T helper that is compatible with the antigen. The final result is the activation of the lymphocyte (internal process P_{ACT}) to start the appropriate immune response.

Space. Some preliminary attempt to code compartments and hence implicitly handle localization of molecules is emerging, but very limited support is available (see Example 6.13).

6.2 Third generation: from calculi to modeling languages

A further step in the design of modeling formalisms is moving from theoretical calculi to modeling languages designed for biology. The most relevant platforms emerged from the two calculi discussed in the previous subsection. A driving principle to move from theoretical descriptions to practical modeling languages is surely to enrich the syntax of the calculi to facilitate the modeling process. This step must be performed by separating concerns as

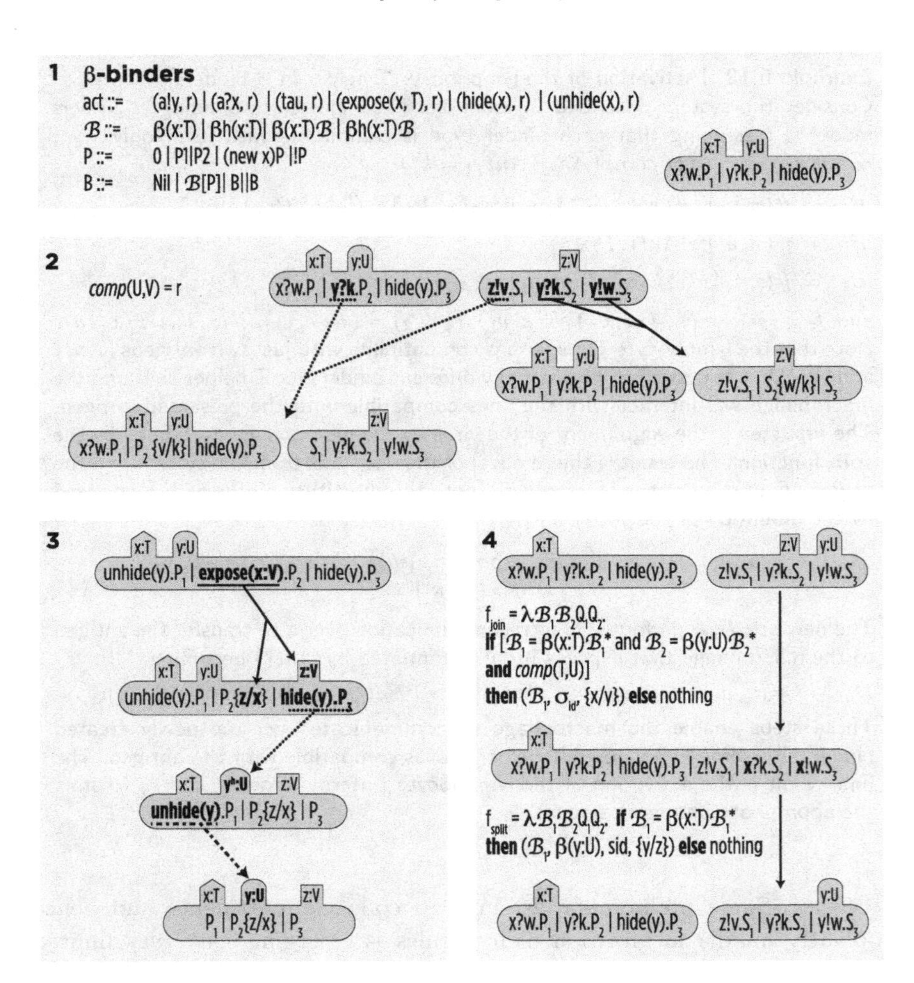

Figure 6.5 β-binders. (1) Syntax. The metavariable T ranges over binder types. $\beta(x : T)$ denotes a binder (interface) named x with type T. P ranges over processes that define the internal behavior of boxes, and B ranges over boxes. (2) Interaction can occur between different boxes through compatible binders (the function *comp* defines the affinity between the types) or internally to a box through complementary channels in π-calculus style. (3) Binder manipulation. Expose creates new binders and may need renaming of bound variables to avoid captures of names. Hide and unhide manipulates the state of binders that can be either visible or hidden. (4) Join and split functions. The join function returns the list of binders of the new box as well as two renaming functions to change the names of the parallel composition of the original boxes to adapt them to the new binders.

Example 6.13. [Encoding compartments in β-binders]
Compartments can be encoded in β-binders by boxes, and interactions between boxes constrained by a naming discipline on binder types. Consider the system

$$B \parallel \beta(y : U)\beta(w : V)[P] \parallel \beta(x : V)[Q]$$

assuming that the type V only occurs in the rightmost box in the system above. The rightmost box acts as a subcompartment of the other box specified above because the binder x can only interact with the binder w. The box enclosing Q can only interact internally or with its enclosing compartment through x.

much as possible, i.e., by keeping qualitative and quantitative descriptions distinct as well as using different syntactic categories and semantic actions to represent different fundamental biological principles. Finally, the run-time of the language should take into account the basic dynamic principles of biological systems as much as possible, such as complexation and decomplexation. We consider here the BlenX language as an example inspired by β-binders because it collects all the features of modeling environments based on process calculi. Note that the coding in BlenX of the β-binders models of the previous section is immediate, being BlenX and extension of β-binders.

6.2.1 BlenX

BlenX is based on boxes that represent biological entities (Fig. 6.6). Boxes have typed and uniquely named interfaces and have an internal behavior specified similarly to β-binders, but with a richer set of primitives.

The dynamic behavior of BlenX models is specified through three classes of actions: *Monomolecular:* actions affect a single box and can be either internal communications (similarly to the rightmost transition in Fig. 6.5(2)) or internal primitives that manipulate interfaces, destroy the box or delay the execution of actions (see Fig. 6.6(1)); *Bimolecular:* actions affect two boxes and can be either communications between boxes or complex/decomplex operations (see Fig. 6.6(2,3)); *Events:* global rewriting rules of the environment that manipulate a set of boxes as well as their structure when global conditions on the structure and cardinality of boxes are satisfied (see Fig. 6.6(4)).

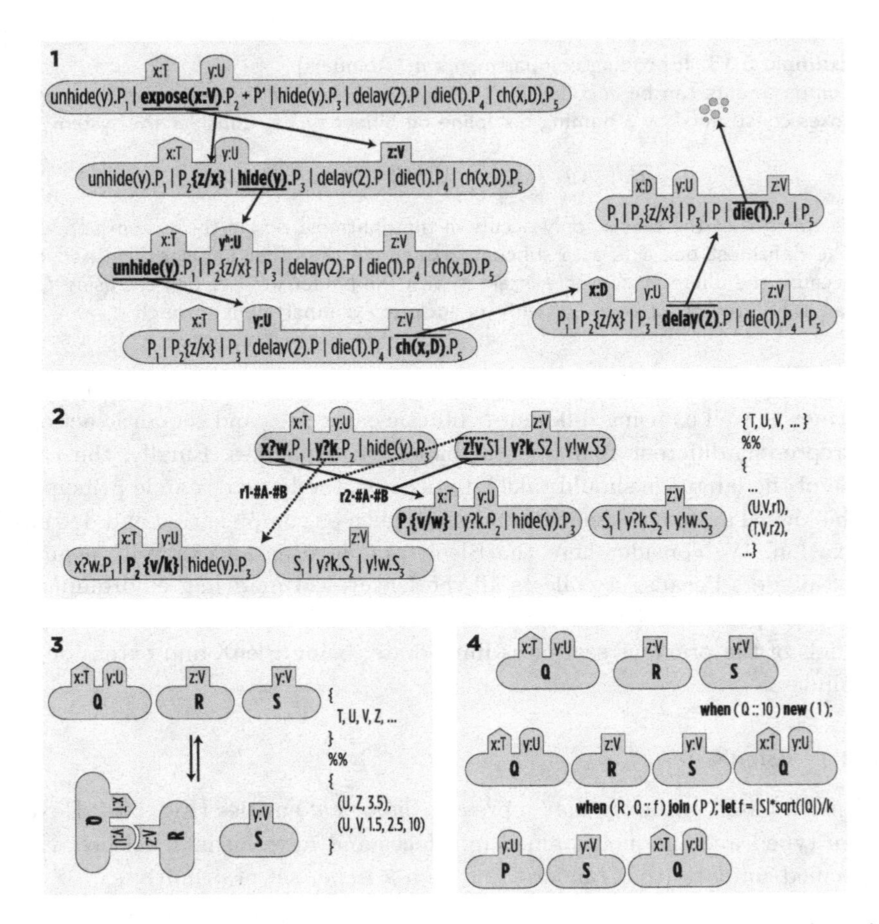

Figure 6.6 BlenX. (1) Monomolecular actions *expose*, *hide* and *unhide* as in β-binders, *ch* changes the type of a binder, *delay* lets time pass and *die* kills a box. (2) Communications. The set on the right lists all the types occurring in the model and the existence of a rate associated with two types in a tuple quantifies the compatibility between the corresponding interfaces. (3) Complex/decomplex operations are generated according to the additional compatibility rates. (4) Events. General conditions on the structure of the system and the cardinality of the boxes trigger *join*, *split*, *new* and *delete* actions.

Events are a very powerful modeling tool that enables the execution of perturbative experiments over the models as it is usually done in wet-labs. It can be specified that whenever a global condition, possibly time-dependent, is satisfied (e.g., a gene can be knocked out, a drug injected in the system to study its interactions or some components can be removed to see how they affect the overall behavior of the system or similarly how robust a system is). Note that any event is associated with a rate and enters the race with all the other actions for selection by the simulation algorithm. Therefore, even if specified differently from the other interaction mechanisms, an event is a normal action of the system that does not require modifications of the stochastic engine. Note that the second event in Fig. 6.6(4) is specified by a general mathematical function that also depends on the concentrations of boxes not affected by the event. BlenX permits the specification of general kinetics by relying on mathematical functions.

The syntax of BlenX is reported in the boxes. Note that optional items are not denoted as usual between square brackets because square brackets are part of the syntax of the language. Therefore we let $\langle A \rangle$ denote optional occurrence of A. We assume as pre-defined the categories of identifiers (Id, sometimes qualified as $boxId$, $varId$, $funcId$, $nodeId$, $binderId$ to specify the object to which they refer) and $numbers$ (either *real* or *decimal*).

A BlenX program is made of an optional *declaration* file for the declaration of user-defined constants and functions (the first group of declarations *dec*), a *binder definition* file that associates unique identifiers to binders of entities used by the program (*affinities*) and a *program* file that contains the program structure (*program*).

A *prog* file is made up of an header *info*, an optional list of rate declarations (*rateDec*), a list of declarations *decList*, the keyword **run** and a list of initial boxes *bp*.

The *info* header contains information needed by the simulator that will execute the program. A stochastic simulation can be considered as a sequence of time-stamped steps that are executed sequentially, in non-decreasing time order. Thus, the duration of a simulation can be specified as a *time*, intended as the maximum time-stamp value that the simulation clock will reach, or as a number of *steps* that the simulator will schedule and execute (see first line in Fig. 6.9(1)). The *delta* parameter can be optionally specified to instruct the simulator to record events only at a certain frequency. For more information on simulation see Chapter 8.

A BlenX program is a stochastic program: every single step that the program can perform has a *rate* associated with it, representing the frequency at which that step can, or is expected to, occur. The *rateDec* specifies the global rate associations for individual channel names or for the four primitives that manipulate interfaces. In addition, a special class $BASERATE$ can be used to set a common basic rate for all the actions that do not have an explicit rate. The special rate *inf* is used to define immediate actions: when enabled, they are always executed before any action with a finite rate.

BlenX

$$
\begin{array}{lcl}
\textit{program} & ::= & \textit{info}\ \langle \textit{rateDec} \rangle\ \textit{decList}\ \textbf{run}\ \textit{bp} \\
\textit{info} & ::= & [\ \textbf{steps} = \textit{decimal}\langle, \textbf{delta} = \textit{number}\rangle\ ]\ |\ [\ \textbf{time} = \textit{number}\ ] \\
\textit{rateDec} & ::= & Id : rate\ |\ \textbf{CHANGE} : rate\ |\ \textbf{EXPOSE} : rate\ |\ \textbf{HIDE} : rate \\
 & | & \textbf{UNHIDE} : rate\ |\ \textbf{BASERATE} : rate\ |\ rateDec, rateDec \\
\textit{rate} & ::= & number\ |\ \textbf{rate}\ (\ Id\ )\ |\ \textbf{inf} \\
\textit{decList} & ::= & dec\ |\ dec\ decList \\
\textit{dec} & ::= & \textbf{let}\ Id\ :\ \textbf{var} = e\ \langle \textbf{init}\ number \rangle;\ |\ \textbf{let}\ Id\ (number) : \textbf{var} = e\ ; \\
 & | & \textbf{let}\ Id\ :\ \textbf{const} = e\ ;\ |\ \textbf{let}\ Id\ :\ \textbf{function} = e\ ; \\
 & | & \textbf{let}\ Id : \textbf{pproc} = process\ ;\ |\ \textbf{let}\ Id : \textbf{bproc} = box\ ; \\
 & | & \textbf{let}\ Id : \textbf{complex} = complex\ ;\ |\ \textbf{let}\ Id : \textbf{prefix} = seq\ ; \\
 & | & \textbf{when}\ (\ cond\ )\ verb\ ; \\
\textit{bp} & ::= & decimal\ Id\ |\ bp\ \|\ bp \\
e & ::= & number\ |\ Id\ |\ |\ Id\ |\ |\ \textbf{log}\ (\ e\ )\ |\ \textbf{sqrt}\ (\ e\ )\ |\ \textbf{exp}\ (\ e\ ) \\
 & | & \textbf{pow}\ (\ e\ ,\ e\ )\ |\ e + e\ |\ e - e\ |\ e * e\ |\ e\ /\ e\ |\ -e\ |\ +exp\ |\ (\ e\ )
\end{array}
$$

Rates for intercommunications over binders are specified in the affinity file. The actual rate of a transition is then computed by considering the possible combinations of different elements of the same species to interact (see Fig. 6.6(3)).

A (state) *variable* is an identifier that can assume real values. A variable is updated when its defining expression *exp* changes or by *update* events. In this case, the function associated with the event is evaluated and the variable is updated with the resulting value. Optionally, variables have initial values specified through the *init* keyword. *Continuous variables* depend on time (Δt specified by the number following the Id) and their value is determined by an expression updated every Δt. A *constant* is an identifier that assumes a value that cannot be changed at run-time and specified through a constant expression (an expression that does not rely on any variable or concentration $|Id|$ to be evaluated). BlenX *functions* are parameterless and

Example 6.14. [BlenX variables, constants, functions]
Examples of variable, constant and function declarations are (the third being a continuous variable with $\Delta t = 0.1$):

```
let v1 : var = 10 * |A|;
let J5 : var = 2 * |X| * log(v1) init 0.1;
let m(0.1): var = mu * m init 0.2;
let pi : const = 3.14;
let e: const = exp(1.0);
let f : function = pow( (J5 / (e * pi * |A|) ) , 4) + 1;.
```

Consider the equation $\frac{dm}{dt} = \mu \cdot m$ that expresses the continuous variation of m over time. Its discretization yields $\Delta m = \mu \cdot m \cdot \Delta t$. To update m every Δt, we write $m_{t(i)} = m_{t(i-1)} + \Delta m$ that amounts to $m_{t(i)} = m_{t(i-1)} + (\mu \cdot m \cdot \Delta t)$, from which the declaration above.

return a real value. Therefore a function is just a named expression used to evaluate a rate or to update the content of a state variable.

Affinities are a peculiar feature of BlenX that drive the interaction between pairs of binder identifiers. For one, this approach avoids any global policy on the usage of names to make components interact; following β-binders it relaxes the exact name pairing, or key-lock, style of interaction typical of the first generation of calculi; it permits a better separation of concerns, as it allows the user to put interaction information in a separate file that can be modified or substituted without altering the program. The usage of affinities in a separate file is similar to program interactions guided by *contracts* or service definitions, like in some web-service models.

An *affinity* is a tuple of three or five elements. The first two elements are binder identifiers declared in the *binderIdList*, while the other elements can either be rates or function identifiers. If the affinity tuple contains a single rate, then the value is interpreted as the base rate of *intercommunication* between binders with identifier equal to the first and second *binderId* respectively. If the affinity tuple contains three rates, then the values are interpreted as the base rate for *complex*, *decomplex* and *intercommunication* between binders with identifiers equal to the first and second *binderId* respectively. When the element after the two *binderId*s is a function identifier, the expression associated with the function will be evaluated to yield a value, then interpreted as the rate of *intercommunication*. Examples of affinity definitions are in Fig. 6.6(2, 3) and Fig. 6.9(2, 3).

BlenX — continued

```
affinities    ::= { binderIdList } ⟨%%{ affinityList }⟩

binderIdList  ::= binderId | binderId, binderIdList

affinityList  := affinity | affinity, affinityList

affinity      ::= ( binderId, binderId, funcId )
                | ( binderId, binderId, ⟨rate, rate,⟩ rate )

box           ::= binders [ process ]

binders       ::= #⟨h⟩( Id ⟨: rate⟩, Id ) | #⟨h⟩( Id, Id ) | binders, binders

process       ::= par | sum

par           ::= Id | rep action . process
                | if condexp then par endif | sum|sum | sum|par
                | par|sum | par|par | ( par )

sum           ::= nil | seq | sum + sum | ( sum )
                | if condexp then sum endif

action        ::= Id ! ( ⟨Id⟩ ) | Id ? ( ⟨Id⟩ ) | delay ( rate )
                | expose ( ⟨rate,⟩ Id : rate, Id ) | ch ( ⟨rate,⟩ Id, Id )
                | hide ( ⟨rate,⟩ Id ) | unhide ( ⟨rate,⟩ Id )

seq           ::= action | action . process | Id . process

condexp       ::= ( Id, Id ) | ( Id, ⟨Id,⟩ hidden ) | ( Id, ⟨Id,⟩ unhidden )
                | ( Id, ⟨Id,⟩ bound ) | condexp and condexp
                | condexp or condexp | not condexp | ( condexp )
```

A box represents an autonomous biological entity that has its own control mechanism (the *process*) and some interaction capabilities expressed by the *binders*.

A *binder* list is made up of a non-empty list of *elementary binders* of the form $\#(Id : rate, Id)$ (active with rate), $\#(Id, Id)$ (active without rate), $\#\mathbf{h}(Id : rate, Id)$ (inactive with rate), $\#\mathbf{h}(Id, Id)$ (inactive without rate), where the first Id is the *subject* (or name) of the binder, *rate* is the stochastic parameter that quantitatively drives the activities involving the binder (hereafter, stochastic rate) and the second Id represents the identifier of the binder. Binder identifiers cannot occur in processes, while subjects of binders can. The subject of an elementary binder is a binding occurrence that binds all its free occurrences in the process inside the box to which the binder belongs. Hidden binders are useful to model interaction, sites that are not available for interaction, although their status can vary dynamically (e.g., an active domain that is hidden by the shape of a molecule and that becomes available if the molecule interacts with/binds to other molecules). Given a list of binders, we denote the set of all its subjects with $sub(binders)$. A box is considered *well formed* if the list of binders has subjects and identifiers all distinct. Well formedness is statically checked and preserved during the program execution.

A process can be a *par* or a *sum*. The non-terminal symbol *par* composes through the binary operator | two processes that run concurrently, while *sum* introduces guarded choices of processes, composed with the operator +. The + operator acts intuitively as an exclusive *or* operator while the | operator acts intuitively as an *and*. Process definitions can contain identifiers, but recursive and mutual recursive definitions are not admitted.

The *rep* operator is used to replicate copies of the process passed as argument. Note that we use only guarded replication, i.e., the process argument of the *rep* must have a prefix *action* that forbids any other action of the process until it has been consumed. Sometimes the *bang* (!) operator is used in place of the *rep* operator. It is defined as $!P = !P \mid P$ that amounts to assume that there are as many copies in parallel of the process P as needed. The *nil* process does nothing (it is a deadlocked process), while the *if–then* statement allows the user to control, through an *expression*, the execution of a *process*. The non-terminal symbol *seq* identifies an action, a process prefixed by an action and a process prefixed by an Id. When a process is defined as *Id.process*, BlenX statically checks that the Id corresponds to a previously defined sequence of prefixes.

The first four *actions* in the definition of BlenX are common to most process calculi. The first pair of actions represent an output/send of a value or a signal on a channel, while the second pair represent the input/reception of value or a signal on a channel. These actions are the ones that model *intracommunication* (communication inside a box) as it happens in the rightmost transition in Fig. 6.5(2). The same input and output prefixes can generate *intercommunications* (communication between boxes) as it

BlenX — species and structural congruence

Free names and *bound names* for processes are defined by assuming that $Id'?(Id).process$, $expose(\langle rate, \rangle Id : rate, Id).process$ are binders for Id in $process$. We extend the definitions to boxes assuming that the subjects of the binders are binding occurrences in $binders[process]$ for $process$. Two processes/boxes are α-equivalent if they only differ in the choice of bound names.
BlenX species are the classes of *structurally congruent* boxes ($\equiv$). The structural congruence for boxes is the smallest relation that satisfies the following laws:

$$process \equiv process', \text{ if } process \ \alpha\text{-equivalent } process'$$

$$process \mid \mathbf{nil} \equiv process$$

$$process_1 \mid (process_2 \mid process_3) \equiv (process_1 \mid process_2) \mid process_3$$

$$process_1 \mid process_2 \equiv process_2 \mid process_1$$

$$sum \mid \mathbf{nil} \equiv sum$$

$$sum_1 \mid (sum_2 \mid sum_3) \equiv (sum_1 \mid sum_2) \mid sum_3$$

$$sum_1 \mid sum_2 \equiv sum_2 \mid sum_1$$

$$!action.process \equiv action.(process \mid !action.process)$$

$$binders[process] \equiv binders[process'], \text{ if } process \equiv process'$$

$$binders, binders'[process] \equiv binders', binders[process]$$

$$\#\langle h\rangle(Id\langle: rate\rangle, Id_1), binders[process] \equiv \#\langle h\rangle(Id'\langle: rate'\rangle, Id_1), binders[process\{Id'/Id\}] \text{ if } Id' \notin sub(binders).$$

Note that the last equation is a schema for all the configurations generated by the optionality parentheses $\langle\ \rangle$.

happens in Fig. 6.6(2) when the link of the communication is bound to an active binder of the box. Note that intracommunications occur on perfectly symmetric input/output pairs that share the same subject, while intercommunication can occur between input/output that have different subjects provided that their binder identifiers are compatible.

The actions *hide*, *unhide* and *expose* are inherited from β-binders (see Figs. 6.5(3) and 6.6(1)).

The remaining actions are peculiar of the BlenX language. The action *ch* changes the type of a binder, *delay* is used to let time pass without performing any concrete action and *die* kills the box that executes it.

As for π-calculus and β-binders, we identify processes and boxes that are syntactically different, but semantically equivalent via a structural congruence.

Example 6.15. [BlenX structural congruence]
Consider the two boxes $b1$ and $b2$ below.

```
let b1 : bproc = #(x:1,A)
   [ ( x!().nil + z?(w).w!().nil ) | x!(z).nil ];
let b2 : bproc = #(y:1,A)
   [ y!(z).nil | ( z?(t).t!().nil + y!().nil ) ];
```

It is $b1 \equiv b2$, hence the boxes belong to the same species. Note that if we have multiple definition of boxes that represent the same species, then at run-time they are collected together and the species name is taken from the first definition (e.g., in the example the name of the corresponding species is $b1$). Hereafter, when we say that in a particular state of execution of a program the cardinality of a box species $b1$ is n we mean that in that state of execution the number of boxes structurally congruent to $b1$ is n.

BlenX offers also the possibility of modeling complexes. A complex is a graph-like structure of box species where boxes are nodes and dedicated communication bindings are edges (see Fig 6.7). Complexes can be instantiated by the user or they can be automatically generated during the program execution by the language.

A complex is created by specifying the list of edges (*edgeList*) and the list of nodes (*nodeList*). Each *edge* is a composition of 4 *Ids*. The first and the third identifier represent node names, while the others represent subject names. Each *node* in the *nodeList* associates the corresponding box name with a node name and specifies the subjects of the bound binders.

A complex can also be generated automatically at run-time. The ability of two boxes to form and break complexes is defined in the binder definition file by specifying triples of stochastic rates (complexation, decomplexation, communication) for pairs of binder identifiers. When two boxes form a complex, the bindings that create the dedicated link are said to be in a *bound* status (denoted by $\#c$). Although the bound status cannot be explicitly specified through the syntax of the language and is used only as an internal representation, a binder in bound status is different from a hidden or unhidden binder and hence the structural congruence definition has to be extended accordingly.

Events specify statements (*verb*) with a rate that are enabled by global conditions *cond*. A condition is made of three parts: *enList*, a list of boxes

BlenX — continued

$$
\begin{array}{lll}
\mathit{complex} & ::= & \{\ (\ \mathit{edgeList}\ )\ ;\ \mathit{nodeList}\ \} \\[6pt]
\mathit{edgeList} & ::= & (\ \mathit{nodeId},\ \mathit{binderId},\ \mathit{nodeId},\ \mathit{binderId}\ ) \\
 & \mid & (\ \mathit{nodeId},\ \mathit{binderId},\ \mathit{nodeId},\ \mathit{binderId}\ ),\ \mathit{edgeList} \\[6pt]
\mathit{nodeList} & ::= & \mathit{node} \mid \mathit{node}\ \mathit{nodeList} \\[6pt]
\mathit{node} & ::= & \mathit{nodeId}\ :\ \mathit{boxId} = (\ \mathit{complBinderList}\ )\ ; \\
 & \mid & \mathit{nodeId} = \mathit{nodeId}\ ; \\[6pt]
\mathit{complBinderList} & ::= & \mathit{Id} \mid \mathit{Id},\ \mathit{complBinderList} \\[6pt]
\mathit{cond} & ::= & \mathit{enList}\ :\ \langle\ \mathit{EvExpr}\ \rangle\ :\ \langle \mathit{rate} \rangle \mid\ :\ \mathit{EvExpr}\ : \\
 & \mid & \mathit{enList}\ :\ \langle\ \mathit{EvExpr}\ \rangle\ :\ \langle \mathit{funcId} \rangle \\[6pt]
\mathit{enList} & ::= & \mathit{boxId} \mid \mathit{boxId},\ \mathit{enList} \\[6pt]
\mathit{EvAtom} & ::= & \mid \mathit{boxId} \mid = \mathit{decimal} \mid \mid \mathit{boxId} \mid < \mathit{decimal} \\
 & \mid & \mid \mathit{boxId} \mid > \mathit{decimal} \mid \mid \mathit{boxId} \mid\ != \mathit{decimal} \\
 & \mid & \textbf{time} = \mathit{real} \mid \textbf{steps} = \mathit{decimal} \\
 & \mid & \mathit{stateOpList} \\[6pt]
\mathit{statOpList} & ::= & \mathit{stateOp} \mid \mathit{stateOp}, \mathit{stateOpList} \\[6pt]
\mathit{statOp} & ::= & \mathit{varId} \leftarrow \mathit{real} \mid \mathit{varId} \rightarrow \mathit{real} \\[6pt]
\mathit{EvExpr} & ::= & \mathit{EvAtom} \mid \mathit{EvExpr}\ \textbf{and}\ \mathit{EvExpr} \\
 & \mid & \mathit{EvExpr}\ \textbf{or}\ \mathit{EvExpr} \mid \textbf{not}\ \mathit{EvExpr} \mid (\ \mathit{EvExpr}\ ) \\[6pt]
\mathit{verb} & ::= & \textbf{split}\ (\ \mathit{boxId},\ \mathit{boxId}\ ) \mid \textbf{join}\ (\ \mathit{boxId}\ ) \\
 & \mid & \textbf{new}\ \langle\langle\ \mathit{decimal}\ \rangle\rangle \mid \textbf{delete}\ \langle\langle\ \mathit{decimal}\ \rangle\rangle \\
 & \mid & \textbf{update}\ (\ \mathit{varId},\ \mathit{funcId}\ )
\end{array}
$$

in the system; *EvExpr*, an expression used to enable or disable the event; a *rate* or rate function, used to stochastically select and include them in the set of standard interaction-enabled actions. Conditions can be specified on simulation time or simulation steps as well. A condition on *simulation time* will be satisfied as soon as the simulation clock is greater or equal to the specified time; a condition on *simulation steps* will be satisfied as

Example 6.16. [BlenX — complexes]
Consider the following BlenX code.

```
let b1 : bproc = #(x:r0,A0),#(y:r1,A1)[ x!().nil ];
let b2 : bproc = #(x:r0,A0),#(y:r1,A1)[ y!().nil ];
let C : complex =
{
  (
    (Box0,y,Box1,x),(Box1,y,Box2,x),
    (Box2,y,Box3,x),(Box3,y,Box0,x)
  );
  Box0:b1=(x,y); Box1:b2=(x,y); Box2=Box0; Box3=Box1;
}
```

The complex C defines a structure equivalent to the one in Fig. 6.7.

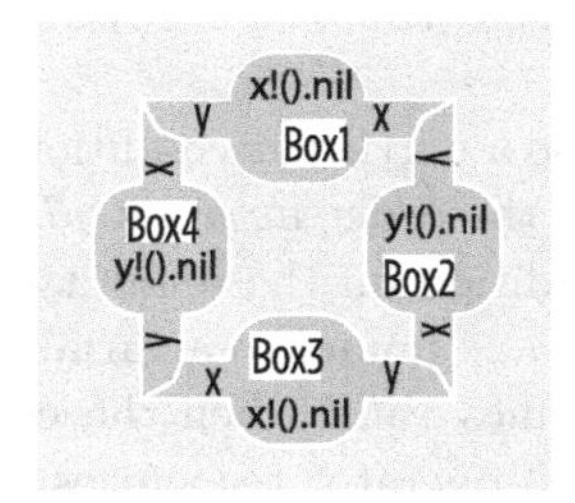

Figure 6.7 The graph of the complex C in Example 6.16.

soon as the step count exceeds the number of steps specified. In both cases, the condition will remain *true* until the event is fired. Events for which the only condition specified is the number of steps or the execution time are guaranteed to fire exactly once.

Events can split a box into two boxes, join two boxes into a single one, inject or remove boxes into/from the system or update the value of variables. Verbs specify how many boxes must occur in the *enList*: split, new and delete require exactly one box, join requires exactly two boxes and update requires the *enList* to be empty. Split substitutes one box of the specified species with two other boxes specified in its $(boxId, boxId)$ argument. Join removes one box for each of the species specified in the list, and introduces one box of the species specified in its $(boxId)$ argument with binders coinciding with the union of the binders of the removed boxes. The verb new will introduce one or more copies of the single box in the list. The verb delete will remove one or more boxes of the species in the list. The verb update modifies the value of a variable in the system. When the event is fired, the result of the evaluation of the function *funcId* is assigned to the variable *varId*. Variables are global *Id*s bound to real values and functions are mathematical expressions on variables and cardinality of species that evaluate to a real values. Note that update has no rate or rate function in its *rate* part: the event is triggered as soon as its *EvExpr* evaluates to *true*.

Example 6.17. [BlenX — events]
Consider the following BlenX code.

```
when(A, B : (|A| > 2 and |B| > 2) : rate(r1)) join (C);
when(A : time = 3.0 : inf) delete;
```

The species involved in the first event are A and B (*enList*). The event will fire with rate $r1$ only when there are more than two As, and two Bs, in the system and will replace one A and one B with C. The second event will fire as soon as the simulation clock reaches 3.0, removing one A from the system.

An update event can use a particular condition based on the traversal of successive states following the order in *stateOpList*; each *stateOp* element in the list expresses a condition on the quantity of an *Id* (i.e. cardinality of boxes for *boxId* or the *value* bound to a variable for *varId* — see Example 6.18). A *stateOp* becomes *valid* when the condition on its *Id* is met for the first time. The "$\rightarrow$" operator recognizes when the quantity bound to *Id* becomes greater than the specified real value, while the "$\leftarrow$" operator recognizes when the quantity bound to *Id* becomes smaller than the specified real value. When a *stateOp* becomes valid, the *EvExpr* evaluates the following *stateOp* of the list until it becomes valid. As soon as the last state in the *stateOpList* becomes valid, the *EvExpr* evaluates to *true* and the update can be executed. Once the event is fired, the *EvExpr* restarts its evaluation from the beginning of the *stateOpList*, waiting for the first *stateOp* to become valid again and so on.

A BlenX model is made up of a static description of the system obtained by listing the entities of the initial configuration with their amount and specificity for interaction. Then the execution of the model and the support of the language are in charge of determining at any step the possible actions to be performed. This approach enables a library-based modeling: the modeler simply selects the components that must be considered in the system and then defines the sensitivity of interaction between them. After this step only a little is needed to equip the boxes with their relevant internal behavior if particular actions need to be implemented. The BlenX modeling process described above allows modelers to build scalable, modular and

Example 6.18. [BlenX — update events]
Consider the oscillatory behavior in Fig. 6.8(a) determined by a deterministic model and assume that we want to count the oscillation with a variable `n`. The first event below increments `n` if the cardinality of `A` first exceeds 20 (first *stateOp* in the list) and then becomes smaller than 20 (second *stateOp*).

```
let n : var = 1;
let f : function = n + 1;
when (: A -> 20, A <- 20 :) update (n, f);
when (: A -> 10, A -> 20, A <- 20, A <- 10 :) update (n, f);
```

The concatenation of an arbitrary succession of states allows us to overcome possible limitations that are often encountered when dealing with stochastic noise. The first event does not correctly capture the oscillations in Fig. 6.8(b), as highlighted in the upper-right corner of the figure. A fine tuning of the states in the condition can correctly capture the behavior of the noisy oscillating system. The second event works properly (see Fig. 6.8(c)).

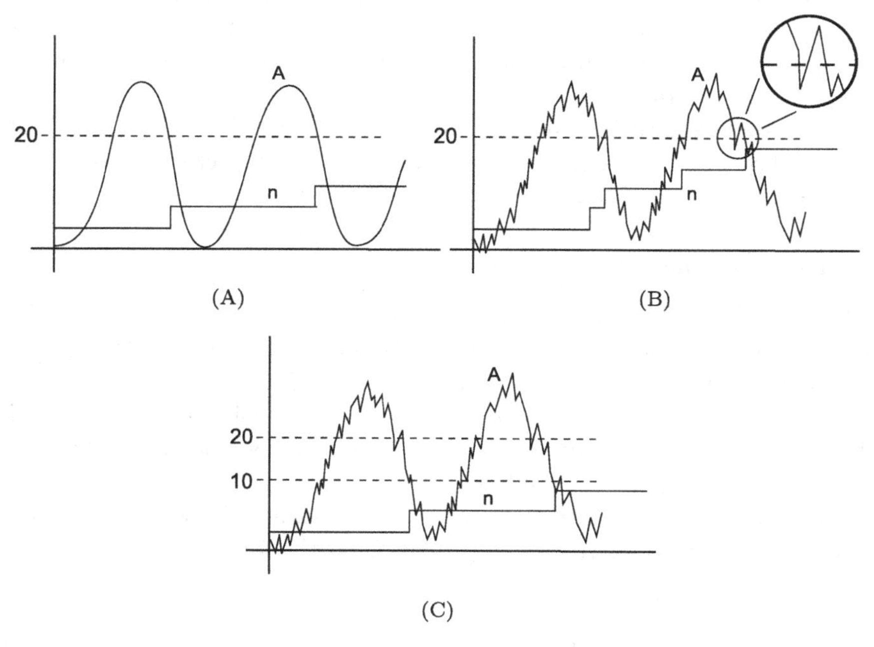

Figure 6.8 Counting the oscillations of species (see Example 6.18). (a) The species A exhibits a deterministic oscillating behavior, captured by the state-list condition in the first event of the example. (b) The species A exhibits a stochastic oscillating behavior and the first event of the example does not work properly. (c) The third event in the example can capture the oscillations correctly.

Example 6.19. [Lotka–Volterra in BlenX]
Consider Fig. 6.9(1). After a directive to the simulator on the number of steps to be simulated, three boxes are declared (see also their graphical representation aside). Box Predator can die with rate 10 or interact over the binder x typed Hunt. The interaction over Hunt is meant to represent the fact that the predator eats a prey. Indeed the box Prey dies with rate *inf* after interacting over Hunt. Prey can alternatively interact over the binder Life to eat some food recursively that is produced without any limitation by the box Nature.

Two additional boxes are defined to name Predator and Prey when ready for duplication (i.e., after a predator eats a prey and and after a prey eats food from nature). The boxes are PredatorRep and PreyRep and both of them trigger an event with rate *inf* that produces two copies of Predator or Prey via a split action.

The last line of the code specifies the initial amount of boxes. Some steps of the system are depicted in Fig. 6.9(2, 3).

compositional models. Finally, all the combinatorial effects are ruled out at modeling level because the very same box represents all the possible states the corresponding biological entity. Example 6.19 shows a BlenX model of the Lotka–Volterra system.

Before moving to explore the relationship between BlenX and ODEs of chemical reaction networks, briefly recall the main characteristics of different generations of process calculi in Fig. 6.10.

6.2.1.1 *BlenX and ODEs*

To map ODEs of chemical reactions into BlenX models, we modify the differential equations to add the species that are appearing implicitly in the system. If a species has a constant total amount, but it can switch between an active and an inactive state, the time evolution of just one of the two states is usually explicitly considered, because the other state can be implicitly derived from the first.

The general mathematical description of a system of n reactant species $S_1, \ldots, S_n$ involved in $R_1, \ldots, R_r$ reactions has the form

$$\frac{d[S_i]}{dt} = \sum_{j=1}^{k} f_j^i([S_1], \ldots, [S_n]) \qquad (6.1)$$

where $i = 1, \ldots, n$, $k \leq r$, and f_j^i is a rate function that can contain constant rate coefficients and other discrete/continuous variables depending on time/species in the model. The complete definition of the system above

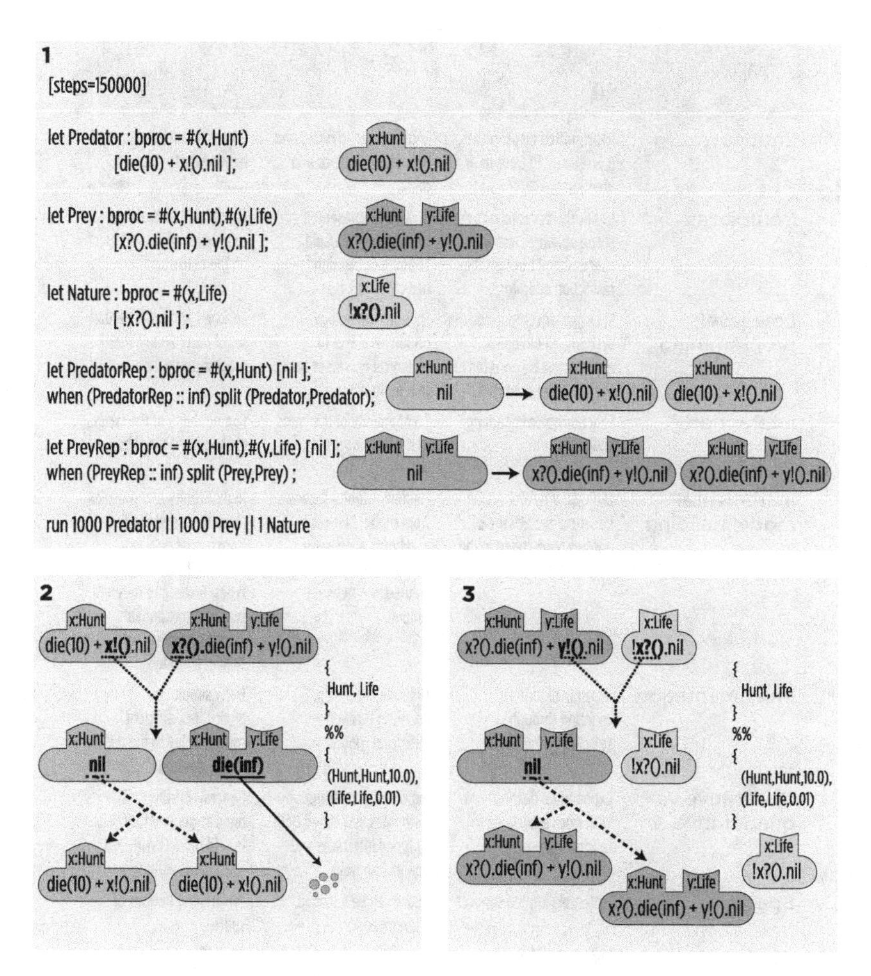

Figure 6.9 Lotka–Volterra in BlenX. (1) Code and graphical representation of boxes. (2) Interaction between predator and prey. The prey dies and the predator duplicates according to the first *when* clause in the code. (3) Interaction between prey and nature. The prey duplicates according to the second *when* clause in the code and nature is still available due to the replication operator in its internal process.

has to be coupled with the set $Var_1, \ldots, Var_t$ of the mathematical expressions for the variation of each discrete/continuous variable. The translation from an ODE system to BlenX is in Algorithm 6.1.

Example 6.20 shows the application of the Algorithm 6.1 to a subset of a budding yeast cell cycle model graphically represented in Fig. 6.11.

	1st generation	2nd generation	3rd generation
Entities	Represented by processes, it is difficult to trace their identity	Encapsulated in boxes and hence easy to trace and count	Encapsulated in boxes and hence easy to trace and count
Complexes	Implicitly represented by scope of names and manipulated through the restriction operator	Explicitly represented through the join/split of boxes or through dedicated links	Improved manipulation of complexes with conditions on their structure and state
Low-level programming	Thought for computing systems, all biological events must be encoded in the available primitives	Thought for biology, primitives closer to biological processes makes model simpler	Enriched and more natural set of primitives. A real programming language
Interaction	Exact complementarity of channel names	Compatibility of the types of the interfaces of boxes	Compatibility of the types of the interfaces of boxes
Incremental model building	Difficult: interconnection structure and affinity of entities hard-coded in the model	Medium: affinity separated from model behavior and boxes allow easy identification of the connections between models	High: Affinity and complex formation separated from model behavior and handled through types. Boxes and different syntactic components identify connections between models
Implementation	Gillespie algorithm, counting through structural congruence	Gillespie algorithm, counting through structural congruence	Many simulation algorithms, efficient counting due to the design of the language
Qualitative vs quantitative	Exponential distributions with rates associated to channel names	Exponential distributions with rates associated both to typed interfaces and channel names	General distributions, rates as general functions, hybrid models with continuous variables
Space	Extremely limited support	Space can be encoded through tricks	Preliminary notions of space
Standards	Very limited import/export capabilities	Some import/export capabilities	Connection with SBML and ODE
Predictive power	Every interaction must be programmed	Many interactions are inferred by the execution engine	Many interactions are inferred by the execution engine

Figure 6.10 The figure summarizes the concerns and the characteristics of the calculi of the three generations described so far.

6.3 Self-assembly

The term *self-assembly* indicates a process in which disordered components form an organized structure or pattern only through local interactions

Algorithm 6.1: From ODEs to BlenX

Input: A system of ODEs of the form $\frac{d[S_i]}{dt} = \sum_{j=1}^{k} f_j^i([S_1], \ldots, [S_n])$ and t variables Var_z, $z \in [1..t]$.
Output: A BlenX program.

begin
  map deterministic rates into stochastic rates (e.g., as in Fig. 5.6);
  declare all the constants, variables and functions in the ODEs;
  for $i \in [1..n]$ **do**
    generate let S_i : bproc = [nil];
    for $j \in [1..k]$ **do**
      case f_j^i **of**

synthesis	$\Rightarrow$	when (S_i :: $\lvert f_j^i \rvert$) new(1)
degradation	$\Rightarrow$	when (S_i :: $\lvert f_j^i \rvert$) delete(1)
change from S_i to S_w	$\Rightarrow$	when (S_i :: $\lvert f_j^i \rvert$) split(Nil,S_w)
changing from S_w to S_i	$\Rightarrow$	when (S_w :: $\lvert f_j^i \rvert$) split(Nil,S_i)

      endcs
    end
  end
  remove event duplicates of reactions appearing in multiple ODEs;
  set the initial state of the model
end

among them, without an external coordination. Debates of how complex structures and functions can emerge from local interactions between simple components occur in many fields (e.g., nanotechnologies, robotics, molecular biology, autonomous computation). Here we concentrate on molecular self-assembly, the process through which molecules assemble in *complexes* without guidance from an outside source, a process which is crucial to the functioning of cells.

Conventional modeling approaches (e.g., ODEs) simply and effectively represent molecules and complexes by associating them with species identifiers or variables. However, there are modeling scenarios in which the explicit definition of all the possible complexes acting in a biological system and all the possible reactions in which they are involved is not always feasible or even possible. Examples are the modeling of processes in which small molecules combine to produce large complexes such as chains or graphs of molecules (e.g., actin polymerization) or in which the number of possible protein complexes and combinations of protein modifications tend to increase exponentially (e.g., multisite phosphorylation — see Example 5.4).

Example 6.20. [ODEs to BlenX]
Consider the ODE that models the activation/inactivation of Cdc20 (Fig. 6.11) and the rate law for the growth of the mass of the cell $dm/dt = \mu * m(1 - m/m^*)$. Cdc20a is the active form of Cdc20.

$$\frac{d[Cdc20]}{dt} = \underbrace{k5p}_{synthesis} + \underbrace{\frac{k5s(m * CycB)^n}{(J5)^n + (m * CycB)^n}}_{(induced)\ synthesis} - \underbrace{k6[Cdc20]}_{degradation} + \underbrace{\frac{k8[Cdc20a]}{J8 + [Cdc20a]}}_{inactivation} - \underbrace{\frac{k7[IEP][Cdc20]}{J7 + [Cdc20]}}_{activation}$$

where ks and Js are the deterministic rates and $CycB$ is an expression that represents the activity of the dimer Cdk:CycB. Synthesis and induced synthesis are zero-order reactions, degradation and inactivation are first-order reactions and activation is a second-order reaction. Algorithm 6.1 produces the output

```
let ca : function = (m*CycB)^n;   (n fixed by the Hill kinetics)
let syn : const = NaV*k5p;
let isyn : function = (k5s*ca)/(J5^n + ca);
let deg : const = k6;
let inact : function = k8/(J8+|Cdc20a|);
let act : function = (k7 * |IEP|)/(J7+|Cdc20|);
let Cdc20in : bproc = [ nil ];
let Cdc20a : bproc = [ nil ];
let IEP : bproc = [ nil ];
when (Cdc20 : : syn) new(1);
when (Cdc20 : : isyn) new(1);
when (Cdc20 : : deg) delete(1);
when (Cdc20 : : act) split(Nil, Cdc20a);
when (Cdc20a : : inact) split(Nil, Cdc20);.
```

Algorithm 6.1 cannot manage the mass updating because it is a variable m updated at discrete time steps. Furthermore, cell division is an event that is not explicitly coded in the ODE system, and it halves the mass value when the concentration of active Cdk:CycB falls below a threshold (here 0.1) after having risen above another threshold (here 0.2). Also, this event is coded by a variable CdkCycB and whenever this variable overcomes the 0.2 threshold and then goes back under the 0.1 threshold, the event that updates the value of the mass with its halved value is executed. In BlenX we can add

```
let m(0.1) : var = mu * m init 0.2;
let mdiv : function = m/2;
when (:CdkCycB -> 0.2, CdkCycB <- 0.1:) update (m, mdiv);.
```

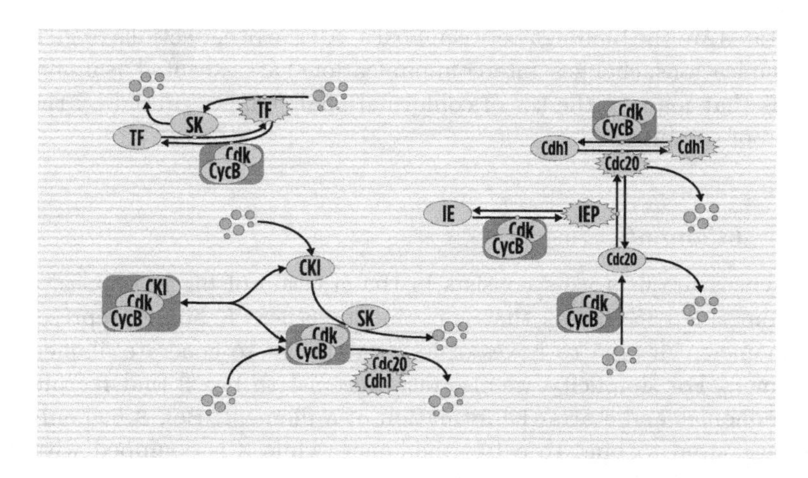

Figure 6.11 Graphical representation of some parts of the cell cycle engine. Solid lines link reactants to products, dashed line represent the mediation effect that some species have on reactions. Cdk proteins bound to CycB work as main regulatory kinases of the process. The complex Cdk:CycB mediates the synthesis of Cdc20 (right part of the figure) during the S-G2-M phase of the cell cycle. Cdc20 is activated by the mitotic process with a signal mediated by an intermediary enzyme IE made active by Cdk:CycB. Deactivation of Cdc20 is generated by mitotic spindle abnormalities through the Mad pathway. The active form of Cdc20 activates Cdh1 that is reverted to its inactive form by Cdk:CycB. The complex Cdk:CycB is inhibited binding to CKI (bottom-left part of the figure) that is abundant in the G1 phase of the cycle. CKI can be driven to fast degradation by a phosphorylation operated by Cdk:CycB (that, however, has little or no activity during G1 phase) or SK (a dimer of Cdk with a cyclin different from CycB). Note also that the active forms of Cdc20 and Cdh1 favor degradation of the complex Cdk:CycB. The cell produces a starter kinase SK to leave G1 phase and enter S phase (top-left part of the figure). The transcription factor TF for the cyclin component of SK is activated by increasing cell mass and its activity is enhanced by SK itself. Instead Cdk:CycB decreases the activity of TF.

The goal of this section is to provide design patterns and examples for using BlenX for programming self-assembly processes. The simplest structure is a *linear* chain, characterized by a single backbone with no branches. A related unbranching structure is a *ring* polymer. A *branched* polymer is composed of a main chain with one or more side chains. Examples of polymers are actin, DNA and microtubules.

For simplicity, the models presented here can only grow, i.e., we do not consider reversible processes in which boxes can detach from a filament. Although this is a simplification, the programs presented give the flavor of the BlenX potential. In all programs the only reactions associated with finite rates are the bindings between boxes. All the other actions (e.g., intra-communications, changes, intercommunications) are executed as immediate actions that manage the set of conformational changes caused by the formation of complexes.

6.3.1 *Filaments*

A polymerization process consists in the creation of big molecules, starting from small molecules that can bind together. A box `M`, representing a monomer, can bind to boxes of its own species to generate filaments of monomers. For simplicity, we also introduce a *seed* box `S` that recruits the first monomer and starts the formation of a filament. Box `S` has only one interface, used to bind to a free monomer, while `M` is equipped with two interfaces, the *left* one used to bind to the last box of a growing filament (can be both a box `S` or `M`) and the *right* one used to bind to a free monomer. The dynamics of the model is depicted in Fig. 6.12.

Note that a filament can only grow on the right side. The seed `S` starts the creation of the filament, while the growing process involves only the binding of `M` monomers. A box `M` can accept the binding of another monomer on its right interface only if it is already part of a growing filament, i.e., its left interface is bound to a filament.

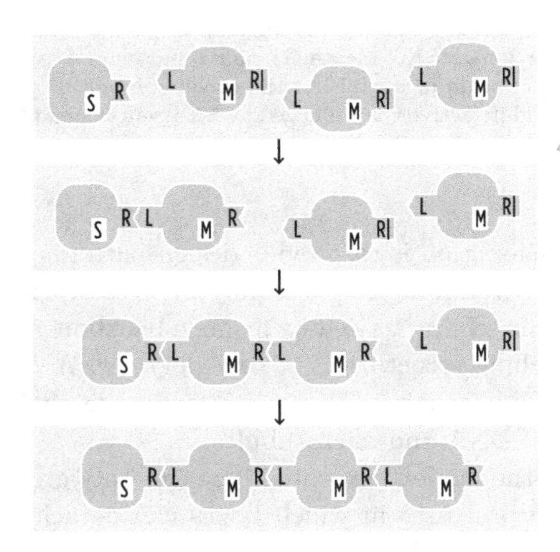

Figure 6.12 Example of filament formation.

The definition of the two boxes, S and M is

```
let S : bproc = #(right,R)[ nil ];
let M : bproc =
   #(left,L), #(right,RI)
   [ if (left,bound) then ch(right,R) endif ];.
```

The binder definition file specifies the compatibility (L,R,1,0,0). Due to the compatibility between L and R, a monomer M can bind to the seed S. After the binding, the internal program of M recognizes that something is bound to the left interface and changes the type of the right interface from RI (R inactive) into R. In this way, the right interface of the bound M gets the capability to bind to the left interface of another free monomer (see Fig. 6.12).

6.3.2 *Trees*

We modify the monomer box M adding a branching interface to create trees of monomers. The branching interface is activated only when M has a monomer bound on its left interface.

```
let M : bproc =
   #(left,L), #(right,RI), #(branching,BI)
   [ if (left,bound) then ch(right,R).ch(branching,B) endif ];.
```

We add also a box T that binds to the branching interface of a monomer in a filament and starts the formation of a branch. This species is introduced to make the rate of branch formation independent of the number of free monomers in the model. T is defined as:

```
let T : bproc =
   #(left,TL), #(right,TRI)
   [ if (left,bound) then ch(right,TR) endif ];.
```

Also for T, the internal program changes the type of the right interface when the box gets bound on its left interface. For trees, we specify two additional affinities: (B,TL,1,0,0) enables the complexation of a box M that is part of a filament with a free box T; (L,TR,1,0,0) enables the complexation of a box T that is bound to a monomer with a free monomer M. A possible run of this program with one box S, three boxes M and one box T is depicted in Fig. 6.13 (arrows decorated with a star represents more then one step).

Example 6.21. [Actin filaments]
Actin is the most abundant small globular protein present in almost all the known eukaryotes. It is one of the most conserved protein among species because it is involved in fundamental cell processes, e.g., muscle contraction, cell motility, cell division and cytokinesis, vesicle and organelle movement, cell signaling, and the establishment and maintenance of cell junctions and cell shape. All these tasks are performed by exploiting the ability of actin to polymerize in long filaments and depolymerize. The main features we model are that actin monomers can generate filaments without a seed; filaments can grow at both ends and filaments can lose monomers from their tips.

Each actin monomers `A` has two interfaces `b` (barbed) and `p` (pointed), initially associated with `PI` (pointed inactive) and `BI` (barbed inactive), respectively. The `b` interface can also associate with `PF` (pointed free) and `PB` (pointed bound), while `p` has symmetric types `BF` and `BB`. The list of compatibilities is `(PF,BF,1,1,0)`,`(PF,BB,1,1,0)`, `(PB,BF,1,1,0)`. The code for `A` is

```
let A : bproc = #(p, BI), #(b, PI)
    [ch(r,b,PF).ch(p,BF).br!().pr!()  | P_REP | B_REP];
let P_REP : pproc = rep pr?().
    if (p, bound) then ch(b, PB).
        if not (p, bound) then ch(b, PF).pr!() endif
    endif
let B_REP : pproc = rep br?().
    if (b, bound) then ch(p, BB).
        if not (b, bound) then ch(p, BF).pr!() endif
    endif.
```

The first process in `A` activates the monomer and starts a copy of `P_REP` and `B_REP` that updates the `b` and `p` interfaces, respectively. The resulting monomer waits for a binding on one of its interfaces. Depending on the interface that becomes bound, there are two symmetric evolutions. For simplicity we describe only the binding on interface `b`. The first condition of `B_REP` is satisfied and `p` is set to `BB`. If the created link is broken, the monomer restores `p` to `BF` and restarts a copy of `B_REP`. If instead the `p` interface binds to a monomer, the first condition of `P_REP` is satisfied and `b` is set to `PB`. Now the monomer is stuck until one of the two interfaces become free again. Unbinding events trigger the second conditions of the `P_REP` and `B_REP` processes that updates the interfaces accordingly.
Note that the filament cannot close into a ring or break in the middle because the pair `(BB, PB)` has no binding and unbinding capabilities.

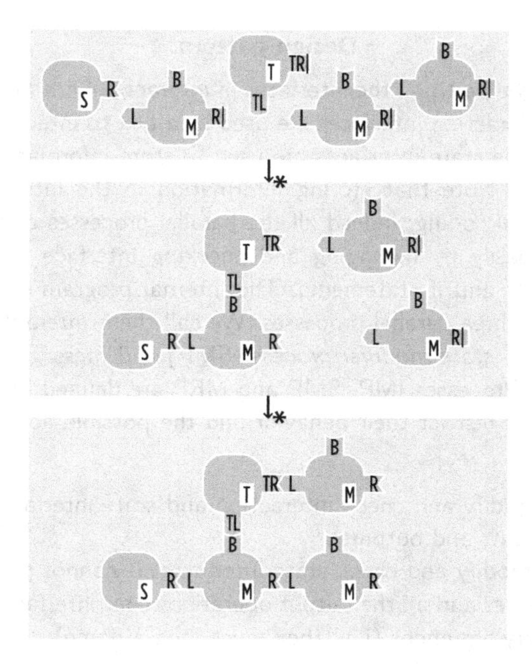

Figure 6.13 Generation of a branching tree.

6.3.3 *Introducing controls*

We introduce programs to build trees with constraints over the depth of the branches and then by imposing a minimum distance among branches that reflects the need for enough physical space between branches to allow the complexation of monomers and branching boxes. Finally, we combine the two models to obtain trees with both characteristics. In order to make the merge procedure of models more general we introduce a BlenX design pattern (see the box).

6.3.3.1 *Controlling branching depth*

A node of a tree has branching depth level n if there are exactly $n - 1$ branches from the node to the root. A tree has branching depth less than n if all its nodes have branching depth level smaller than n. We define a BlenX model that generates trees with branching depth of less than 5. The boxes involved in this model are monomers, branches and seeds. The BlenX graphical and textual representation of monomers is in Fig. 6.14

Design pattern

For each box we partition the interfaces in *interaction interfaces* and *state interfaces*. Interaction interfaces are used by a box to interact with other boxes, while the state interfaces are used to store information regarding the box state. Note that storing information in the interfaces simplifies the internal coding; indeed all the parallel processes can access this information easily by managing and checking interface types through change actions and if statements. The internal program of each box is composed by three parallel processes. We call them *interaction modifier process* (IMP), *state modifier process* (SMP) and *message receiver process* (MRP). Processes IMP, SMP and MRP are defined following a list of criteria that restrict their behavior and the possible actions they can perform:

- IMP can modify and check interaction and state interfaces. It cannot perform inputs and outputs;
- SMP can modify and check state interfaces. It cannot perform inputs over interfaces and all the output operations over interfaces must have a non-empty argument (i.e., they must send a name);
- MRP receives incoming messages from interaction interfaces and sends a signal over each received names.

Figure 6.14 Monomers for controlling branching depth. L, BI and RI are interaction interfaces, while DU is a state interface.

The state interface has subject `depth` and can be associated with five different types: `DU`, `D1`, `D2`, `D3` and `D4`. Type `DU` is used when the box is free, while the others are used when the box is bound and represent the node branching depth level. We refer to this interface as *depth* interface.

The interaction modifier process `MD_IMP` activates the right interface when the box binds on its left interface and activates the branching interface only if the strucure exposed on the depth interface is one among D1, D2 and D3. Hence, MD_IMP is defined as follows:

```
let MD_IMP : pproc =
   if (left, bound) then ch(right,R) endif
   | if not ((depth,D4) or (depth,DU)) then ch(branch,B) endif;.
```

The state modifier process MD_SMP modifies the type associated with the depth interface in order to make it represent the number of branches from the root to the current box. This task is performed through an exchange of messages among the state modifier processes of different boxes. Thus MD_SMP is a parallel composition of two processes, one receiving and using messages for updating its state interfaces, and another sending messages to bound boxes for triggering their update:

```
let MD_SMP : pproc = MD_r_SMP | MD_s_SMP.
```

MD_r_SMP is a sequential process made of an input on depth_msg, which guards a sum of inputs over specified channels whose firing enables proper modifications of the depth interface:

```
let MD_r_SMP : pproc =
   depth_msg?().(
      d1?().ch(depth,D1)
      + d2?().ch(depth,D2)
      + d3?().ch(depth,D3)
      + d4?().ch(depth,D4)
   );.
```

The interaction partner of MD_r_SMP is:

```
let MD_s_SMP : pproc =
     right!(depth_msg).(
      if (depth,D1) then right!(d1) endif
      + if (depth,D2) then right!(d2) endif
      + if (depth,D3) then right!(d3) endif
      + if (depth,D4) then right!(d4) endif
     )
   | branch!(depth_msg).(
      if (depth,D1) then branch!(d1) endif
      + if (depth,D2) then branch!(d2) endif
      + if (depth,D3) then branch!(d3) endif
     );.
```

This process has two parallel subprocesses guarded by outputs on `right` and `branch`, respectively. The output on `right` can be consumed only when the box binds to another box on the corresponding right interface. When consumed, the name `depth_msg` is sent to the `MD_MRP` of the bound box, and it is propagated inside that box by an output on `depth_msg` without object (code `right?(channel).channel!()` in `MD_MRP`). The output `dept_msg!()` synchronizes with the `MD_r_SMP` process of that box and enables a sum of input processes on channels `d1`, `d2`, `d3` and `d4`. The choice operator is resolved according to the next message sent by `MD_s_SMP` and forwarded to the bound box by `MD_MRP`. The initial output of `MD_s_SMP` enables a sum of mutually exclusive `if`. Each alternative sends through the right interface a different channel depending on the depth interface type. `MD_MRP` sends a message on this channel that synchronizes with one of the inputs of `MD_s_SMP`, causing the property update of the depth interface (code is the same as before, but the instantiation of `channel` is with `d1,..,d4` depending on the depth). Note that this communication protocol propagates the depth information to the newly attached box and is executed atomically with a sequence of immediate actions.

The `MD_s_SMP` process performs a sequence of immediate actions on the branch interface, similar to those described for the right interface, also to propagate the depth information along the branches. In this case the depth `D4` case is not considered because a branch cannot grow from a monomer with branching depth equal to four.

The *message receiver process* `MD_MRP` is a parallel composition of similar processes, one for each interaction interface:

```
let MD_MRP : pproc =
   rep right?(channel).channel!()
   | rep left?(channel).channel!()
   | rep branch?(channel).channel!();.
```

The branching box `T` is in Fig. 6.15.

The interaction modifier process `TD_IMP` activates the right interface when the box binds with another box on the left:

```
let TD_IMP : pproc = if (left, bound) then ch(right,TR) endif;.
```

The state modifier process `TD_SMP` propagates the depth information to the new filament nodes. In this case, given that it is a branch node, we increase the depth level and we do not consider the depth level one (`D1`),

Figure 6.15 Branching box for controlling branching depth. TL, BI and TRI are interaction interfaces, while DU is a state interface.

because the branch node contributes to its depth level and so it has depth level at least two:

```
let TD_SMP : pproc =
   depth_msg?().(
      d1?().ch(depth,D2)
      + d2?().ch(depth,D3)
      + d3?().ch(depth,D4)
   )
   | right!(depth_msg).(
      if (depth,D2) then right!(d2) endif
      + if (depth,D3) then right!(d3) endif
      + if (depth,D4) then right!(d4) endif
   );.
```

The message receiver process TD_MRP follows the same pattern of MD_MRP.

The seed box S does not have any state interface and its BlenX graphical and textual description is in Fig. 6.16.

The interaction capabilities of this box never change and therefore the interaction modifier process is the empty process `nil`. The state modifier process SD_SMP has to propagate the initial depth information when a filament is initiated, i.e., the first box it binds to is informed that its depth level is one:

```
let SD_SMP : pproc = right!(depth_msg).right!(d1);.
```

SD_IMP|SD_SMP|SD_MRP
S
R

let S : bproc =
#(right,R)
[SD_IMP | SD_SMP | SD_MRP];

Figure 6.16 Seed box for controlling branching depth. R is an interaction interface.

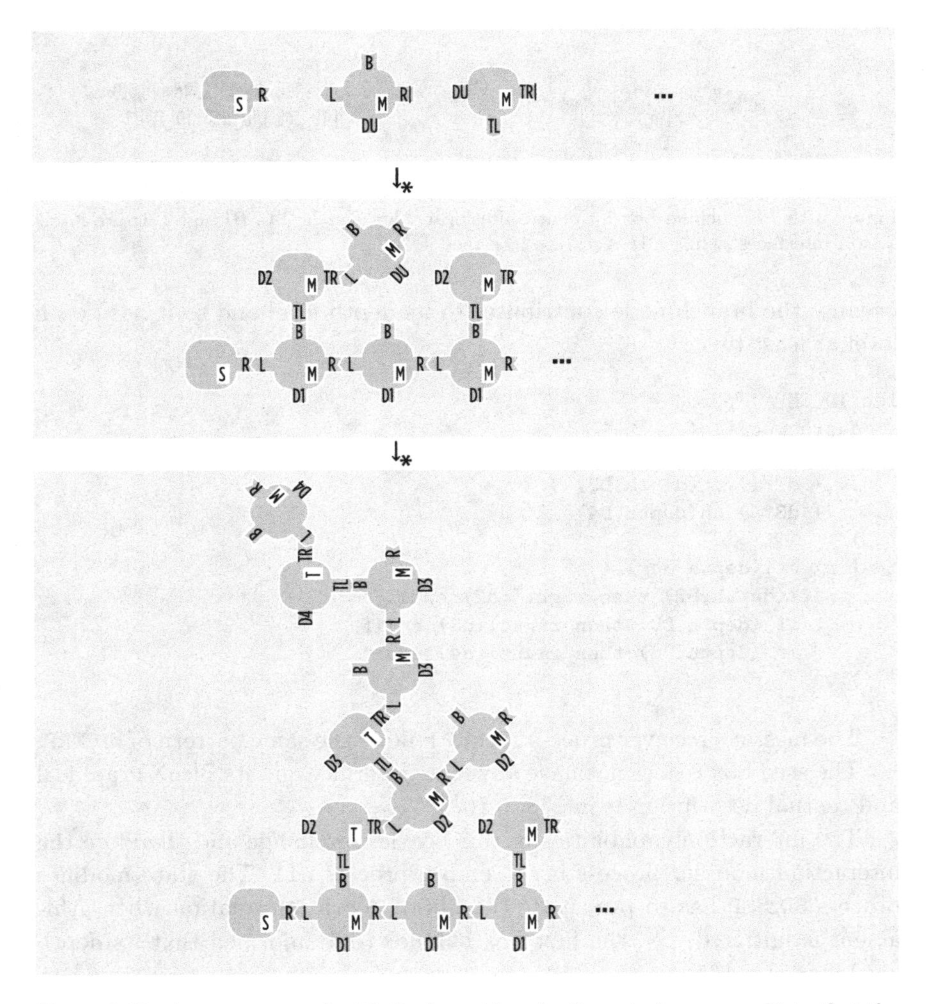

Figure 6.17 A tree generated with the branching depth control program. Note that the monomer branching interfaces are inactive at level four. The dots indicate the presence of other boxes in the system.

Figure 6.17 shows an example of computation of the branching depth control program.

6.3.3.2 *Controlling branching distance*

Here we consider the generation of trees having at least four monomers between their branches. Each monomer stores its distance from the nearest

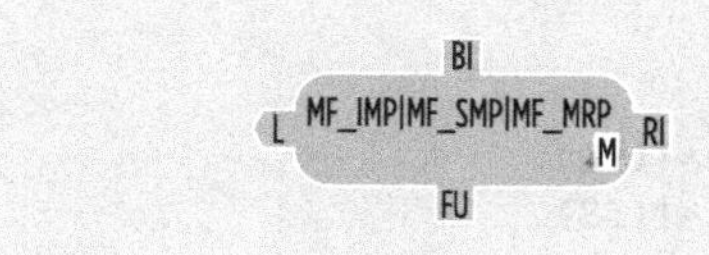


```
let M : bproc =
   #(right,RI), #(left,L), #(branch,BI), #(distance,FU)
      [ MF_IMP | MF_SMP | MF_MRP ];
```

Figure 6.18 Monomer for controlling branching distance. L, BI and RI are interaction interfaces, while FU is a state interface.

branch (the seed node is considered to be a branch) in a distance interface associated with the types FU, F1, F2, F3 and F4. The FU type represents the state of a free monomer, while the other types represent bound monomers. The number indicates the distance from the nearest branch with the exception of F4 that means the distance is greater or equal to four. The BlenX code for M is in Fig. 6.18.

The interaction modifier process MF_IMP permits the polymerization on the right interface when the box is bound on the left and permits branching formation only if the nearest branch is at least four nodes far away. If the distance interface has type F4, the process activates the branch interface. Then, an if checks for modifications of the distance interface, and when the distance is becoming smaller than F4, it disables branching.

```
let MF_IMP : pproc =
     if (left, bound) then ch(right,R) endif
   | if (distance,F4) then ch(branch,B).
          if not (distance,F4) then ch(branch,BI) endif
     endif;.
```

The state modifier process MF_SMP changes the distance interface according to the binding events. The BlenX code for the state modifier process MF_SMP is

```
let MF_SMP : pproc =
   ( if (distance,F1) then right!(distance_msg).right!(f2) endif
   + if (distance,F2) then right!(distance_msg).right!(f3) endif
   + if (distance,F3) or (distance,F4) then
        right!(distance_msg).right!(f4)
     endif)
   |rep distance_msg?().(
        if(distance,F1) then
           f1?() + f2?() + f3?() + f4?()
        endif
      + if(distance,F2) then
           f1?().ch(distance,F1).propagate!(f2)
```

```
          + f2?() + f3?() + f4?()
        endif
      + if(distance,F3) then
           f1?().ch(distance,F1).propagate!(f2)
         + f2?().ch(distance,F2).propagate!(f3)
         + f3?() + f4?()
        endif
      + if(distance,F4) or (distance,FU) then
           f1?().ch(distance,F1).propagate!(f2)
         + f2?().ch(distance,F2).propagate!(f3)
         + f3?().ch(distance,F3)
         + f4?().ch(distance,F4)
        endif
    )
    | rep propagate?(distance).(
         ( if (left,bound) then
              left!(distance_msg).
              left!(distance)
           endif
         + if not (left,bound) then kill!() endif)
         |
         ( if (right,bound) then
              right!(distance_msg).
              right!(distance)
           endif
         + if not (right,bound) then kill!() endif)
      )
    | rep kill?().nil;.
```

The first parallel component propagates the distance information to the box that binds to the right interface. If its distance is n, the distance of the newly attached box will be $n + 1$. In order to communicate this information, the process sends first a message containing the channel `distance_msg` to make the receiving box aware that a message regarding its distance from a branch is incoming. After that, depending on the type of its distance interface, it can send three different channels: `f2`, `f3` and `f4`. They represent distance two, three and distance greater or equal to four respectively. In this case the process cannot send distance one (represented by the channel `f1`) because a monomer is at least at distance one from a branch and so the distance of a neighbor is at least two. The channel `f1` is only sent by branch and seed nodes when they bind to a monomer.

The second parallel component manages the incoming distance messages and updates the distance interface, if required. After an input operation on `distance_msg`, four `if` processes made up of alternatives selected according to the distance interface become active. These four processes wait for an output action over one of the `fn` channels (where $\mathtt{n} \in \{1,2,3,4\}$). If an intercommunication happens then the distance interface is updated only if the new distance is smaller than the current one. If the interface is updated, the distance, increased by one, is propagated by the third parallel component that is activated by an output operation over the `propagate` channel. Note that the second parallel component is under a replication operator because it is executed each time another process in parallel performs an output over the `distance_msg`. The process performing the propagation ignores the origin of the received information (left or right interface). Thus, it sends the distance message both over the right and the left interfaces with two processes in parallel. These processes only differ for the interface on which they operate. They consist of two processes composed with a choice operator. The first process is triggered if the interface is bound; in this case it communicates the received distance over it. The other is executed if the interface is not bound and it triggers an output operation over a channel called `kill`. Its role consists in avoiding pending communications in the case the interface is not bound.

The process `MF_MRP` that dispatches messages appropriately is defined as in the branching depth model.

The state of a branching box `T` does not depend on the bound boxes and hence no state interface is needed (see Fig. 6.19).

The interface modifier process `TD_IMP` is the same of the branching depth model. The state modifier process `TF_SMP`, even without state interfaces, has to cooperate with the state modifier processes of the other boxes to communicate the distance information to them. When a monomer binds to the box `T`, `TF_SMP` communicates to the new neighbor that its new distance is one. The communication of the distance is done as previously and it is managed through the `TF_MRP` dispatcher.

Figure 6.19 Branching box for controlling branching distance. TL and TRI are interaction interfaces.

Figure 6.20 Seed box for controlling branching distance. R is an interaction interface.

```
let TF_SMP : pproc =
   left!(distance_msg).left!(f1)
   | right!(distance_msg).right!(f1);.
```

Also, the seed box S has no state interface (Fig. 6.20).

The interaction modifier process SF_IMP is empty (nil) because the interaction capabilities of the box never change. The state modifier process SF_SMP has the same role as the branching box TF_SMP process, with the difference that there is only one interface available for binding. Thus, the information regarding the distance is sent only over it:

```
let SF_SMP : pproc = right!(distance_msg).right!(f1);.
```

Figure 6.21 shows an example of a tree generated by the branching distance control program.

6.3.3.3 *Merging controls*

We merge the controls of the models of the previous subsections to generate trees with depth level less than five and with a branching distance of at least four monomers. We create a new seed box S, a new branching box T and a new monomer box M by starting from their corresponding definitions in the two previous models. We denote with S^d, T^d and M^d the box definitions for the branching depth control model and with S^b, T^b and M^b the box definitions for the branching distance control model. Each box B (with B ranged over by S, T and M) has the same interaction interfaces of B^d and B^b and a set of state interfaces, which is the union of the state interfaces of B^d and B^b. The state modifier process of each box B is obtained by composing in parallel the state modifier processes of B^d and B^b. Moreover, the message receiver process is the same as B^d and B^b because the two boxes have the same interaction interfaces. Finally, we define a new interaction modifier process. To merge the two models we only need to define a new IMP. This intuitive procedure is enough general to allow the merging of

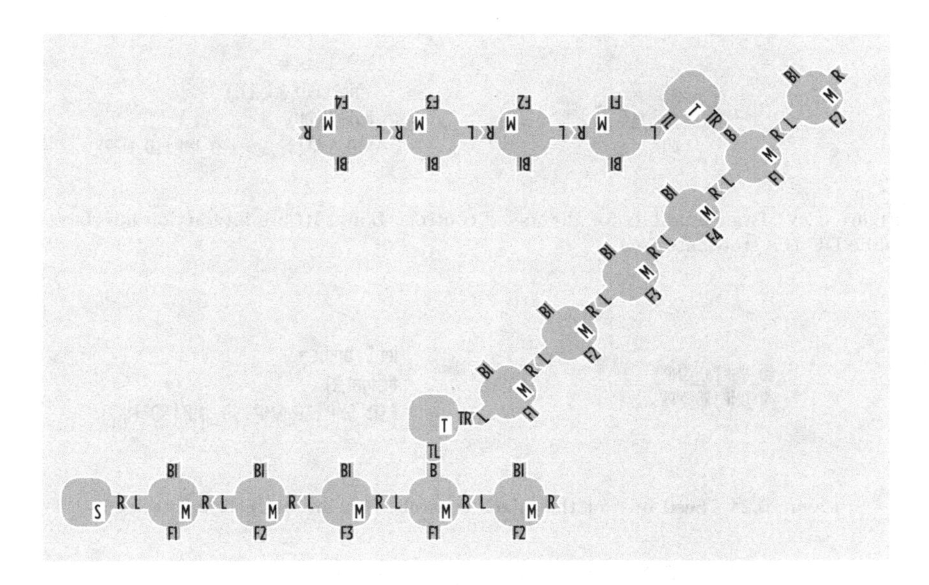

Figure 6.21 A tree generated by the branching distance control program.

Figure 6.22 Monomer for the mixed controls. L, BI and RI are interaction interfaces, whereas DU and FU are state interfaces.

more than two different control models. The box `M` of this merged model is in Fig. 6.22.

We define an `M_IMP` that allows the binding on the right interface only if the box is bound on the left interface and allows the growth of a branch only if the box has level less then five and it is far away at least three monomers from a branch:

```
let M_IMP : pproc =
   if (left, bound) then ch(right,R) endif
   |
   if (distance,F4) and not (depth,D4) and not (depth,DU) then
      ch(branch,B).
      if not (distance,F4) then ch(branch,BI) endif
   endif;.
```

Figure 6.23 Branching box for the mixed controls. L and RI are interaction interfaces, while DU is a state interface.

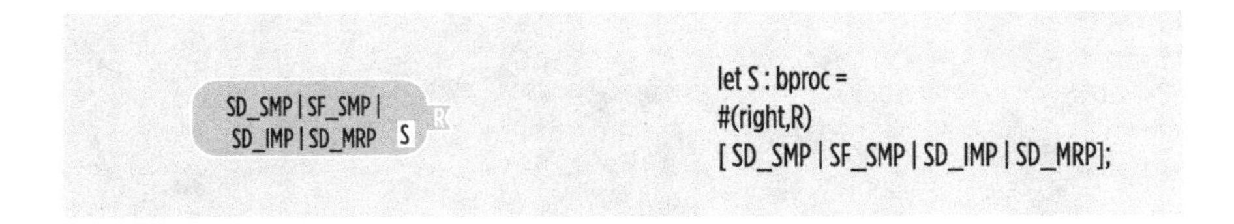

Figure 6.24 Seed box for the mixed controls. R is an interaction interface.

The definition of the merged branch box T is easier because the interaction modifier processes of the branch boxes of the models we are merging are the same. Thus we can use one of them (see Fig. 6.23).

Similarly, the merged seed box S of this model is in Fig. 6.24.

Note that in these models we used interface types as counters. Since interface types are not numbers, no arithmetic on them is possible (e.g., we cannot increase or decrease the values represented by interface types with plus or minus operators). Hence we have to represent each possible number with a different interface type and implement addition and subtraction operations with processes that changes interface types in the appropriate way (e.g., the previous definition of TD_SMP).

Example 6.22. [Actin filaments continued]
Actin filaments (see Example 6.21) interact with a multitude of molecules; in particular, they can bind with the Arp2/3 complex, a seven subunit protein. This complex has the capability to bind existing actin filaments and become a nucleation site for actin monomers. This leads to the formation of tree-like structures that are important for biological processes such as cell locomotion, phagocytosis and intracellular motility of lipid vesicles. The introduction of controls over the development of the structure has parallels in the biological world. In fact, when the Arp2/3 complex binds to a filament, because of structural constraints, this makes the complexation of other branching proteins around its proximity impossible. The intuitive model described here can be improved in order to mimic the behavior of the actin molecules more realistically.

6.3.4 *Rings*

We model the formation of ring polymers of length `N` by relying on a different design pattern to underline the flexibility of the language in allowing different programming styles.

Assume an initial population of boxes `A`, each containing the same process. Each box has an interface with type `L` (left) and an interface with type `R` (right), which are compatible, i.e., `(R,L,1,0,inf)`. Stochastic simulations can produce various kinds of complexes as depicted in Fig. 6.25.

The figure reveals the difficulties we have in controlling the formation of non-trivial complexes by only acting on the specification of interface capabilities.

To control ring formation, we equip each box with a state interface used to store the information regarding the length of the chain to which the box belongs (initially, the type of that interface is `S1`). Boxes `A` can bind together, forming chains that grow until they reach a size in which they close, forming a ring. We consider the ring as the stable form. A characteristic of self-assembly is that rings must be the result of local interaction between boxes, avoiding any global coordination.

Each box can be in one of three states: not bound to any other box (`Free`); bound either on the left or on the right (`Bound1`) or bound both on the left and on the right (`Bound2`). The process in the boxes implements a state machine that controls transitions from one state to another through

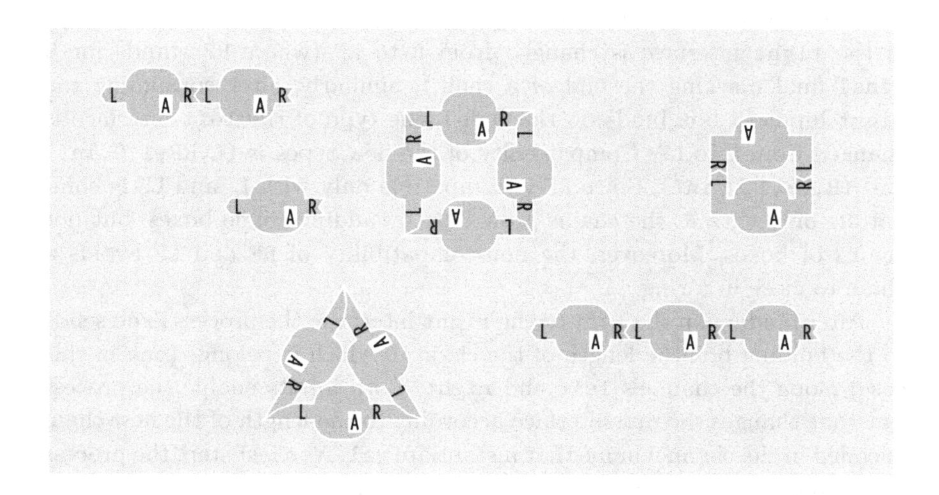

Figure 6.25 Examples of ring complexes.

the type of the box interfaces. After binding, the involved boxes start a protocol made up of a sequence of immediate actions to change the internal program state and their interface types. The definition of A is

```
let A : bproc =
    #(left,L), #(right,R), #(num,S1)
    [   b0!()
        | rep b0?().Free
        | rep b1?().Bound1
        | rep b2?().Bound2
    ];.
```

Each monomer starts by synchronizing on b0 and enters the state Free. Processes Free, Bound1 and Bound2 implement the three different states of our state machine. In order to implement a recursive behavior, we use the replication operator.

The process Free is

```
let Free : pproc =
   (left?(val).ch(right,RF).Set + right?(val).ch(left,LF).Set)
    | left!(one) + right!(one);

let Set : pproc =
    val!() |
    one?().ch(num,S2).b1!() ... + n-1?().ch(num,SN).b1!();.
```

After consuming the left input (a box binds on the left), the type of the right interface is changed from R to RF (where RF stands for R Final final marking the end of a chain). Similarly, after consuming the right input (a box binds on the right), the type of the left interface is changed from L to LF. Compatibility of the new types is (L,RF,1,0,inf) and (R,LF,1,0,inf). Since RF is compatible only with L, and LF is compatible only with R, the chains grow only by adding single boxes, but not chains of boxes. Moreover, the non-compatibility of RF and LF avoids a chain to close in a ring.

After binding on the left or the right interface, the process Free sends to the binding box the length of the chain to which it belongs (one in this case) along the channels left and right. Both inputs enable the process Set that changes the num interface according to the length of the new chain encoded in the channel name that instantiate val. As a last step, the process Set changes the state of the box from Free to Bound1 (activated by a signal along b1). The steps described so far are depicted in Fig. 6.26.

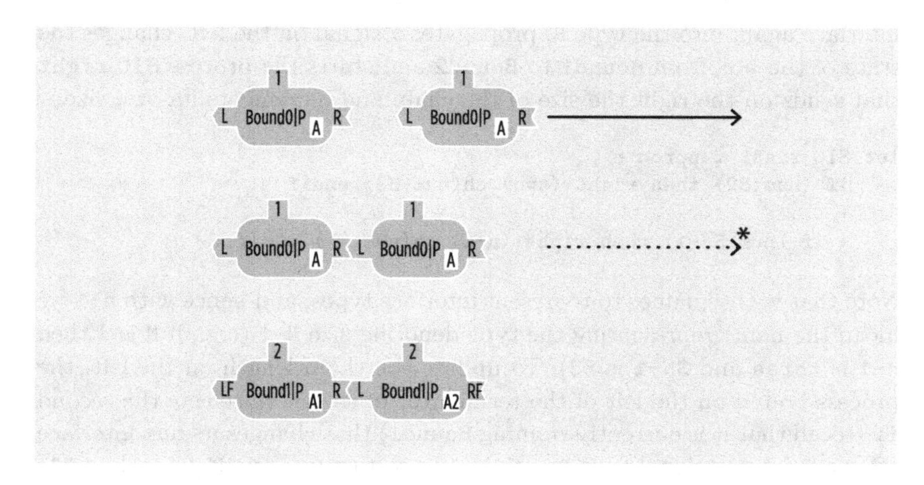

Figure 6.26 Starting a ring formation. Two boxes bind, free L and R interfaces are changed to LF and RF and the state interface is updated to store the new chain length.

The two boxes are now in the state encoded by the process Bound1.

```
let Bound1 : pproc =
    if (right,RF) then
        right?().ch(right,R).left!().b2!().SIC_right
    endif
    +
    if (left,LF) then
        left?().ch(left,L).right!().b2!().SIC_left
    endif
    +
    if (right,R) then right?().b1!().IncreaseCounter endif
    +
    if (left,L) then left?().b1!().IncreaseCounter endif
    +
    if (right,RF) and (num,SN) then ch(right,RC).
        if (right,bound) then ch(right,R).b2!() endif
    endif
    +
    if (left,LF) and (num,SN) then ch(left,LC).
        if (left,bound) then ch(left,L).b2!() endif
    endif.
```

The first two if conditions are satisfied when a box is a terminator of a forming chain. After consuming the right input (a free box binds on the right, becoming the new end of the chain), the process changes the right interface again into the type R, propagates a signal on the left, changes the state of the box from Bound1 to Bound2 and starts the process SIC_right that sends on the right the size of the chain and increments its own size

```
let SIC_right : pproc =
    if (num,S2) then right!(two).ch(num,S3) endif
    + ...
    + if (num,SN-1) then right!(n-1).ch(num,SN) endif;.
```

Note that we use names to represent interface types, and hence with n-1 we mean the name representing the type denoting size N-1 (e.g., if N is 4 then n-1 is three and SN-1 is S3). To update the chain length on the left, the process bound on the left of the terminator of the chain selects the second if (recall that it is currently running Bound1) that changes its num interface after receiving a signal from the right and restarting Bound1 (signal on b1) by activating IncreaseCounter.

The process IncreaseCounter changes the num interface with a type that represents the actual size incremented by one

```
let IncreaseCounter : pproc =
    if (num,S2) then ch(num,S3) endif
    + ...
    + if (num,SN-1) then ch(num,SN) endif;.
```

The steps described so far are depicted in Fig. 6.27. If a box binds on the left of a chain, the specular control processes for the left interface are used (second and fourth if) where SIC_left coincides with SIC_right with all occurrences of right replaced by left. The other two control processes are part of Bound1 to control the growing of the chain on both sides and to close the chain in a ring when it reaches the appropriate length.

If the right interface has type RF and the num interface has type N, then the right interface is changed into RC and the process is blocked until the right interface is bound. Since the process controlling the left side is specular, when the chain reaches a length of N the right and left side interfaces assume structures RC and LC, respectively, and remain blocked until they bind together. However, since the compatibility of these two structures is defined as (RC,LC,inf,0,0) they immediately close and the two blocked processes can continue their execution, leading to a complex made up of structurally equivalent boxes and hence representing a completely symmetric ring. As an example, if N is equal to 3, when a chain of size 3 is formed, it immediately closes in a symmetric triangle (see Fig. 6.28).

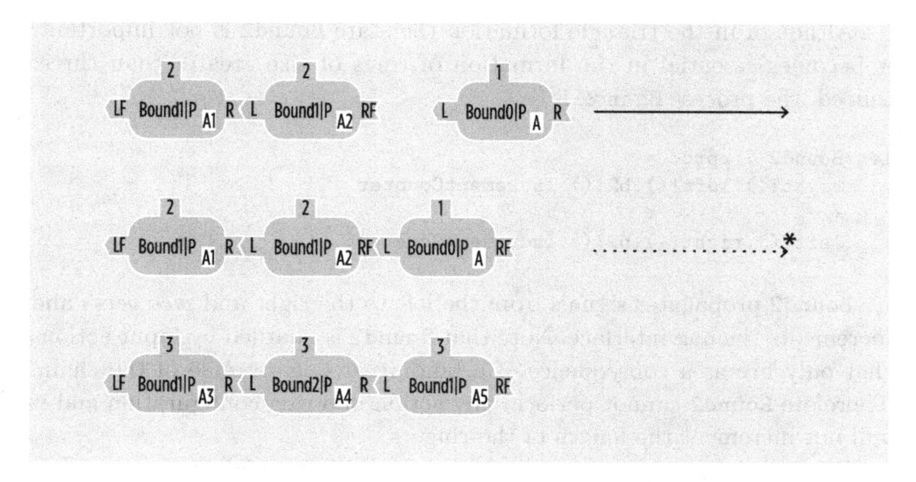

Figure 6.27 Prolonging the chain. After a new monomer binds to the right of a chain, the old terminator is reset to R, the new terminator RF is generated and the chain length is updated.

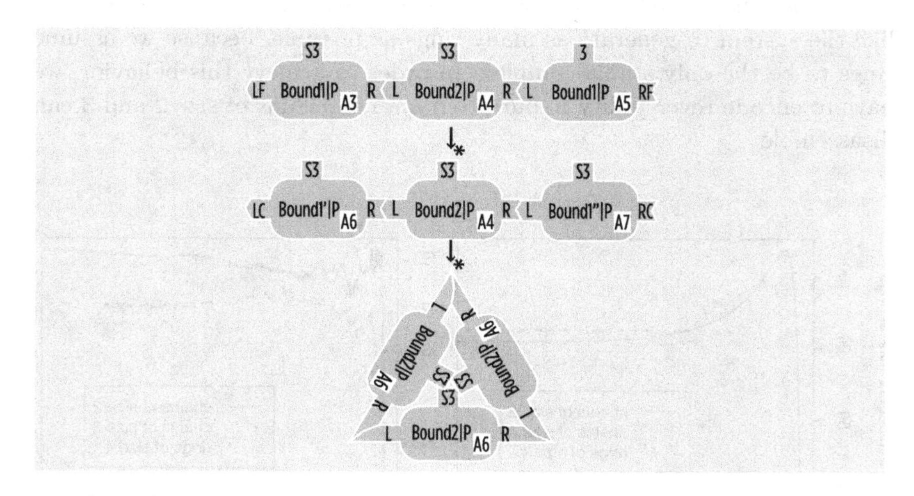

Figure 6.28 Closing a chain. When the pre-defined length is reached, the terminators of the chain are changed into LC and RC, thus enabling immediate closing of the chain. After closing, LC and RC are reset to L and R, respectively.

Although in the triangle formation the state Bound2 is not important, it becomes essential in the formation of rings of size greater than three. Indeed, the process Bound2 is

```
let Bound2 : pproc =
    right?().left!().b2!().IncrementCounter
    +
    left?().right!().b2!().IncrementCounter;.
```

Bound2 propagates signals from the left to the right and *vice versa* and increments the num interface. Note that Bound2 is guarded by input actions that only fire as a consequence of a binding at one extreme of the chain. Therefore Bound2 cannot perform any action in a ring configuration and it will not increment the length of the ring.

We run the previous program 100 times with N equal to 4 and an initial population of 1000 boxes A. The average dynamics and the standard deviation are in Fig. 6.29.

The results show that given an initial population of boxes, we reach a final configuration in which we have almost the same number of chains of size 3 and rings of size 4, and some chains of length 2. We would instead like the system to generate as many rings as possible, because we assume rings to be the only stable complex. In order to achieve this behavior, we have to encode reversibility in our program, i.e., chains of size 2 and 3 can disassemble.

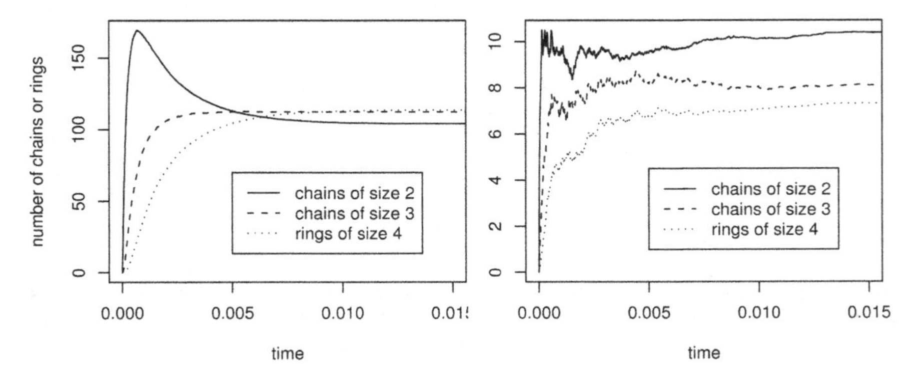

Figure 6.29 Average dynamics and standard deviation of 100 stochastic simulations regarding the formation of rings of size 4. Each simulation takes in average around two seconds. Reproduced with permission from [Larcher *et al.* (2010)].

6.3.4.1 *Adding reversibility*

We add reversibility by allowing the unbinding of boxes located at the left and right sides of a chain. To implement this protocol we work on the process `Bound1`, because it is the state characterizing the boxes located at the sides of a chain. We add the following alternatives

```
if (left,LF) then ch(r,right,RU).b1!() endif
+ if (right,RF) then ch(r,left,LU).b1!() endif
+ if (not (left,bound)) and (not (right,bound)) then
    ch(left,L).ch(right,R).ch(num,S1).b0!() endif
```

where `r` is a rate describing the speed of the reversible reaction. Each terminator of a chain can decide to unbind by changing the type of the bound interface (left interface for the right side and *vice versa*). The compatibilities of the new types are `(LU,R,0,inf,0)` and `(RU,L,0,inf,0)`, that implement immediate detaching of the terminator from the chain. After the unbinding, for both sides, the condition of the third `if` becomes true and hence the interfaces and the process of the unbound boxes return to the initial configuration (recall the recursive behavior of box A; see Fig. 6.30).

The last complex of the picture shows how the remaining chain also returns to a consistent configuration, where the boxes have the correct internal state and their interfaces `num` have the correct types. In order to obtain this result, we modify `Bound2` by adding alternatives that control the binding status of the interfaces and recognize when there is an unbinding,

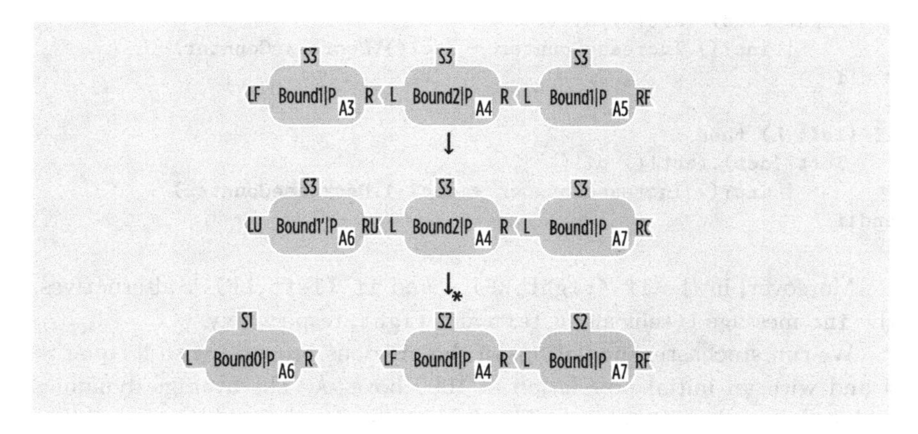

Figure 6.30 Reversibility. The left box of the chain is detached and interfaces are reset to correct types.

send a decrement signal to the rest of the chain and reset the state to `Bound1`.

```
if not(right,bound) then
    ch(right,RF).left!(dec).b1!().DecreaseCounter endif
+
if not(left,bound) then
    ch(left,LF).right!(dec).b1!().DecreaseCounter endif.
```

The `DecreaseCounter` process is very similar to that of the `IncreaseCounter` but decrements the type of the `num` interfaces. The process propagates on the left/right a name `dec` that instantiates `act` in the additional alternatives added.

```
right?(act).left!(act).(act!().b2!()
     | inc?().IncreaseCounter + dec?().DecreaseCounter)
+
left?(act).right!(act).(act!().b2!()
     | inc?().IncreaseCounter + dec?().DecreaseCounter).
```

Note that the increasing or decreasing processes are enabled depending on the received name and process `Bound2` is enabled again. Similarly, `Bound1` has to manage the `inc` and `dec` messages. This is done by replacing the `if (right,R)..` and `if (left,L)..` alternatives with

```
if (right,R) then
    right?(act).(act!().b1!()
        | inc?().IncreaseCounter + dec?().DecreaseCounter)
endif
+
if (left,L) then
    left?(act).(act!().b1!()
        | inc?().IncreaseCounter + dec?().DecreaseCounter)
endif.
```

Moreover, in the `if (right,RF)..` and `if (left,LF)..` alternatives, the `inc` message is sent along `left` and `right`, respectively.

We run stochastic simulations of the previous program with `N` equal to 4 and with an initial population of 1000 boxes `A`. The average dynamics and standard deviation are in Fig. 6.31. Note that by adding reversibility we have that the number of rings continues to increase, until no more rings can be formed.

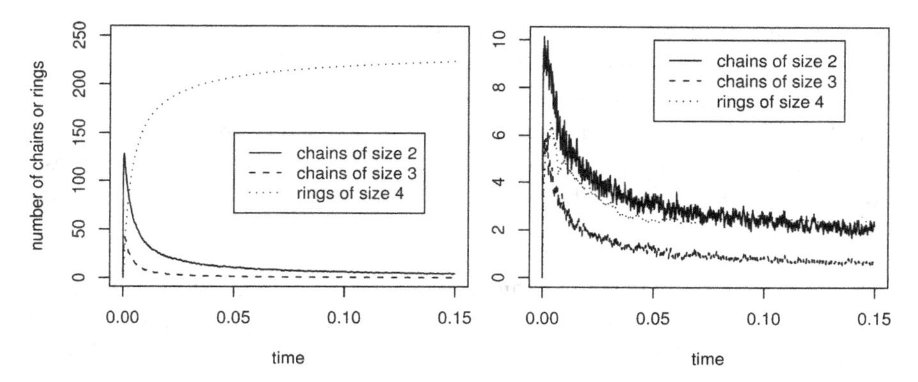

Figure 6.31 Average dynamics and standard deviation of 100 stochastic simulations regarding the formation of rings of size 4 with reversibility. Each simulation takes on average around ten seconds. Reproduced with permission from [Larcher *et al.* (2010)].

6.4 An evolutionary framework

We discuss here how to simulate the evolution of networks by relying on a language-based modeling approach. The nodes of the network are the individuals in a population and the arcs represent interactions. Evolution proceeds through selection acting on the variance generated by random mutation events. Individuals replicate in proportion to their performance, referred to as *fitness*. This process is described by Algorithm 6.2.

Algorithm 6.2: Population evolution

Input: Number of generations.
Output: A population of evolved individuals according to a fitness measure.

begin
    Population := GenerateInitialPopulation();
    for *i=0 to generations* **do**
        for *each Individual in Population* **do**
            output := Simulate(Individual);
            fitnesses[Individual] := ComputeFitness(output)
        end
        NewPopulation := ReplicateAndMutate(fitnesses, Population);
        Population := NewPopulation
    end
end

This algorithm differs slightly from the generic evolutionary algorithms used in computer science, being closer to real biological observations made for the asexual reproduction of organisms. Each individual in the population is codified as a BlenX program, and the boxes in each program are the components of each individual. The interaction among these boxes is the behavior of the network. There are four procedures in the algorithm:

(1) `GenerateInitialPopulation`: the initial population can be generated randomly from a pre-defined network configuration, or it can be a network with no interactions. All the individuals in the initial population can be identical at the beginning, as they will be differentiated later by the mutation phase.
(2) `Simulate`: each individual in the population is simulated stochastically and separately.
(3) `ComputeFitness`: the output of the simulation is used to compute the fitness value of the current individual. The fitness value depends on the observed phenotype.
(4) `ReplicateAndMutate`: this is the most important part of the algorithm; individuals with the highest fitness values are more likely to survive, replicate and produce a progeny that resembles them, without, however, being completely equal to them.

The `ReplicateAndMutate` algorithm (6.3) creates a new population with the same number of individuals of the current generation, using the current individuals as a base. At each step it chooses one individual, with probability proportional to its fitness (`ChooseOne` in the code above). This is achieved by constructing a cumulative probability array a from the *fitness* array, generating a random number in the range 0..`Population.Size`, and then finding the index into which the random number falls.

The selected individual will replicate and pass to the next generation. During the replication, each protein in the individual is given the chance to mutate, according to a probability.

A mutation is selected among all the possible types by the `GetRandomMutation` function, and this mutation is applied. Finally the individual, which can be either identical to its predecessor or mutated, is added to the new population.

The next subsection describes some possible mutations.

Algorithm 6.3: Replication and mutation of individuals

```
Input: A population, the fitness levels of its individuals, a duplication
probability, a mutation probability.
Output: A new population.

begin
    for i = 0 to Population.Size do
        Individual := ChooseOne(fitnesses, Population);
        for each Protein in Individual.Proteins do
            if Random() < DuplicationProb then
                Individual.Proteins.Add(Protein.Duplicate())
            end
            for each Domain in Protein.Domains do
                if Random() < MutationProb then
                    MutType := GetRandomMutation();
                    if IsMutFeasible(Domain, MutType) then
                        D2 := Individual.PickCompDom(Domain,
                        MutType);
                        Individual.Mutate(Domain, D2, MutType)
                    end
                end
            end
        end
        NewPopulation.Add(Individual)
    end
end
```

6.4.1 *Mutations*

Here, we consider the end effects of mutations occurring at the DNA level. These mutations ultimately affect the dynamics of the molecular networks abstracting individuals. For example, mutations in a DNA sequence can change the protein amino-acid sequence, leading to changes in its tertiary structure with implications for the affinity of this protein with other proteins or substrates. Similarly, events at DNA level such as gene duplication or domain shuffling can alter network structure and dynamics.

Figure 6.32 introduces the toy model we use in this subsection to illustrate mutations.

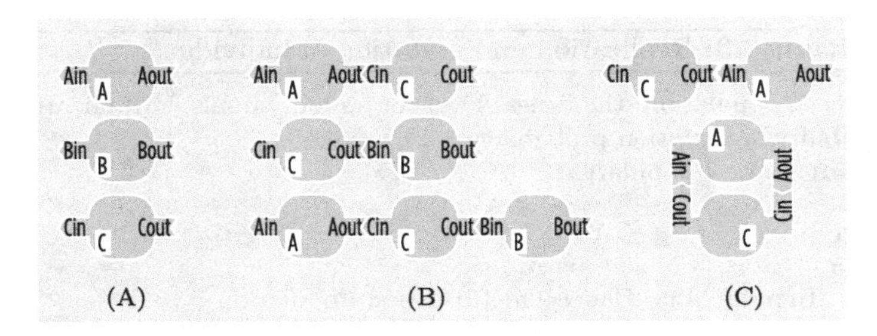

Figure 6.32 The toy model of a three-protein network: (A) the proteins in the system; (B) the complexes that can be observed assuming that *Aout* can bind to *Cin* and that *Cout* can bind to *Bin*; (C) the complexes that can be observed after a mutation that does not allow any more *Cout* to bind to *Bin* and enables the binding between *Cout* and *Ain*.

6.4.1.1 *Duplication and deletion of proteins*

Gene duplication at DNA level is implemented with a duplication of the box of the protein codified by the gene. The new box will have the same internal structure with new binder identifiers that, however, inherit the same interaction capabilities of the corresponding binders of the original box. Duplication of binder identifiers is needed because subsequent mutations on one of the binders of the duplicated protein must not affect the original one. Furthermore, since the new protein is a new distinct entity, it must not be structurally equivalent to the original one. The same considerations hold for the internal processes: duplication and deletion of domains may lead to a modification of the internal structure (see below); the internal processes must be duplicated so that each box has its own, distinct internal behavior. This is implemented by replacing all the occurrences of binder identifiers in the processes with the new identifiers.

Deletion of a protein is accomplished by deleting its box and all the internal processes that refer to the deleted protein.

6.4.1.2 *Mutation of domains*

Point mutations in DNA can change the protein amino-acid sequence, and consequently lead to the mutation of a domain and to changes in the interaction capabilities of the protein to which it belongs. This is achieved by changing the rates of the interactions (affinities) in which the mutated domain (binder) takes part. More specifically, the mutation on a domain

Example 6.23. [Duplication of genes]
Assume that a mutation duplicates the gene that codifies protein C. We will then add a box C' (with active domains $C'in$ and $C'out$) to the system. The interaction capabilities of C' are inherited from C. Therefore $C'out$ can bind to Bin. The complexes that can be observed are the ones in Fig. 6.32(b) and other 3 complexes that are obtained by replacing C with C' in the figure.

can be specified as:

(1) a *change of interaction*, for which we modify the affinity, adding a number sampled from a normal distribution;
(2) an *addition of an interaction* between two domains $d1$ and $d2$, modeled as the addition of an affinity $(d1, d2, x, 0, 0)$, with $x > 0$; and
(3) a *removal of an interaction* between two domains $d1$ and $d2$ setting $(d1, d2, 0, 0, 0)$.

We are assuming for simplicity that decomplexation never occurs.

Example 6.24. [Mutation of domains]
Assume that the domain $Cout$ of protein C is no more able to bind to Bin after a modification, but it can bind to Ain. We then set $(Cout, Bin, 0, 0, 0)$ and $(Cout, Ain, x, 0, 0)$ with x sampled from a normal distribution. As a consequence, the complexes that can be observed in the system are depicted in Fig. 6.32(c).

6.4.1.3 *Duplication and deletion of domains*

Duplicating or removing domains can be done by acting on the binders list and on their affinities; however, for these domains to act as sensing or effecting domains in cooperation or in antagonism with the existing ones, the internal behavior of the processes must also be changed.

Consider a sensing domain: when a signal arrives — by means of a ligand binding, or by phosphorylation of a residue — the internal behavior of the protein changes, bringing it to a different *state*. If that domain is duplicated, the internal behavior must be changed accordingly. The second domain may act *concurrently* with the former domain, with the result that the activation of this second domain will bring the protein into the same state as the old one, acting in parallel. This is the case, for example, of a receptor that can bind to two different signal molecules. Alternatively, the duplicated domain can affect the capability of the protein to reach that state, and so must act *in coordination* with the original one (e.g., kinases that must be phosphorylated twice to activate).

These mutations are obtained by manipulating the structure of the internal process to *transform* its behavior. In both cases, we assume that the internal process has a standard structure: a parallel composition of different processes, each representing a different *state*. The set of processes in parallel is a set of *mutually exclusive* ones: at any given time only one of the processes can be active (e.g., not blocked waiting for a communication). Moreover, each process in the set *enables* another one by issuing a communication immediately before blocking itself. This way, we are coding a finite state automaton for the control of the box.

Cooperative domains require a signal on each of them to reach the desired internal configuration. The transformation can be accomplished by substituting the process codifying for the current active state with a new process, adjusting the channel names used for the intracommunications and binder identifiers accordingly.

Example 6.25. [Duplication and deletion of domains]
Consider the code below that represents the activation of a domain p when a signal is received on a domain $r1$. Note that the signal on sig in $state1$ is an abstract message that inactivates p and restores $state0$ in the box.

```
rep state0?().(r1?().state1!())
rep state1?().(unhide(p).sig?().hide(p).state0!())
```

Assume a duplication of the domain $r1$ and the the new domain $r2$ is acting competitively with $r1$. To manage the activation of p, the code for $state0$ becomes

```
rep state0?().(r1?().state1!() + r2?().state1!()).
```

If the duplication of the domain is cooperative the code for $state0$ is changed as follows:

```
rep state0?().(r1?().r2?().state1!() + r2?().r1?().state1!()).
```

The case of *concurrent* or *competitive* domains, where each of the signals can lead to the desired internal configuration, can be handled in a similar way.

Deletion of a domain requires undoing the steps done while duplicating it. This task is accomplished again by transformation of the internal process, restoring the behavior to the original one.

Note that mutations can be implemented in BlenX, because language-based models are not meant to be solved (like other formalisms, for example those based on ODEs); instead, the model code is compiled into a format that is executed by the simulator *virtual machine*. The transformations corresponding to the mutations are implemented by manipulating the code of the compiled model.

6.4.1.4 *Measure of fitness*

When analyzing the evolution of specific biological systems, one needs to consider the fitness benefit of that system to the organism (i.e., to its reproductive success). There are three main approaches to measuring fitness (see Fig. 6.33).

The fitness of an individual can be computed analytically by a metric of the structure — its representation in the model — of the individual or by measuring an external parameter related to the time-dependent behavior of the network. The analytical solution does not require simulation: the presence or absence of a particular feature in the individual influences its fitness in a positive or negative way, which is computed by evaluating a formula on each individual.

Measuring an external parameter to compute the fitness requires simulation. Consider for example a model of bacterial chemotaxis: measuring how much food an individual can reach in a fixed number of steps is indicative of its ability to swim towards food. In this case, the measure of fitness is independent of any characteristic of the network, and depends only on the individual's high-level behavior. Unfortunately, this kind of measurement

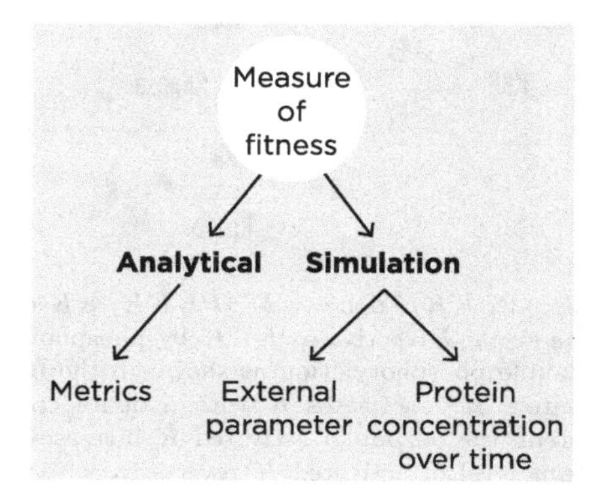

Figure 6.33 Different approaches to measure fitness.

can only be applied in a very limited number of networks: the measure of fitness is problem-dependent, and it varies with the kind of network and the characteristics a scientist wants to investigate. A behavior at the level of a single network seldom influences the whole organism in a direct and measurable way.

The third way to measure fitness takes direct measures of the trace of protein and molecule concentrations generated over time by the simulation. This measure can be done in various ways, including stability analysis, integration of the signal or measure of the derivative.

While it is usually complicated to define and measure fitness, network dynamics can provide a good proxy in case of biological networks. The concentration of the proteins involved in such networks defines the proper functioning of the network, and how these concentrations fit a specific time course determines how well the network operates.

The next subsection reports an example of this framework's application.

6.4.2 *A case study: MAPK*

The mitogen-activated protein kinase cascade (MAPK cascade) is a series of three protein kinases which is responsible for cell response to growth factors. A model for the MAPK cascade is in Fig. 6.34. Its ODE analysis shows that the production of activated Ks as a consequence of the input signal $E1$ is an ultra-sensitive switch and the response curves are steeply sigmoidal.

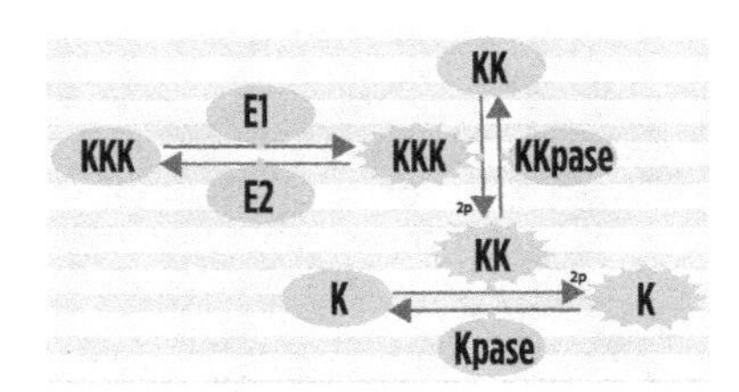

Figure 6.34 MAPK cascade. KKK denotes $MAPKKK$, KK denotes $MAPKK$ and K denotes $MAPK$. The signal $E1$ activates KKK by phosphorylation, which in turn activates KK with a double phosphorylation as shown in the figure by the tag $2p$ on the arrow. When activated, KK activates K with a double phosphorylation. When $E1$ is added to the system, the output of activated K increases rapidly. By removing the signal $E1$, the output level of activated K reverts back to zero. The transformations in the reverse direction are the result of the signal $E2$, the $KKpase$ and the $Kpase$.

We simplify all the enzymatic reactions in the MAPK cascade

$$E + S \underset{k_2}{\overset{k_1}{\rightleftharpoons}} ES \xrightarrow{k_3} EP \xrightarrow{k_4} P + E$$

with reactions of the form

$$E + S \xrightarrow{K_{EP}} E + P.$$

The simplified model still behaves as an ultra-sensitive switch and the code is described in Examples 6.26 and 6.27.

We analyze the evolution of a population of networks according to a fitness function, which captures the essential behavior of the MAPK cascade model. We generate an initial population of 500 individuals containing the network, shown in Fig.6.35a. We set up initial conditions with a single kinase $K1$, a single phosphatase $P1$, an activation signal $E1$ and a deactivation signal $E2$; the model has no interactions among entities. In other words, we consider an ancestral organism that has all the base proteins, but lacks a signaling system similar to the MAPK cascade. The dynamic of each individual is then simulated for 7000 simulation steps and we remove the signal $E1$ at the step 1500 using a time-triggered *delete* event. By using the output of the simulation, we measure for each individual how rapidly the output of an active kinase increases and how much the output of the same kinase persists after removing the signal $E1$ before returning to the initial condition. Let $out = \{n_0, n_1, ..., n_{7000}\}$ be the tuple representing the active $K1$ dynamics of an individual, then the fitness for out is

$$fitness(out) = \mu + \Big(\frac{\sum_{j=i_1}^{e_1} n_j}{K_M^* * (e_1 - i_1)} - \Big(\gamma * \frac{\sum_{j=i_2}^{e_2} n_j}{K_M^* * (e_2 - i_2)}\Big)\Big).$$

The two sums, which we denote respectively with $A1$ and $A2$, represent discrete integrals and are normalized with respect to their possible maximum values (see Fig. 6.36). The values i_1, e_1, i_2 and e_2 are parameters that define the boundaries for the computation of the two discrete integrals in the fitness formula, and the value K_M^* represents the maximum value for the active K response.

Moreover, μ represents the minimum fitness and γ controls the relative importance to responding to a signal and turning the response off after its removal.

In order to maintain a biological validity for the new individuals, we only accept mutations that satisfy the following constraints: (1) signals $E1$ and $E2$ cannot be removed; (2) a kinase can only activate other kinases or itself; (3) kinases are specific (e.g., they do not phosphorylate multiple

Example 6.26. [MAPK in BlenX]
We model in BlenX the MAPK system in Fig. 6.34. We assume all the rates are equal except for interface manipulation primitives that are assumed to have rate `inf`. The peculiarity of this model is that we do not use complexes, but just interactions between boxes. The boxes for the two signals E1 and E2 behave similarly by continuously offering an output over their interface.

```
let E1 : bproc = #(e1, start)[ rep e1!().nil ];
let E2 : bproc = #(e2, stop)[ rep e2!().nil ];.
```

E1 activates KKK by interacting with its `recv` interface when KKK is not active (interface type of `recv ikkk`). Symmetrically, E2 deactivates KKK by interacting with its `recv` interface when KKK is active (interface type of `recv akkk`). To allow the above interactions we include the tuples `(start, bkkk, r)` and `(stop, akkk, r)` in the binder definition file. Note that we have just one rate in the affinities because we only rely on interaction between boxes.

```
let KKK : bproc = #h(p,okkk)#(recv, ikkk)[
    recv?().s1!().nil | kkk0 | kkk1 | rep p!(plus).nil ].
```

The KKK box is a two-state machine that unhides and hides its `p` interface over which continuously offers an output with a `plus` argument. After changing the type of the `recv` interface the process encoding the current state waits for a new signal on `recv` to move to the other state. The first process on the left of the parallel composition within KKK is needed for the very first signal to initialize the machine to state 1. The processes encoding the states are:

```
let kkk0 : pproc = rep s0?().hide(p).ch(recv,ikkk).
                         recv?().s1!().nil;
let kkk1 : pproc = rep s1?().unhide(p).ch(recv,akkk).
                         recv?().s0!().nil;.
```

The KK needs two signals to be fully activated (see the 2 on the arrow in Fig. 6.34). Therefore, we use an intermediate state in which KK enters after the first signal from KKK. The encoding is (`kk0` activates the state machine):

```
let KK: bproc = #h(p, oKK), #(recv, iKK)
          [ kk0 | rep s0?().kk0 | rep s1?().kk1 | rep s2?().kk2
          | rep p!(plus).nil ];.
```

proteins) and (4) phosphatases are not specific but can only deactivate kinases.

The variation of fitness during a simulation is depicted in Fig. 6.37. Note the "steps" in the fitness. We observed this typical behavior in almost all our runs. In the first generations, individuals have to find the correct signal: the *jump* in (A) is realized when the the activation signal $E1$ hits one of

Example 6.27. [MAPK in BlenX — continued]
The processes in KK are:

```
let kk0 : pproc = recv?(x).ch(recv,intKK).s1!().nil;
let kk1 : pproc = recv?(what).what!() |
            ( plus?().unhide(p).ch(recv,aKK).s2!()
              + minus?().ch(recv,bKK).s0!() );
let kk2 : pproc = recv?(x).hide(p).ch(recv,intKK).s1!();.
```

The process kk0 receives a signal on recv from the interface p of an active KKK (we need the affinity (oKKK, iKK, r)), and moves the state machine to an intermediate state kk1 by changing the type of recv to intKK and sending a signal along s1. The state kk1 can revert back to kk0 via a phosphatase interaction forwarding a minus signal (see the definition of the phosphatase below) or move to the active state kk2 if the signal plus is received from KKK. When kk1 moves to kk0, the type of recv is changed into iKK and a signal along s0 is sent. When kk1 evolves to kk2, the effecting interface p of KK is unhidden so that KK can start signaling plus along it and the type of recv is set to aKK. For this last transition to happen we need the affinity (oKKK, aKK, r). The phosphatase that inactivates KK is

```
let KKpase: bproc = #(x, pase2)[ rep x!(minus).nil ];.
```

Since it has to signal the intermediate or active state of KK on recv, we have to add the affinities (pase2,intKK,r) and (pase2,aKK,r).
The behavior of K is similar to the one of KK and hence its encoding follows the same pattern.

```
let K: bproc = #h(p, oK), #(recv, bK)
         [ k0 | rep s0?().k0 | rep s1?().k1 | rep s2?().k2
         | rep p!(plus).nil ];.
```

The processes in K and the phosphatase are:

```
let k0 : pproc = recv?(x).ch(recv, intK).s1!().nil;
let k1 : pproc = recv?(what).what!() |
         ( plus?().unhide(p).ch(recv,aK).s2!()
           + minus?().ch(recv,bK).s0!() );
let k2 : pproc = recv?(x).hide(p).ch(recv,intK).s1!();
let Kpase: bproc = #(x, pase1)[ rep x!(minus).nil ];.
```

The affinities that we have to add to include the dynamics of K are (oKK, iK, r), (oKK, aK, r), (pase1,intK,r) and (pase1,aK,r). The system is

```
run 2 E1 || 2 E2 || 2 KKpase || 2 Kpase || 20 KKK
                                        || 200 KK || 200 K.
```

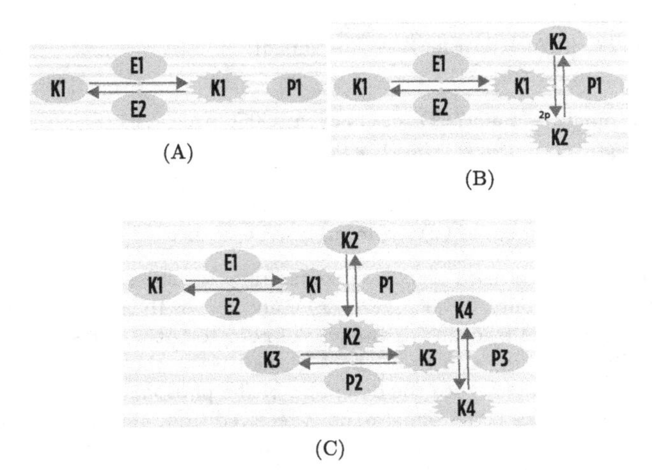

Figure 6.35 (A) Basic individual of the initial configuration. (B) A particular individual we obtained, with a two-level phosphorylation. (C) An alternative evolution, with single phosphorylation kinases but a longer cascade.

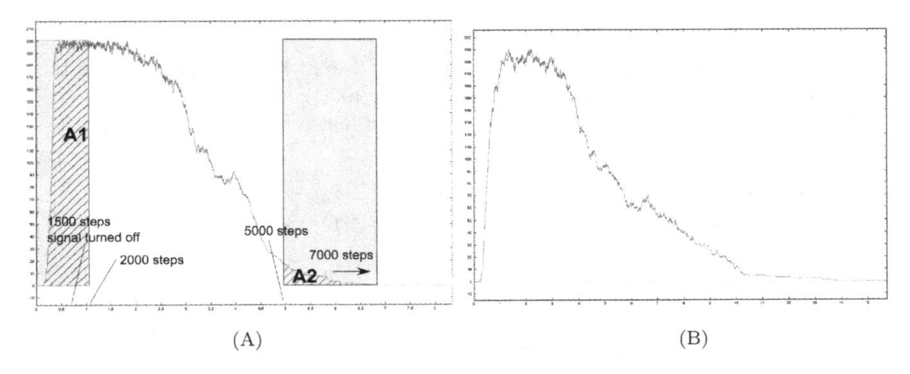

Figure 6.36 (A) Time course of the active K concentration over the simulation time, superimposed to the integral areas for the fitness function we implemented. The fitness parameters are $i_1 = 0$, $e_1 = 2000$, $i_2 = 5000$, $e_2 = 7000$, $K^*_M = 200$, $\mu = 0.1$ and $\gamma = 0.75$. (B) Time course of K for a network with high fitness originated over 2000 generations. Reproduced with permission from [Dematté *et al.* (2008c)].

the kinases. In (B) instead, we have the slow adaption to the introduction of the deactivation signal: the presence of the signal allows the cascade to be switched off, but reduces the gain of the switch in response to an activation signal. The second *jump*, in (C), is where double phosphorylations or more kinases are added to the cascade, allowing the network to regain the lost efficacy and react in a steep way to the activation signal. The last

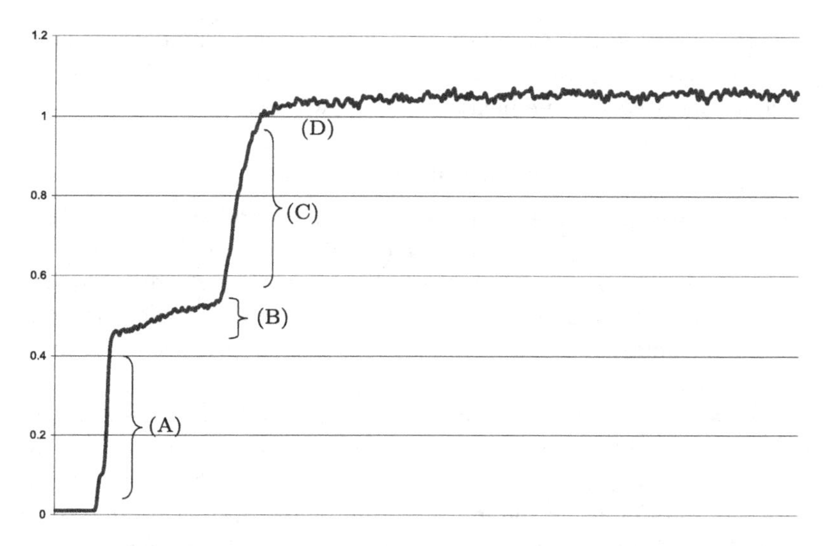

Figure 6.37 Changes in fitness during a typical evolutionary simulation. (A) and (C) are jumps in fitness, (B) is a slow adaptation and (D) is a steady state. Reproduced with permission from [Dematté *et al.* (2008c)].

phase, (D), is where more phosphatases are added in order to switch off the response in a quicker way.

In the next subsection we introduce domain-specific languages: a further step ahead that solves some of the drawbacks that are still affecting the modeling languages studied in this subsection.

6.5 Domain-specific languages

Some of the main issues of the languages described in the previous sections are that they rely on

(1) a difficult programming paradigm (process calculi and concurrent programming);
(2) graphs to represent complexes that slow down simulations;
(3) *ad hoc* abstract machines with limited optimization, resulting in performance and maintenability drawbacks.

An internal domain-specific language (ℓ) compiled to $C^{\#}$ exploits the most widely used imperative programming paradigm and overcomes these limitations. Optimizing $C^{\#}$ compilers provide the grounds for improving performance and hence usability of language-based modeling. Further improvements come from the design of new stochastic simulation algorithms (see Algorithm 8.13). In the next subsection we present ℓ.

Domain-specific languages

A *domain-specific language* (DSL) is defined to bridge the applicative domain and the programs needed to model the problem. The bridge consists of a common vocabulary between the various stakeholders (domain users and experts, analysts, developers) and a set of domain rules that describe the domain collaborations/interactions that give rise to the main domain processes.
A DSL can be either *internal/embedded* or *external*. An embedded DSL is a program built in the language that implements the DSL and that extends the descriptive power of the host language with domain vocabulary and rules by using the infrastructure of the hosting language. An external DSL is built from scratch without any hosting support.
A good design choice for an internal DSL is to use a linguistic abstraction of the applicative domain on top of a semantic model (basic abstractions and rules) implemented in the hosting language. It is essential to identify the right abstractions to make the implementation of a DSL for a specific domain cost-effective. Furthermore, the interface of the DSL with the hosting language, as well as with the developers, should be as limited as possible. If the design is good we gain in expressiveness and in the ability to better interact with the domain experts during development.

6.5.1 ℓ *design*

The experience gained in stochastic π-calculus and *BlenX* for modeling biological systems and the issues that emerged are the ground for five main design goals:

(1) *Performance.* The use of modeling and simulation techniques for real biological systems is calling for better performance than process algebra-derived formalisms.
(2) *Local states.* Process algebra-derived languages and agent-based systems usually manage the change of state of a component of the system by exploiting message passing and differentiation of processes. This

mechanism is most of the time too complex to express an element's change of state from inactive to active for which a Boolean flag would work fine.

(3) *Reactions vs processes.* Chemical reactions are a well-known formalism to represent biochemical pathways and allow the modeler to discuss models with biologists in an easy way. Process-based representations are not so intuitive when the size of the system is growing because of the synchronization. Message passing between components makes it difficult to involve biologists in the modeling activities.

(4) *Standard programming techniques.* Process algebra is an advanced topic even for computer scientists and their programming technique is not familiar for most of the programmers and modelers. An imperative style of programming is instead familiar to most scientists.

(5) *Space.* It is becoming increasingly evident that cell compartments cannot be ignored to account for the mechanistic details of biological processes.

The main performance bottleneck of process algebra-based formalisms is checking the structural congruence of processes to count the number of elements of a given type in the system. This number is used in computing the transition rates of the system and an optimization here would bring a substantial time saving in simulations. A challenge is identifying whether two complexes (usually represented as graphs) are the same species. It amounts to computing graph isomorphism to distinguish complexes made of the same components bound together differently (e.g., the triangle complex and the three-component linear complex in Fig. 6.25 are different).

Most of the time it is not known how the proteins in a complex are bound together. The graph representation of complexes forces assumptions on their structure that are not supported by experimental evidence. Therefore ℓ releases the ability of representing the binding structure of complexes by using multisets (a complex may contain more copies of the same element) of boxes (biological components) to gain in performance.

The advantage of process-based and agent-based systems over chemical reactions is that an interaction can be associated with operations that modify the state of the system depending on the context, possibly modifying agents/components not involved in the firing reaction. The choice of ℓ is to represent components as boxes with an internal state that can be manipulated by reaction rules equipped with pieces of code that run as side conditions of the reaction. Thus ℓ keeps the simplicity of reaction-based systems

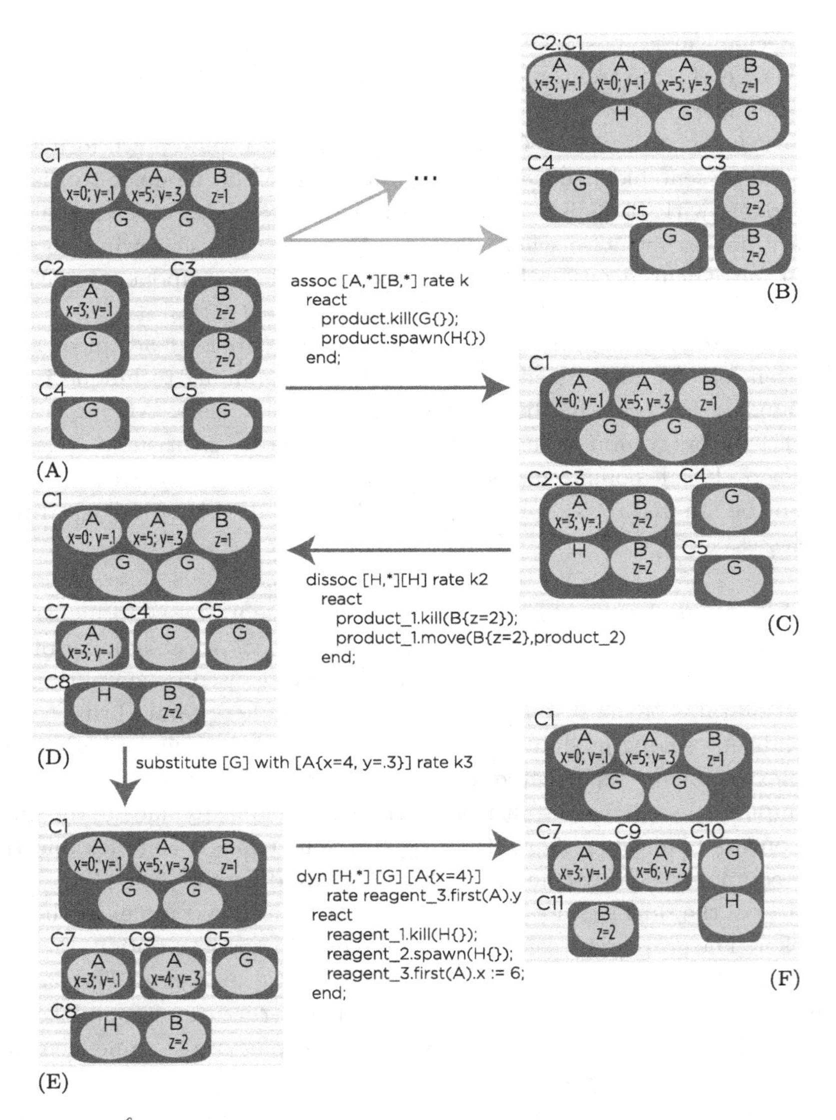

Figure 6.38 ℓ intuition. (A) A system is a multiset of complexes ($C1$–$C5$). Boxes may have local variables (x, y, z in the figure). Reaction rules are defined on patterns that identify complexes to which they can be applied and have code associated with to modify the system when the reaction is fired. Special variables (product and reagent) are predefined to access the elements of the complexes identified by the patterns. More than one complex can satisfy a pattern: (B) and (C) are two possible target states of the first assoc rule. The main types of rules are assoc to form complexes, dissoc to split complexes (from (C) to (D)), substitute to replace an element with another one (from (D) to (E)) and dyn for general rules (from (E) to (F)).

and their computational model extended with the ability of implementing modifications to the system as a consequence of the reaction selected.

Finally, to have the ability to exploit common expertise to develop models, ℓ is imperative and embedded into C# to exploit the optimizing compilers and tools developed for C#.

6.5.2 ℓ *intuition*

We refer to Fig. 6.38 to describe the modeling intuition of ℓ and its dynamics. A system is a multiset of complexes (Fig. 6.38a) — the round rectangle layouts. A single component is represented as a complex of just one element (e.g., the complexes $C4$ and $C5$ in the figure — the elements/boxes are the ellipsis within the complexes). The type of boxes and complexes in Fig. 6.38(a) are:

```
A{x:int,y:real}; B{z:int}; G{}

C1[A{x=0,y=.1}, A{x=5,y=.3}, B{z=1}, G{}, G{}],
C2[A{x=3,y=.1}, G{}], C3[B{z=2}, B{z=2}], C4[G{}], C5[G{}].
```

Note that the names of the boxes are interpreted as the types of the boxes. Values of fields are from basic types (int, bool, real) and can be the evaluation of arithmetic or logical expressions.

Patterns are used to define families of rules that apply to complexes with similar characteristics such as *all the complexes that contain at least an* A — this is what the pattern $[A, *]$ expresses in the assoc rule mapping the state (A) in the figure to either (B) or (C).

A complex matches a pattern if there is a bijection between the boxes in the pattern and in the complex. If the pattern contains the special character $*$ that matches any box, it is enough to find an injection of the boxes in the pattern into the ones of the complex (Fig. 6.39).

The dynamics are defined by reaction rules (multiset rewriting rules) with associated code to modify the system after the reaction is performed. There are four reaction rules: assoc (merge two complexes that match the patterns in the rule); dissoc (remove from a complex that matches the first pattern a subcomplex that matches the second pattern); substitute (replace the complex that matches the first pattern with the complex second argument) and dyn (apply the code in the reaction to the complexes that match the patterns). The rules assoc, dissoc and substitute are special cases of the

Pattern	Matching complexes	Pattern	Matching complexes
[A, *]	C1, C2	[A, G]	C2
[B, *]	C1, C3	[B, G, *]	C1, C2
[B]	∅	[A{x=3}, *]	C2

Figure 6.39 Pattern matching of the complexes in Fig. 6.38(A).

rule dyn and have been introduced only to simplify model writing. The reaction has stochastic rates that can be either constants or user-defined functions. The code associated with reactions can access the reagents and the products through pre-defined variables called $reagent_1, .., reagent_n$ and $product_1, .., product_n$ where the index refers to the position of the pattern in the rule. The actions that can be performed are deletion of a box from a complex, spawning of some new boxes within a complex, movement of a box from one complex to another or update of the fields of boxes. One rule at time is applied and multiple choices are solved stochastically via a race.

Consider the rule assoc leading from (A) to either (B) or (C) in Fig. 6.38. The first computational step is the identification of the complexes in (a) that match the patterns in the rule. The first pattern is matched by $C1$ and $C2$ and the second pattern is matched by $C1$ and $C3$ (Fig. 6.39). $C1$ matches both patterns, but its multiplicity in the system is just 1, hence dimerization is not allowed. Therefore, the possible complexes resulting from the assoc rule are $C1 : C3$ (not depicted in the figure), $C2 : C1$ (depicted in (B)) and $C2 : C3$ (depicted in (C)).

The complex $C2 : C1$ is obtained by merging $C2$ and $C1$ and then applying the code within the syntactic brackets react and end. The complex obtained by the merge can be accessed by the variable *product* and the associated actions are performed on the product. Hence a copy of the box G is removed and 2 copies of H are added. Similarly, the complex $C2 : C3$ is generated.

The stochastic rate and the number of the complexes matching the patterns determine which reaction to fire among the enabled ones via a race condition (see Sect. 8.3). Assume that the reaction leading to (C) is selected.

Consider the rule dissoc leading from (C) to (D) in Fig. 6.38. The only complex matching the first pattern is $C2 : C3$ from which a subcomplex made of one H box exactly is removed (to avoid ambiguities no $*$ is allowed in the second pattern of a dissoc). The application of the rule generates a complex containing A and two Bs and a complex containing one H. The code associated with the rule removes a B from the first product complex and moves the remaining B from the first product to the second one, resulting in the complexes $C7$ and $C8$.

Consider the rule substitute leading from (D) to (E) in Fig. 6.38. The complexes that match the first pattern are $C4$ and $C5$. Assume that the stochastic simulation algorithm selects $C4$. We then replace $C4$ with the complex specified in the the second position of the rule, yielding $C9$.

Consider the rule dyn leading from (E) to (F) in Fig. 6.38. $C8$ matches the first pattern, $C5$ matches the second pattern and $C9$ matches the last pattern. The rule dyn just runs its associated code and removes one H from $C8$ yielding $C11$, adds an H to $C5$ yielding $C10$ and sets x to 6 in $C9$.

Example 6.28. [Enzymatic reaction in ℓ]
Consider the enzymatic reaction in Example 5.2. Assuming that K_1, K_2, K_3, K_4 are the stochastic constants corresponding to k_1, k_2, k_3, k_4 and the complexes are $E\{\}, S\{\}, P\{\}$, the ℓ code is

assoc $[\mathrm{E}]$ $[\mathrm{S}]$ rate K_1
dissoc $[\mathrm{E}, S]$ $[\mathrm{S}]$ rate K_2
substitute $[\mathrm{E}, S]$ with $[\mathrm{E}\{\}, \mathrm{P}\{\}]$ rate K_3
dissoc $[\mathrm{E}, P]$ $[\mathrm{P}]$ rate K_4.

We end this section by considering space that is a primitive notion of ℓ. The type location of ℓ assigns boxes to compartments and the reaction rule move translocates complexes from a compartment to another one. Rates in rules can be associated with locations to model space-dependent kinetics as in

```
substitute  [A,B,*]  [A] with [1:B,2:C]  in cytoplasm rate 2
                                         in nucleus rate 3.
```

Example 6.29. [Space in ℓ]
Consider Fig. 6.40. The first rule translocates A from the extracellular space into the cytosol. The second move rule has a code associated with it: the complex containing F is moved from the cytosol to the nucleus and the x field of its D component is updated to 3. The last move translocates the complex containing C from the organelle to the cytosol. After the third move in Fig. 6.40 the system on the left is mapped into the system on the right.

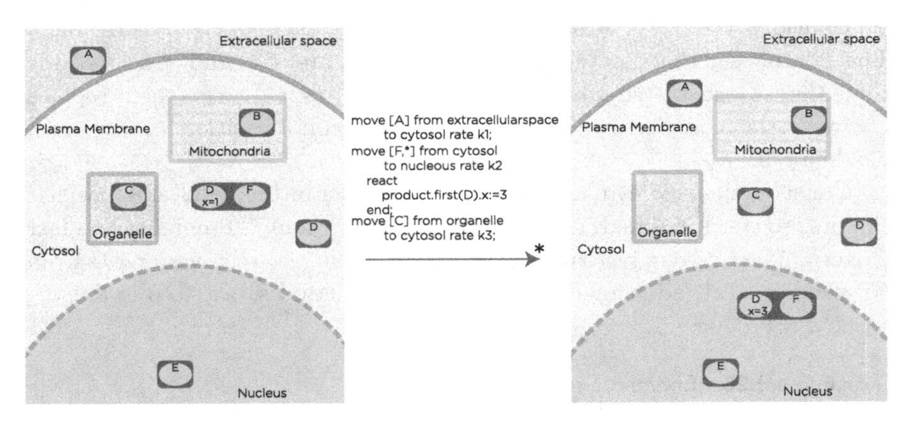

Figure 6.40 ℓ space and movement. The basic type location is used to denote cell compartments and the move reaction (that can also have a react part) is used to implement the translocation of complexes from one compartment to another. The star on the arrow means that it represents more than one step (indeed, three in this case).

6.5.3 ℓ *definition*

The syntax of ℓ in BNF form is introduced in the following boxes in this book. The formal definition of ℓ is in Appendix C.6. Note that optional items are not denoted as usual within square brackets because square brackets are tokens of ℓ. Therefore, we let $\langle A\rangle^*$, $\langle A\rangle^+$ and $\langle A\rangle^?$ denote at least zero, at least one and at most one occurrence of A, respectively. Furthermore, we assume pre-defined the category of identifiers (Id, sometimes qualified as $fieldId$ to specify the object they refer to.

BasicType stands for a primitive type used in ℓ(bool, int, real and location), whereas *BasicLiteral* ranges over their values. The meaning of bool, int and real is the usual one, whereas location indicates compartment names.

ℓ boxes and complexes

BasicType ::= bool | int | real | location
BasicLiteral ::= *BoolLiteral* | *IntLiteral* | *RealLiteral* | *LocationLiteral*
BoxType ::= Id
BoxDecl ::= *BoxType*$\{\langle fieldId:$ *BasicType*$;\rangle^*\}$
BoxExp ::= *BoxType*$\{\langle filedId =$ *Exp*$;\rangle^*\}$ (each field initialized)
BoxPattern ::= *BoxType*$\{\langle fieldId =$ *Exp*$;\rangle^*\}$
CplxExp ::= [*Exp* : *BoxExp* $\langle$, *Exp* : *BoxExp*$\rangle^*$]
Run ::= run *Exp* : *CplxExp*@*Exp* $\langle$; *Exp* : *CplxExp*@*Exp*$\rangle^*$ end

BoxDecl is the declaration of a *box*: it specifies a name for the box (Id–interpreted as the type of the box, hence *BoxType*) together with a possibly empty sequence Id : *BasicType* that are its *fields* and their types. The name of the fields is unique inside the box. *BoxExp* represents a box having all its declared fields instantiated. A *BoxPattern* shares the same syntax of a *BoxExp*; however, it may not provide a value for all fields; this will be used for counting and searching for boxes.

CplxExp represents complexes as multisets of boxes, i.e., it is a non-empty sequence of the form *Exp* : *BoxExp*, where *Exp* is an integer expression that denotes the number of instances of *BoxExp* in the complex. When *Exp* = 1, we can omit it.

The whole initial system state is then defined via the *Run* clause, which specifies a list of complexes having the form Exp_1 : *CplxExp*@Exp_2, where Exp_1 is the initial population of the complex in the system that is being modeled, and Exp_2 is the compartment the complex belongs to. Note that isolated boxes in a system are actually represented by a singleton complex, made of just that box.

The dynamics of the system are specified by multiset rewriting rules based on complex patterns (see the boxes). Example 6.31 illustrates pattern matching.

The rewriting rules specify stochastic rates rate *Exp* in their *RateClause* to be applied stochastically. The rate expression can inspect the boxes in the

Example 6.30. [Box and complexes in ℓ]
$A\{\mathtt{x} : \mathtt{int};\ y\ :\ \mathtt{real}\}$ is A declaration of a box type A containing fields x and y of type int and real and one possible instantiation.

```
A{x : int; y : real}          A{x = 3; y = 1.0}
```

A complex made of 2 A boxes as defined above and a B box with no field is

```
[2: A{x = 3; y = 1.0}, B{}].
```

Finally, a system made of 10 copies of the complex just declared in the cytoplasm and 5 copies of a similar complex but with an additional C box in the nucleus is

```
run 10 : [2: A{x = 3; y = 1.0}, B{}]@cytoplasm;
     5 : [2: A{x = 3; y = 1.0}, B{}, C{z = 1}]@nucleus;
end;.
```

Patterns and pattern matching

A *Pattern* is a sequence of *BoxPattern*, possibly followed by a wildcard $*$. A pattern without $*$ matches with complexes having exactly the specified boxes, whereas the wildcard allows the pattern to match with complexes including other boxes as well. The order in which boxes appear in patterns is irrelevant: patterns are handled according to associativity and commutativity. A *BoxPattern* of the form $B_1\{f_1 = v_1, .., f_n = v_n\}$ matches with a box $B_2\{g_1 = h_1, .., g_m = h_m\}$ if $B_1 = B_2$, $n \leq m$ and for each $i \in [1..n]$ there exists $j \in [1..m]$ such that $f_i = g_j$ and $v_i = h_j$. A complex c matches with a pattern p ($p \vdash c$), if one of the following conditions holds:

- p does not end with $*$, and there is a *bijective* mapping θ between box patterns in p and boxes in c, where correspondent elements match;
- p does end with $*$, and there is an *injective* mapping θ between box patterns in p and boxes in c, where correspondent elements match.

When we want to point out the mapping θ, we write $p \vdash_\theta c$ instead of $p \vdash c$.

reagents via special variables $\mathtt{reagent}_\mathtt{i}$, where i ranges from 1 to the number of reagents. By default, rate expressions follow the *mass action kinetics* law so that rates are implicitly multiplied by the abundance of the reagents. Another common kinetics law is Michaelis–Menten and it is expressed by

Example 6.31. [Pattern matching in ℓ]
Consider complexes $c_1 = [A\{x = 1\}]$, $c_2 = [B\{\}]$, $c_3 = [A\{x = 0\}, A\{x = 1\}]$, $c_4 = [A\{x = 1\}, B\{\}]$ and $c_5 = [A\{x = 1, y = 4\}]$ and patterns $p_1 = [A]$, $p_2 = [A, A]$, $p_3 = [A, *]$, $p_4 = [B]$, $p_5 = [B, *]$ and $p_6 = [A\{x = 1\}]$. Then, only the following relations hold: $p_1 \vdash c_1$, $p_1 \vdash c_3$, $p_1 \vdash c_6$, $p_2 \vdash c_4$, $p_2 \vdash c_5$, $p_3 \vdash c_2$, $p_3 \vdash c_3$, $p_4 \vdash c_3$, $p_4 \vdash c_5$, $p_5 \vdash c_1$, $p_5 \vdash c_3$, $p_5 \vdash c_6$. Reconsider also Fig. 6.38.

ℓ patterns, rate clause, expressions, rewriting rules

Pattern^{no*} ::= *BoxPattern* | *BoxPattern*, *Pattern*^{no*}

Pattern ::= *Pattern*^{no*} | *Pattern*^{no*}, $*$

RateValue ::= *Exp* | custom *Exp* | mm *Exp* : *Exp*

RateClause ::= $\langle$in *Exp*$\rangle^?$ $\langle$when *Exp*$\rangle^?$ rate *RateValue*

Exp ::= *BasicLiteral* | *Id* | (*Type*) null | *Exp* && *Exp* | ¬*Exp*
| *Exp* = *Exp* | *Exp* < *Exp* | *Exp* + *Exp* | *Exp* − *Exp*
| *Exp* $*$ *Exp* | *Id*($\langle$*Exp*$\langle$, *Exp*$\rangle^*\rangle^?$) | *BoxOps* | *CplxOps*

BoxOps ::= *Exp*.*Id* | *Exp*.first(*BoxPattern*) | *Exp*.count(*BoxPattern*)
| *Exp*.spawn(*BoxExp*)

CplxOps ::= spawn(*CplxExp*@*Exp*) | count(*Pattern*@*Exp*)

AssocDecl ::= assoc [*Pattern*] [*Pattern*] *RateClause* $\langle$react *Block*$\rangle^?$

DissocDecl ::= dissoc [*Pattern*] [*Pattern*^{no*}] *RateClause* $\langle$react *Block*$\rangle^?$

DynDecl ::= dyn $\langle$[*Pattern*]$\rangle^*$ *RateClause* react *Block*

SubDecl ::= substitute $\langle$[*Pattern*]$\rangle^*$ with $\langle$*CplxExp*$\rangle^*$ *RateClause*
$\langle$react *Block*$\rangle^?$

MoveDecl ::= move [*Pattern*] from *Exp* to *Exp* rate *RateValue*
$\langle$react *Block*$\rangle^?$

RateValue mm. All the other kinetic laws (e.g., Hill kinetics) can be defined by *RateValue* custom. Both keyword mm and custom prevent the implicit rate multiplication with abundance of molecules implemented by the mass action kinetics. The optional part *in* Exp_1 of a *RateClause* specifies the compartment in which the rewriting rule applies (i.e., the pattern matching in the rule is done only with the complexes in Exp_1). The optional part *when* Exp_2 specifies a Boolean condition that must be true to apply the rule. When in *Exp* is missing, the association is performed in every possible compartment. Similarly, the absence of when *Exp* causes the association to always be performed.

Expressions *Exp* are built from constants (*BasicLiteral*, Id, null) through logical operators, relational operators and arithmetic operators. It is also possible to call functions. The expressions more peculiar to ℓ are *BoxOps* to access fields of boxes (*Exp*.Id), to search for boxes within a complex (first), to count boxes matching a pattern (count) or to spawn a new box (spawn). There are also expressions acting on complexes (*CplxOps*) to spawn new complexes or count existing ones.

Example 6.32. [Rate clauses and expressions in ℓ]
Consider the rule assoc $[A]$ $[B, *]$ rate $5.2 *$ reagent$_1$.first$(A\{\})$.mass. The rule firing rate is proportional to the mass of the first reagent. If there are 10 complexes $[A]$ and 20 complexes $[B, C]$, the rate is implicitly multiplied by $10 \cdot 20$.
The expression spawn$([2 : B\{x = 3, y = 4\}, 1 : A])$ will spawn a new complex with two B boxes and one A box, and evaluate to a reference for such a complex. The expression count$([A, *])$ will evaluate to the number of all the complexes in the current system having at least one A box, including the one which has just been created. Furthermore, if c is a variable referring to the complex we just spawned, c.first(B).y evaluates to 4, while c.count(B{x = 3}) evaluates to 2.

The rules assoc, dissoc, dyn, move and substitute control the evolution of a system. A rule

$$\textsf{assoc}\ p_1\ p_2\ \textit{RateClause}\ \textsf{react}\ \textit{Block}$$

allows pairs of reacting complexes matching with p_1 and p_2 to associate. When that happens, the two reagent complexes merge their boxes and form a new larger complex, mimicking the association of two macromolecules. The *RateClause* specifies the conditions and speed under which association happens (see above). When an assoc rule is fired and the complexes in the patterns are merged, the code block specified in the react part is run. This can access the newly formed product (via a special `product` variable) and modify it further, e.g., by changing box fields, or adding/removing boxes, or spawning entirely new complexes.

A rule

$$\textsf{dissoc}\ p_1\ p_2\ \textit{RateClause}\ \textsf{react}\ \textit{Block}$$

ℓ commands, declarations and model

Block	::=	var Id := *Exp*; *Block* | *Cmd* ; *Block* | end
Cmd	::=	skip | Id := *Exp* | if *Exp* then *Block* else *Block* | return *Exp*
		| while *Exp* do *Block* | $Id(\langle Exp\langle, Exp\rangle^*\rangle^?)$ | *BoxCmd*
BoxCmd	::=	*Exp*.Id := *Exp* | *Exp*.kill() | *Exp*.kill(*BoxPattern*)
		| *Exp*.assoc(*Exp*) | *Exp*.move(*BoxPattern*, *Exp*)
		| foreach Id : *BoxType* in *Exp* *Block* | *BoxOps* | *CplxOps*
Type	::=	*BasicType* | *BoxType* | cplx | void
FunDecl	::=	*Type* $Id(\langle Type\ Id\ \langle, Type\ Id\rangle^*\rangle^?)$ *Block*
LetDecl	::=	let Id := *Exp*
Decl	::=	*AssocDecl* | *DissocDecl* | *DynDecl* | *MoveDecl*
		| *SubDecl* | *BoxDecl* | *FunDecl* | *LetDecl*
SpaceDecl	::=	location $\langle LocationLiteral\rangle^*$
Model	::=	$\langle SpaceDecl;\rangle^? \langle Decl;\rangle^*$ *Run*

specifies the dual operation, namely the dissociation of a complex into two subcomplexes. Here, p_1 specifies the complex to break up, whereas p_2 matches with a subcomplex to separate. No wildcard $*$ is allowed in p_2, since that would cause an arbitrary random subcomplex to be detached. This restriction is represented in the syntax by $Pattern^{no*}$. In the case where p_2 has multiple matches inside the reagent, we let all of them define an equally probable dissociation, hence effectively dividing the rate among all possible splits. After the rule triggers and the split is performed, the react code block is run, and can access the new complexes using the two special variables $\texttt{product}_1$ and $\texttt{product}_2$.

A rule

$$\text{dyn } p_1 \,..p_n \ \textit{RateClause} \text{ react } \textit{Block}$$

is similar to assoc, except that no complex merge is performed, and the react code block still has access to the not-yet-merged complexes $\texttt{reagent}_1, .., \texttt{reagent}_n$. This rule can produce association/dissociation by programming them manually in the react code block (e.g., by using assoc and move commands — see also Appendix C.6.3). While dyn is a very general-purpose mechanism, associations and dissociations are so common as to deserve their own constructs in the language. The modeler has to use dyn only for, e.g., monomolecular reactions or reactions involving more than two reagents and not modeled with substitute.

A rule

substitute p_1 .. p_n with $CplxExp_1$ .. $CplxExp_m$ *RateClause* react *Block*

substitutes the complexes matching the patterns $p_1, .., p_n$ with the concrete complexes corresponding to $CplxExp_1, .., CplxExp_m$. After that, the code *Block* is run and can access the new complexes via product_i.

A rule

move p from Exp_1 to Exp_2 rate *RateValue* react *Block*

moves the complexes matching the pattern p from the compartment identified by Exp_1 to the one identified by Exp_2. The code *Block* is then run, possibly altering the moved complexes via the special variables product_i. It is worth noting that both move and substitute can be modeled using the dyn rule, as we show in Appendix. C.6.3, but they simplify model writing.

Examples of reaction rules are in Fig. 6.38. To generate the enabled reactions in a configuration of the system, the ℓ semantics restricts the set matches it considers to a *canonical* subset obtained by collapsing matches that only differ by permutations of their components. This is needed to avoid unwanted rate amplification as described in Example 6.33.

Example 6.33. [Rate amplification in ℓ]
Consider the rule

dyn $[A,*]$ $[B,*]$ rate custom 1.0 react reagent_1.assoc(reagent_2); end.

Intuitively, this rule should cause the association of any two complexes, provided one includes an A while the other includes a B. Consider now a system formed by complexes $C_1 = [A,B,X]$ and $C_2 = [A,B,Y]$. Here, both patterns match with both complexes. This gives rise to *two* matches, namely $p_1 \mapsto C_1, p_2 \mapsto C_2$ and $p_1 \mapsto C_2, p_2 \mapsto C_1$. Hence, we would generate two transitions, effectively making the association rate 2.0 instead of the expected 1.0. This problem arises whenever two (or more) patterns share matches in the current system.

The code blocks in rules are written in a simple statically typed imperative language. Variable types (*Type*) include all the basic ones (*BasicType*), boxes (each *BoxDecl* declares a new type) and complexes (cplx). The value of a variable having a box or complex type is a reference to the box or the complex. A set of primitives allows one to freely access and modify boxes and complexes. For instance, adding boxes in a complex is done via the expression *Exp*.spawn(*BoxExp*), which creates a box *BoxExp* inside complex *Exp* and evaluates by a reference to the new box. Removing a box instead is done with the command *Exp*.kill(*BoxPattern*), which searches for a box matching the pattern within complex *Exp* and deletes it. New complexes are created by the expression spawn(*CplxExp*@*Exp*) which creates a complex *CplxExp* in compartment *Exp*, evaluating by a reference to the new complex. Existing complexes are removed via *Exp*.kill(), where *Exp* evaluates to a reference to the complex. It is also possible to merge two complexes, as it happens for association (assoc). More generally, one can loop over all the boxes of a given type in a complex using the command foreach b : $BoxType$ in $complex$. Furthermore, the expression *Exp*.first($BoxPattern$) returns a box matching $BoxPattern$ among those in complex *Exp*. Note that ℓ includes many common imperative constructs such as assignment, conditional, while loops and function calls, whose meaning is standard.

Finally, a model is made of space declarations, followed by any other declaration and a run clause.

Example 6.34 shows some reactions and some rate calculation. We refer to Appendix C.6 for the formal definitions underlining the reasoning and results in the examples.

6.5.4 *Relationships with other formalisms*

We describe here an algorithm that translates chemical reaction models into equivalent ℓ models. An application of the algorithm is in Example 6.35.

Algorithm 6.4 allows us to translate Style specifications into ℓ models by mapping graphical specifications into chemical reactions according to the algorithm in Fig. 5.20 and then applying the above translation of chemical reactions into ℓ (Algorithm 6.5). An example of application of this algorithm is in Example 6.36.

The relationships between some of the formalisms introduced in the book and their translation algorithms are in Fig. 6.41.

Example 6.34. [Reactions and their rates in ℓ]
Consider the rule

dyn $[A, *]$ rate custom 1 react
    foreach $b : B$ in reagent$_1$ $b.y := b.y + 1;$ end;
end.

In a system $\mu_1 = \{[A\{x = 1\}, A\{x = 2\}, B\{y = 4\}]\}$, the rule above causes a transition to $\mu_1' = \{[A\{x = 1\}, A\{x = 2\}, B\{y = 5\}]\}$ with stochastic rate 1.

Now, consider the rule

dyn $[A, *][B, *]$ rate custom 1.32 react
    foreach $b : B$ in reagent$_1$.$y := b.y + 1;$ end;
end.

In a system $\mu_2 = \{5 : [A\{x = 1\}, A\{x = 2\}, B\{y = 4\}]\}$, the rule above causes a transition to $\mu_2' = \{4 : [A\{x = 1\}, A\{x = 2\}, B\{y = 4\}], [A\{x = 1\}, A\{x = 2\}, B\{y = 5\}]\}$ with stochastic rate 1.32. There is only one match $C_1 = C_2 = [A\{x = 1\}, A\{x = 2\}, B\{y = 4\}]$, so the rate evaluates to 1.32. Note that we disregard the fact that there are five such complexes in the system, since we are using custom rates. By comparison, removing the custom keyword would cause mass action law to be applied, changing the rate into $1.32 \cdot 5 \cdot (5 - 1)/2 = 13.2$, i.e., 1.32 times the number of (unordered) molecule pairs.

Finally, consider the rule

dyn $[A]$ rate reagent$_1$.first$(A).x$ react
    var $a :=$ reagent$_1$.first(A);
    if $a.x < 3$ then spawn$([C]$@a.loc$)$; else spawn$([D]$@a.loc$)$; end;
end.

In the system $\mu_3 = \{2 : [A\{x = 1\}], 3 : [A\{x = 2\}], 4 : [A\{x = 3\}]\}$, the pattern $[A]$ is matched by all the three complexes, which differ by the value of x. In the cases $x = 1$ and $x = 2$, a new complex $[C]$ is spawned, leading the system to $\mu_3' = \mu_3 \cup \{[C]\}$. The rate for this transition can be computed applying the mass action law as follows. The $x = 1$ subcase contributes $1 \cdot 2$, while the $x = 2$ one contributes $2 \cdot 3$, hence the overall rate is 8. By contrast, the $x = 3$ case leads the system to $\mu_3'' = \mu_3 \cup \{[D]\}$ with a rate of $3 \cdot 4 = 12$.

Example 6.35. [Lotka–Volterra in ℓ]
Consider the chemical reactions in Example 5.13. The application of Algorithm 6.4 generates first the boxes R{}, G{} and F{} corresponding to rabbit, grass and fox. Then, the reactions in Example 5.13 are translated into

```
substitute [R] [G] with [2 : R{}] [G{}] rate sh1;
substitute [R] [F] with [2 : F{}] rate sh2;
substitute [F] with rate sh3;
```

where $sh1$, $sh2$ and $sh3$ are the stochastic rates corresponding to the deterministic rates in Example 5.13. Note that the with argument is a complex and not a pattern; hence we have to fully specify the boxes in the complex (i.e., we have to include the curly brackets and fields if defined). Note also that the third substitute has the with argument empty and it is used to model the death of a fox. The intuition is that the first substitute makes a rabbit reproduce (the 2 abundance in the complex) when it eats grass and leave grass available for other rabbits. The second substitute makes fox reproduce when it eats a rabbit. The last substitute models a fox's death.

Algorithm 6.4: From chemical reactions to ℓ

Input: A set CR of chemical reactions on species $\{X_1, .., X_n\}$.
Output: A ℓ program.

begin
  generate $X_i\{\}, i \in [1, .., n]$ boxes;
  for $n_i X_i + .. + n_{i+k} X_{i+k} \xrightarrow{r} m_j X_j + .. + m_{j+h} X_{j+h} \in CR$ **do**
    generate
      substitute $[n_i : X_i]..[n_{i+k} : X_{i+k}]$ with $[m_j : X_j]..[m_{j+k} : X_{j+h}]$
        rate sr
    where sr is the stochastic rate constant corresponding to r
  end
end

Algorithm 6.5: From Style to ℓ

Input: Style specifications.
Output: An ℓ program.

begin
  The translation in Fig. 5.20 generates a set CR of chemical reactions;
  Apply Algorithm 6.4 to CR
end

Example 6.36. [MAPK in ℓ]
Consider the MAPK cascade depicted with the graphical language Style in Fig. 6.34. According to Algorithm 6.5 we first produce the chemical reactions corresponding to the graphical model following the translation in Fig. 5.20. The chemical reactions are the following, where we let $aKKK, aKK$ and aK denote the active forms of KKK, KK and K, respectively. We also let iKK and iK denote the intermediate state of KK and K after one phosphorylation (recall that the $2p$ label on the arrow is a shorthand hiding a reaction – see also the BlenX model in Example 6.26). We assume all the reaction having rate r.

$$KKK + E1 \xrightarrow{r} aKKK + E1 \qquad aKKK + E2 \xrightarrow{r} KKK + E2$$
$$aKKK + KK \xrightarrow{r} aKKK + iKK \qquad aKKK + iKK \xrightarrow{r} aKKK + aKK$$
$$aKK + KKpase \xrightarrow{r} iKK + KKpase \qquad aKK + K \xrightarrow{r} aKK + iK$$
$$iKK + KKpase \xrightarrow{r} KK + KKpase \qquad aKK + iK \xrightarrow{r} aKK + aK$$
$$iK + Kpase \xrightarrow{r} K + Kpase \qquad aK + Kpase \xrightarrow{r} iK + Kpase$$

Algorithm 6.4 applied to the above set of chemical reactions produces the ℓ code below.

```
KKK{}; aKKK{}; KK{}; iKK{}; aKK{}; K{}; iK{}; aK{}; E1{}; E2{};
KKpase{}; Kpase{};
substitute [KKK] [E1] with [aKKK{}] [E1{}] rate sr;
substitute [aKKK] [E2] with [KKK{}] [E2{}] rate sr;
substitute [aKKK] [KK] with [aKKK{}] [iKK{}] rate sr;
substitute [aKKK] [iKK] with [aKKK{}] [aKK{}] rate sr;
substitute [aKK] [KKpase] with [iKK{}] [KKpase{}] rate sr;
substitute [aKK] [K] with [aKK{}] [iK{}] rate sr;
substitute [iKK] [KKpase] with [KK{}] [KKpase{}] rate sr;
substitute [aKK] [iK] with [aKK{}] [aK{}] rate sr;
substitute [iK] [Kpase] with [K{}] [Kpase{}] rate sr;
substitute [aK] [Kpase] with [iK{}] [Kpase{}] rate sr;
```

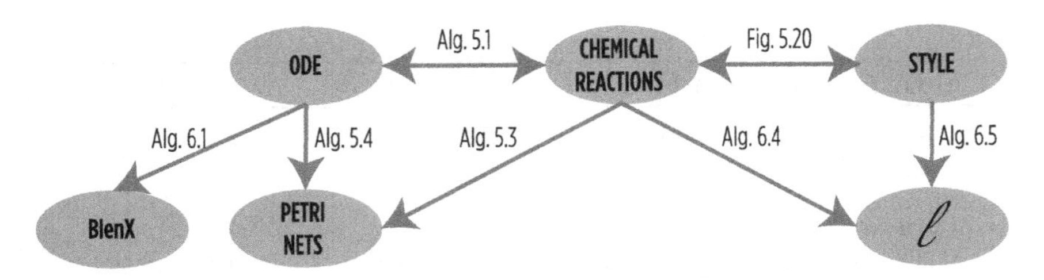

Figure 6.41 Relationships between the main formalisms introduced in the book. The arrows are labeled with references to the corresponding algorithms.

6.6 Summary

We introduced the algorithmic modeling (language-based) approach. We surveyed the main languages developed and focused on BlenX and ℓ. BlenX is a process algebra-based language. We reported its syntax and an intuitive semantics for it as well as a set of examples of applications (including models, self-assembly and evolutionary approaches). We then moved to the last generation of languages by introducing the domain-specific language ℓ with its syntax and intuition. We also reported some examples.

6.7 Further reading

Algorithmic systems biology has been introduced in [Priami (2009a)]. Similar concepts are also expressed in [Fisher *et al.* (2011)]. A general description of algorithmic modeling is in [Priami (2009b, 2012); Errampalli *et al.* (2004); Priami (2006)], while general surveys of process calculi for biological modeling are [Prandi *et al.* (2005); Guerriero *et al.* (2009b)]. Some general descriptions of process-algebra based modeling techniques are in [Priami and Quaglia (2004); Guerriero *et al.* (2009b)]. Stochastic π-calculus has been introduced in [Priami (1995, 1996, 2002)], and its biological version has been introduced in [Priami *et al.* (2001)]. Some applications are in [Curti *et al.* (2004); Lecca *et al.* (2004); D'Ambrosio *et al.* (2004); Lecca and Priami (2007)]. A connection between stochastic π-calculus and SBML is described in [Eccher and Priami (2006)]. Variants of the stochastic π-calculus have been introduced in [Phillips and Cardelli (2004); Phillips *et al.* (2006); Kuttler *et al.* (2007); Versari and Busi (2007); John *et al.* (2008)].

β-binders references are [Priami and Quaglia (2005a,b); Prandi *et al.* (2008); Degano *et al.* (2006)] and some applications are in [Ciocchetta *et al.* (2005); Prandi *et al.* (2006); Ciocchetta and Priami (2007); Guerriero *et al.* (2007b)]. A relation between β-binders and SBML is discussed in [Ciocchetta *et al.* (2008)]. The κ-calculus is studied in [Danos and Laneve (2003)], its evolution in rule-based modeling is in [Danos *et al.* (2007)] and the similar approach BNG is in [Hlavacek *et al.* (2006)]. Other process calculi defined for modeling biological systems are Bioambients [Regev *et al.* (2004)], Brane calculi [Cardelli (2005); Danos and Pradalier (2005)], CCS-R [Danos and Krivine (2004)], Bio-PEPA [Ciocchetta and Hillston (2007)].

A BlenX implementation is described in [Dematté *et al.* (2008b)]. Additional information on BlenX are in [Dematté *et al.* (2008); Dematté *et al.*

(2008a); Dematté *et al.* (2010)]. The formal semantics of BlenX has been defined in [Priami *et al.* (2009b); Romanel (2010)]. The complexity of computing structural congruence in process algebra-based languages is in [Romanel and Priami (2008)] and additional theoretical foundations on BlenX are in [Romanel and Priami (2010)]. The use of BlenX in an evolutionary framework is introduced in [Dematté *et al.* (2007b, 2008c)], while self-assembly is discussed in [Larcher *et al.* (2010); Larcher (201)] (self-assembly in stochastic π-calculus is in [Kahramanoğulları *et al.* (2009, 2013)]). The translation from ODEs to BlenX programs is in [Palmisano *et al.* (2009)] and highlights of the compositional nature of BlenX are in [Zámborszky and Priami (2010); Zámborszky (2010)] and an application to drug discovery is in [Dematté *et al.* (2007a)]. The possibility of managing non-exponential rates in language-based models is in [Mura *et al.* (2009)]. BlenX simulations reported in the book are performed with the BetaWB and COSBI LAB tools available at the url http://www.cosbi.eu/research/cosbi-lab and http://www.cosbi.eu/research/cosbi-lab, respectively. At these urls there are also collections of models. Besides BNG mentioned above, there are other approaches similar to BlenX such as the rule-based languages with nested structure [Carsten *et al.* (2011); Plotkin (2013)]. Additional papers describing mappings between ODEs and language-based approaches are [Calder *et al.* (2005); Cardelli (2009)].

Some intuitions on the computational idea underlying the design of ℓ are in [Priami *et al.* (2012); Nikolić *et al.* (2012)]. An implementation of ℓ and a collection of models is available at the url http://www.cosbi.eu/research/L. A relationship between language-based modeling and visualization is in [Cardelli and Priami (2009)].

The formal semantics of the languages extensively discussed in this chapter is in Appendix C where we use the structural operational semantic definitions introduced in [Plotkin (1981)].

Additional information on computational modeling of cell cycle are in [Novak *et al.* (2003); Sible and Tyson (2007)] and a complete BlenX model is in [Palmisano (2010)]. The MAPK pathway can be found in [Orton *et al.* (2005)] and its complete BlenX model is in [Dematté (2010)].

Chapter 7

Dynamic modeling process

The modeling process starts from the definition of the objective of the model and the criteria for the model's being accepted by the users. The acceptance criteria are fundamental because they set the termination condition for the modeling process.

The modeler must acquire and organize knowledge through literature and databases search, discussions with the domain experts, acquisition of data from the system to be modeled (e.g., high-throughput experiments, time series of concentrations — see Sect. 2.6). This knowledge must be documented.

Then the modeler has to define the logical structure of the model in a very high-level formalism that should be comprehensible to the domain experts. Domain experts must participate in the modeling process to ensure credibility of the model. The outcome of this step is a *model schema* that is translated into a *concrete model* (i.e., a model that can be simulated). We need to select a technology for the concrete model among those presented in the previous chapters and define a mapping from the schema that must be as automatic as possible.

A relevant part of the modeling process is the *model calibration*. Calibration is the systematic adjustment of the model parameters to make the model behavior more accurately reflect a given data set. A risk here is overfitting, that will be checked in the validation phase (see Sect. 3.2.2.1).

Finally, we check whether the model passes the acceptance criteria we defined at the beginning of the process.

Figure 7.1 reports the workflow, which is an iterative process just like software development. At each step we may need to backtrack and refine the construction to match the expectations of the user and the acceptance criteria. Model evaluation or acquisition of new knowledge on the way are

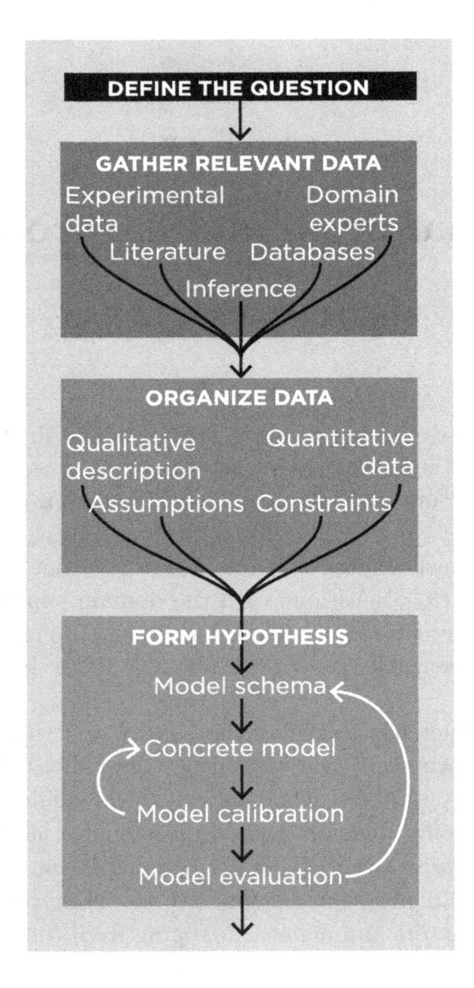

Figure 7.1 Modeling process.

the main causes of reworking steps already performed. Therefore the modeling process has both synthesis steps (the black arrow in the figure) and refinement steps (the white arrows in the figure) until the acceptance criteria are met.

7.1 Setting the objectives and the acceptance criteria

No model is good for all. This means that we have to make some choices even before starting the modeling process. The definition of the model objectives

Modeling objectives

— What do you want to model?
— What do you investigate with the model?
— What do you expect from the model?
— Why do you need a model?

will drive all the modeling process. To avoid ambiguities in this crucial phase of modeling, it is helpful to write down a precise statement answering the questions in the above box. The statement should be as precise as a theorem statement because it defines the mission of the modeler for the current problem. We will use as an example for illustrating the modeling process the metabolism of an anti-cancer drug called gemcitabine (Example 7.1) for which we set the objectives in Example 7.2.

Example 7.1. [Simplified gemcitabine metabolism]
Gemcitabine (2-2-difluorodeoxycytidine, dFdC) is a nucleoside analog used in oncology to block DNA replication in tumor cells. Gemcitabine is transported from plasma into the cell through the cell membrane. Gemcitabine can be deaminated by cytidine deaminase (CDA) in the cytoplasm and in the extracellular environment leading to the metabolite 2′-2′-difluorodeoxyuridine (dFdU). Both dFdC and dFdU can be phosphorylated by the deoxycytidine kinase. The monophosphorylated gemcitabine is phosphorylated two more times with the intervention of nucleoside monophosphate kinase (NMPK) and nucleoside diphosphate kinase (NDPK), respectively. The active versions of gemcitabine to be incorporated in the DNA are the triphosphate (dFdC-TP and dFdU-TP) metabolites. Gemcitabine triphosphate is incorporated into DNA causing chain termination and blocking cell proliferation. Monophosphorylated gemcitabine can be deaminated as well by deoxycytidylate deaminase (dCMPD), whereas dCMPD is inhibited by the aminated gemcitabine triphosphate dFdC-TP. The gemcitabine triphosphate competes with the natural nucleoside triphosphate dCTP for incorporation into nascent DNA chain and inhibits DNA synthesis, thus blocking cell proliferation in the early DNA synthesis phase. A graphical description of the steps above is in Fig. 7.3.

The perfect model does not exist. According to George Box, all models are wrong, but some are useful; the practical question is how wrong they can be before they are not useful. Therefore acceptance criteria are needed to stop the modeling process when the model is useful enough. The model must be suitable to carry out the investigations identified in the objectives

Example 7.2. [Objectives of the gemcitabine model]
We follow here the structure of the questions in the box on the previous page to set the objectives (see italicized words). We *want to model* the metabolism of the drug gemcitabine within a cell until DNA binding for understanding the effect of inhibitory mechanisms and other components that affect drug efficacy. We *want to investigate* the mechanisms of action of gemcitabine. We *expect* a better understanding of the functioning of gemcitabine and production of hypotheses for a better dosage of the drug and the mechanism that regulates the efficacy of the drug. We *need a model because* it is difficult and expensive to perform *in vivo* or *in vitro* experiments.

and match the user's expectations (properties of interest). Correspondence-related properties and the model's measures of goodness (see Sects. 3.2.2.1 and 3.2.2.2) must be considered. The model must be credible and have a good match in the I/O validity without overfitting. It is next to impossible to have models that are 100% accurate, sensitive and specific. Therefore we must set thresholds above which we accept the model.

Acceptance criteria

— Does the model address the questions identified in the objectives?
— Is the model credible and valid?
— Does the model match the expected accuracy, sensitivity and specificity?
— Is the model tractable?
— Does the model have the expected level of performance?
— Does the model have the expected structural properties?

The model must be tractable and have properties such as efficiency, efficacy, effectiveness and scalability. Structural properties are considered if the model has to be integrated in a larger model or if it has to interoperate with other models. Note that in practice not all the properties of a model are considered in the definition of the acceptance criteria (see Example 7.3).

7.2 Building the knowledge base

The knowledge base to build a model is populated by collecting experimental data, literature descriptions, databases content and domain expert knowledge.

Example 7.3. [Acceptance criteria of the gemcitabine model]
We want a model to describe gemcitabine's mechanism of action and reveal hints to determine the dosage of the drug for efficacy with respect to conditions that can vary in patients. We want a model trusted by experimentalists and that has a correspondence of at least 90% in the I/O validity check. We want a model that is extensible and modular to add further molecular details in subsequent stages.

To guide the assembly of the knowledge needed, the box below reports some guiding questions for this phase of the modeling process. A descriptive starting point for our running example is in Example 7.4. We can classify all the information we need into four main classes: qualitative descriptions,

Assembling knowledge

— Which are the components of the system?
— Which are the attributes of the components?
— Which are the interactions between the components?
— Which are the dynamic modifications of the attributes?
— Which are the conditions governing attribute changes and interactions?
— Which is the spatial structure of the system?
— Which is the quantitative data avaialble on the aspects above?
— Which are your assumptions?
— Which are the constraints imposed on the model?

quantitative data, assumptions and constraints (see Fig. 7.1). Note that the same fact can be included in different classes case by case. For instance *A interacts with B* is a qualitative statement if it is reported in the literature or if it emerges from an experiment; it is an assumption if we model this interaction but we do not have evidence of it in experiments or literature; it is a constraint if we restrict the model space by superimposing the interaction to all the possible models of the system.

Qualitative knowledge of the system can be further classified as structural and dynamic knowledge. Structural knowledge includes the description of the components and their attributes plus the spatial structure of the system. The initial allocation of components onto the spatial structure is a structural description of the system.

Example 7.4. [Knowledge base for dFdC of the gemcitabine model]
We associate with dFdC its attributes, characterized by the set of values that they can assume (e.g., their type). The type of the attributes define the modification capabilities. The main attributes of dFdC are *phosphorylated* that can assume values from 0 to 3 and the Boolean (*yes, no*) value *deaminated.* A particular attribute is *location* that associates with each component the set of locations in the spatial structure of the system in which it can be located. When a component can be observed in more than one spatial compartment, it means that it can translocate from one to the others and we are implicitly both listing the translocation capabilities and defining the spatial structure of the system. dFdC can be in the extracellular compartment, in the cytoplasm or in the nucleoplasm. When listing the attribute of a component, we also explicitly associate with each attribute its initial value. dFdC initially is not phosphorylated, it is not deaminated and it is located in the extracellular compartment.

The modifications that dFdC can undergo are a triple phosphorylation first by dCK, then by NMPK and finally by NDPK. This happens both in its deaminated and non-deaminated form. Moreover, dFdC can be deaminated by CDA, both in the extracellular matrix and in the cytoplasm, and dFdC-MP can be deaminated by dCMPD if dCMPD is not inhibited by dFdC-TP.

The inhibition of deamination of the phosphorylated form of dFdC can be modeled by a binding between dFdC-TP and the deamination agent dCMPD. Moreover, the binding of dFdC with DNA is enabled only by the three phosphate forms.

The translocations that dFdC can undergo are from the extracellular compartment to the cytoplasm and from the cytoplasm to the nucleoplasm. The translocation from cytoplasm to nucleoplasm of gemcitabine can only happen when it is phosphorylated three times.

Example 7.5 reports a template for recording the information related to the dFdC component as discussed above. Note that conditions could be implictly contained in the name of the component used in the modification. Similarly, modification 1(c)iv applies to dFdC both in the extracellular matrix and in the cytoplasm. For instance, modification 1(c)iii is only possible if dFdC is phosphorylated twice. Moreover, the condition 1(c)i is satisfied by both the deaminated and non-deaminated version of dFdC. An alternative way of writing translocation 1(e)ii would have been Cytoplasm $\rightarrow$ Nucleoplasm [dFdC.phosphorylated $= 3$], where the sentence in the square brackets after the translocation is the enabling condition. The rest of the knowledge base is in Example 7.6. Note the initial value of the location attribute of CDA. It means that initially CDA is in both the extracellular matrix and in the cytoplasm.

Qualitative knowledge on the dynamics of the system includes the modification capabilities that determine the set of possible values of the attributes of a component, the description of the binding capabilities identifying the interconnection network of components and the translocation

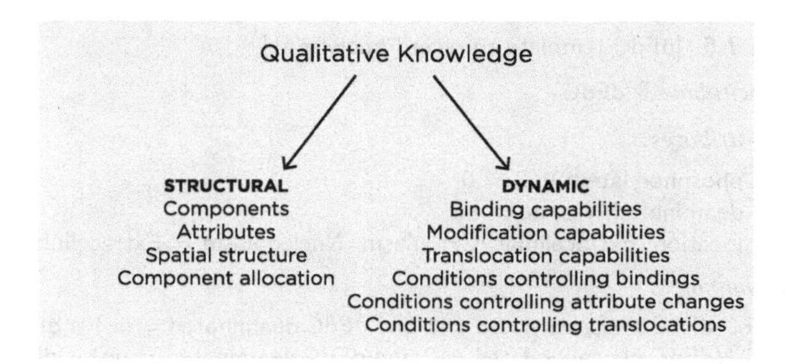

Figure 7.2 Qualitative knowledge to be included in the knowledge base for modeling.

capabilities that determine the potential movements of a component. The dynamics of the system may be controlled by conditions of the system or of the components.

Figure 7.2 reports a diagram of the qualitative knowledge needed to build the knowledge base for a model.

It is sometimes useful to have different names for identifying the same component with different attribute values (e.g., in Example 7.1 we identify gemcitabine with dFdC when it is not deaminated, and with dFdU when it is deaminated. Similarly, we used different names to identify different levels of phosphorylation). It is necessary to register all these naming conventions in the knowledge base (see Example 7.5).

Quantitative data is extremely relevant to prepare the model for simulation. Indeed, we need the initial concentration or amount of the components as well as the relevant kinetic information about the speed and probability of interactions, translocations and modifications. Most of the time the complete set of information is not available, but other quantitative data (e.g., time series of concentrations coming from macroarray, proteomic data coming from mass spectrometry, cell imaging data) can be used to derive the parameters for the model. Sometimes we have only some measurement that the model simulation has to match and we may use fitting procedures for the not available parameters. Even worse, sometimes no quantitative data is available and we have to resort to semi-quantitative or relative measures such as *protein A increases its concentration by 10% with respect to protein B* or *reaction A is much faster than reaction B*. At this stage we only need to register all this information in our knowledge base.

Hereafter we will adopt the convention *write what you need* (e.g., if some component has no attribute relevant for the investigation at hand, we will

Example 7.5. [dFdC template to record knowledge]

(1) *Gemcitabine* is dFdC

(a) *Attributes*:

i. phosphorylated: 0..3 = 0;
ii. deaminated: yes, no = no;
iii. location: Extracellular, Cytoplasm, Nucleoplasm = Extracellular;

(b) *Naming*:

i. (dFdC.phosphorylated = 1 and dFdC.deaminated = no) is dFdC-MP;
ii. (dFdC.phosphorylated = 2 and dFdC.deaminated = no) is dFdC-DP;
iii. (dFdC.phosphorylated = 3 and dFdC.deaminated = no) is dFdC-TP;
iv. (dFdC.phosphorylated = 1 and dFdC.deaminated = yes) is dFdU-MP;
v. (dFdC.phosphorylated = 2 and dFdC.deaminated = yes) is dFdU-DP;
vi. (dFdC.phosphorylated = 3 and dFdC.deaminated = yes) is dFdU-TP;

(c) *Modifications*:

i. dCK phosphorylates dFdC, [dFdC.phosphorylated = 0];
ii. NMPK phosphorylates dFdC, [dFdC.phosphorylated = 1];
iii. NDPK phosphorylates dFdC, [dFdC.phosphorylated = 2];
iv. CDA deaminates dFdC;
v. dCMPD deaminates dFdC-MP, [dCMPD not bound to dFdC-TP];

(d) *Bindings:*

i. dFdC-TP binds dCMPD;
ii. dFdC[phosphorylated = 3] binds DNA;

(e) *Translocations*:

i. dFdC: Extracellular → Cytoplasm;
ii. dFdC-TP, dFdU-TP: Cytoplasm → Nucleoplasm.

simply write nothing about the attributes of that component). Note that the attributes that can assume a single value do not need the initialization value (e.g., the location attribute of dCK in the Example 7.6). It is good practice to avoid duplication of information, thus we include references to descriptions already stored in the knowledge base (e.g., bindings of dCMPD in Example 7.6).

We stressed already that the modeling process is an iterative process that starts from a draft model and is refined until it captures the properties of interest without violating assumptions and constraints, or contradicting experimental and literature data. Moreover, the refinement process will continue until the outcome of the model fits with an high degree of precision with the experimental observations. Therefore it is very important to

Example 7.6. [Templates for the components other than dFdC]

(2) *Cytidine deaminase* is CDA;
   (a) *Attributes*:
      i. location: Extracellular, Cytoplasm = {Extracellular, Cytoplasm};
   (b) *Modifications*:
      i. see 1(c)iv, Example 7.5;

(3) *Deoxycytidine kinase* is dCK;
   (a) *Attributes*:
      i. location: Cytoplasm;
   (b) *Modifications*:
      i. see 1(c)i, Example 7.5;

(4) *Nucleoside monophosfate kinase* is NMPK;
   (a) *Attributes*:
      i. location: Cytoplasm;
   (b) *Modifications*:
      i. see 1(c)ii, Example 7.5;

(5) *Nucleoside diphosfate kinase* is NDPK;
   (a) *Attributes*:
      i. location: Cytoplasm;
   (b) *Modifications*:
      i. see 1(c)iii, Example 7.5;

(6) *Deoxycytidylate deaminase* is dCMPD;
   (a) *Attributes*:
      i. location: Cytoplasm;
   (b) *Bindings*:
      i. see 1(c)v, Example 7.5.

(7) *DNA*
   (a) *Attributes*:
      i. location: Nucleoplasm;
   (b) *Bindings*:
      i. see 1(d)ii, Example 7.5.

Example 7.7. [References for the knowledge base of the gemcitabine model]
We record here papers, experiments and databases used to build the knowledge base. We also add for any item a pointer to the qualitative and quantitative information on the systems recorded in the knowledge base.
An example of references follows.

(1) V. Heinemann, Y-Z. Xu, S. Chubb, A. Sen, L. W. Hertela, G. B. Grindey and W. Plunkett. Cellular Elimination of $2',2'$-Difluorodeoxycytidine $5'$-Triphosphate: A Mechanism of Self-Potentiation. *Cancer Research*, **52**:533–539, 1992.
This paper has been used to develop the initial, simplified model.

(2) S. A. Veltkamp, D. Pluim, M. A.J. van Eijndhoven, M. J. Bolijn, F. H.G. Ong, R. Govindarajan, J. D. Unadkat, J. H. Beijnen and J. H. M. Schellens. New insights into the pharmacology and cytotoxicity of gemcitabine and $2',2'$-difluorodeoxyuridine. *Molecular Cancer Therapeutics*, **7**(8):2415–2425, 2008.
This paper has been used to infer the kinetics from time series data. It is further used to refine the structure of the model.

document choices and solutions throughout the modeling process. We must clearly mark assumptions and constraints, if any, in the knowledge base.

To complete the knowledge base of a system, we must collect the bibliographic references that we have used to assemble knowledge on the qualitative and quantitative information, as well as the experiments we refer to and the supporting databases that we have consulted. The enumeration that we used in the listing of the knowledge on the system should be used to link papers, experiments and databases to specific items in the knowledge base. Example 7.7 is not intended to be a complete set of references for the gemcitabine model, but just a template of how reference information can be recorded.

7.3 From the knowledge base to a model schema

We start with a graphical representation of the biological processes and we then add the missing knowledge with a tabular approach.

There are many formats for producing diagrams of biological processes. None of the proposed notations is emerging as superior to the others and each of them has advantages and disadvantages. We use here the graphical language introduced in Sect. 5.6. A zen approach in which *the less,*

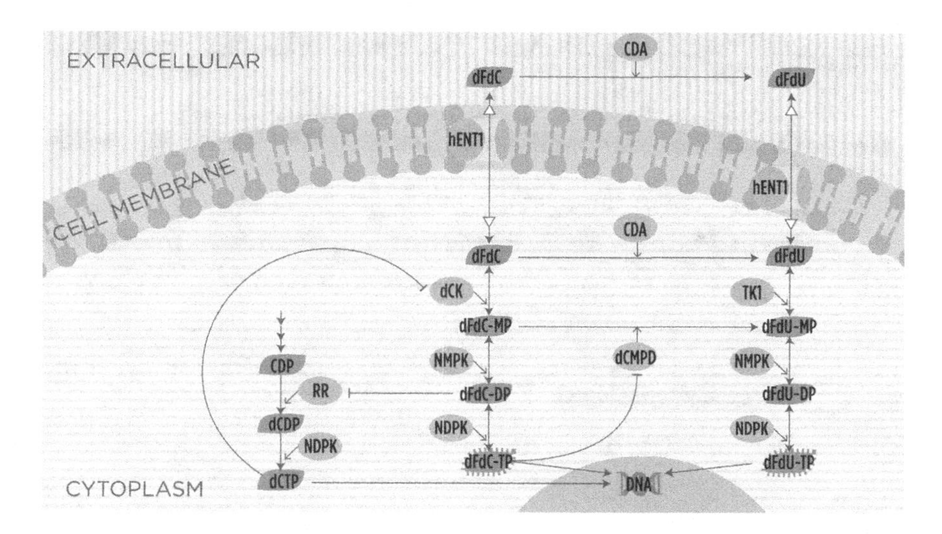

Figure 7.3 Simplified model of the gemcitabine metabolism (see Example 7.1).

the better in a single visualization layer is the key principle driving our choices. Figure 7.3 depicts the basic visualization layer of the metabolism of gemcitabine as explained in Example 7.1).

An alternative representation of the knowledge base (which also works as a model schema) is based on tables and constrained natural language statements. We consider structure tables (Components, Complexes, Compartments), dynamics tables (Global Dynamics, Binding Dynamics, Bimolecular Dynamics, Translocations) and parameter tables (Parameters, Initial State). Figures 7.4–7.13 show how the tables look and what they store (all the tables in the figures are screenshots of the COSBI LAB Model tool available at http://www.cosbi.eu/research/cosbi-lab). The exact syntax of constrained natural language statements is defined in the online help of COSBI LAB Model.

Example 7.8 describes the content of the tables in Figs. 7.4–7.13 that refers to the gemcitabine example of this chapter. Note that this tabular representation of the knowledge base works as a model schema as well because the constrained natural language is not ambiguous and can be automatically translated into executable representations (indeed, the tool automatically translates tables into a BlenX model). The quantitative information in the tables feeds the model for simulation.

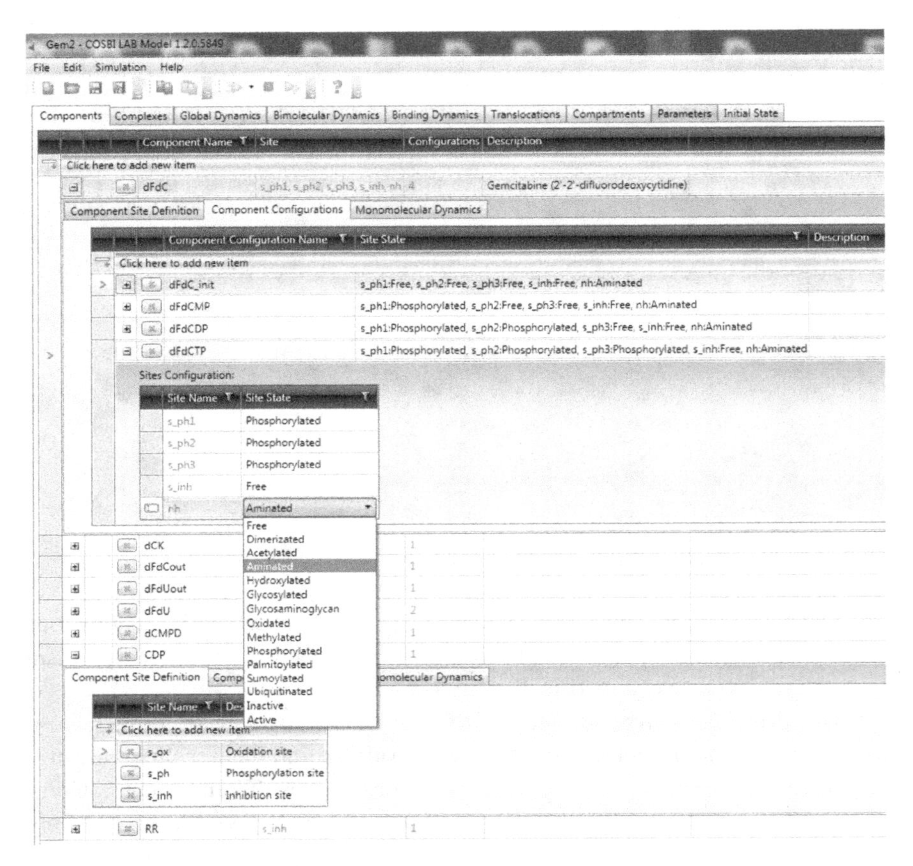

Figure 7.4 Components are characterized by a name, some interaction sites, some configurations and a description (e.g., dFdC that has four interaction sites and can be in the system with four different configurations). Sites have states that can be selected among a list of pre-defined states. Components configurations are defined through the state of its sites. Component configurations are used to collect all the different states of the same entities under a common name for analysis purposes.

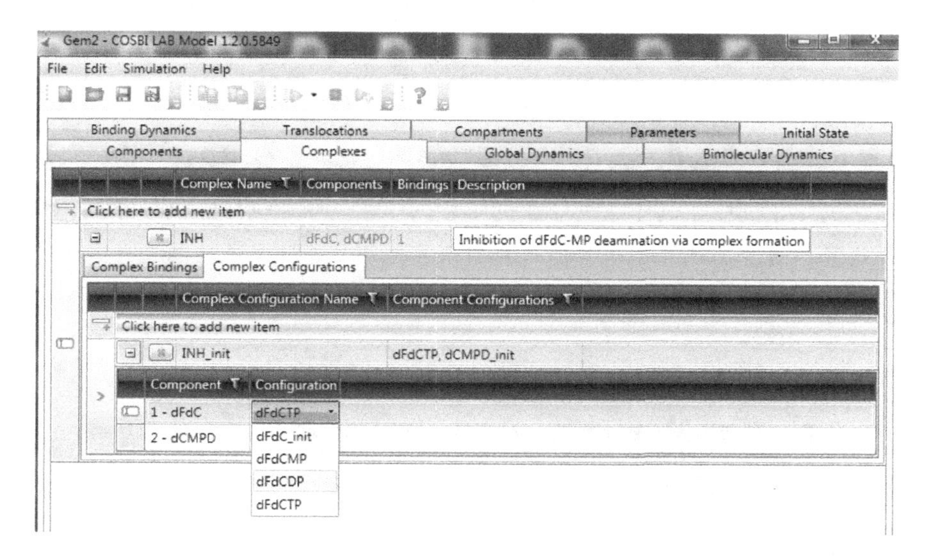

Figure 7.5 Complexes are characterized by a name, the list of components and their bindings forming the complex. Since components may have different configurations, we specify the configurations that allow complex formation (e.g., only configuration dFdC-TP of gemcitabine is allowed to form a complex with dCMPD). This table is needed to specify the complexes that are available in the initial state of the system; those that are created dynamically are implicitly represented by binding dynamics (Fig. 7.10).

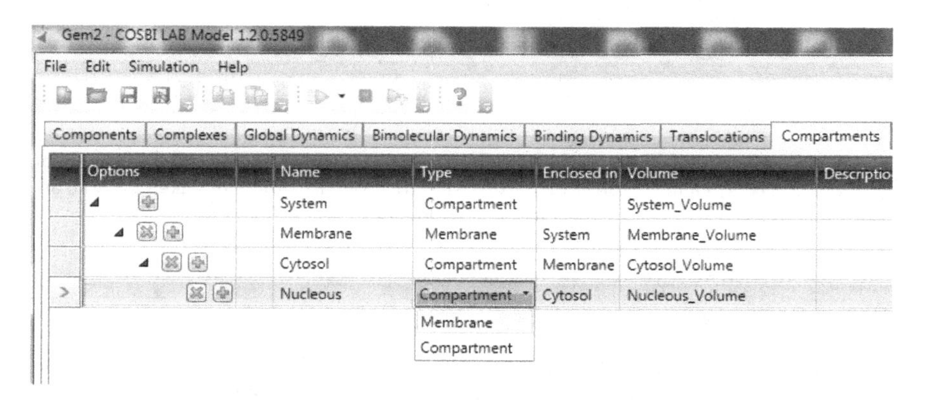

Figure 7.6 Compartments define the spatial structure of the system as a tree, where the relation *child of* is interpreted as *contained in*. The root of the tree is always system. The type of elements are compartment or membrane and for each of them it is possible to specify a volume and a description. The volume is defined here as a variable that is then instantiated in the Parameter table (see Fig. 7.13).

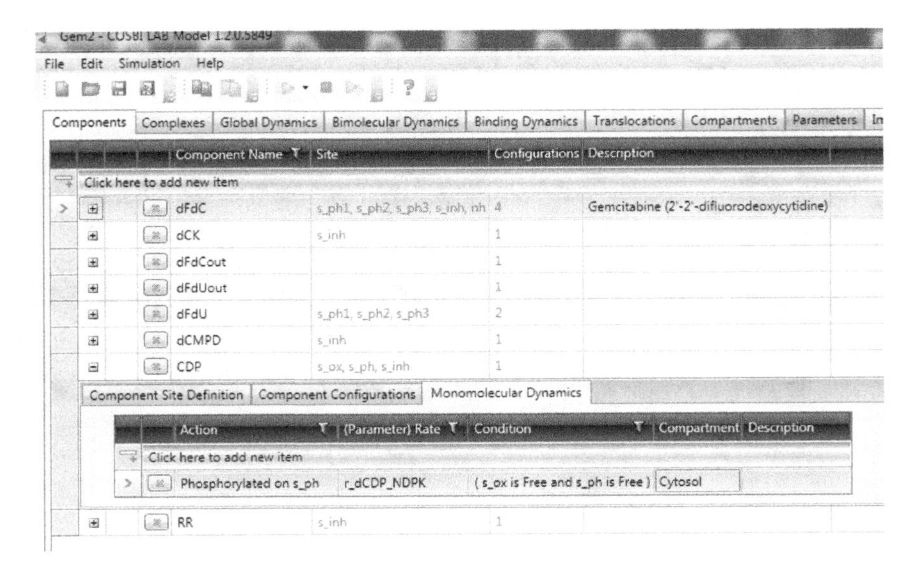

Figure 7.7 Monomolecular dynamics affect a single component, hence it is defined within the component definition. It is characterized by a constrained natural language statement, a rate constant for the kinetics, an enabling condition and a compartment.

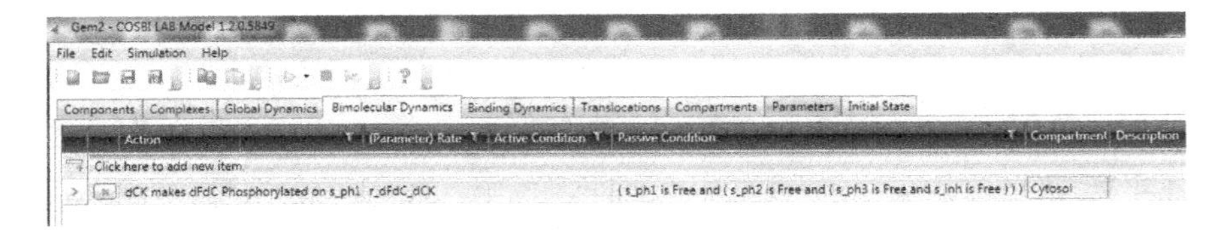

Figure 7.8 Bimolecular dynamics describe the actions in which an active component (subject of `makes`) changes the state of a passive component (object of `makes`). There are conditions referring to active and passive components.

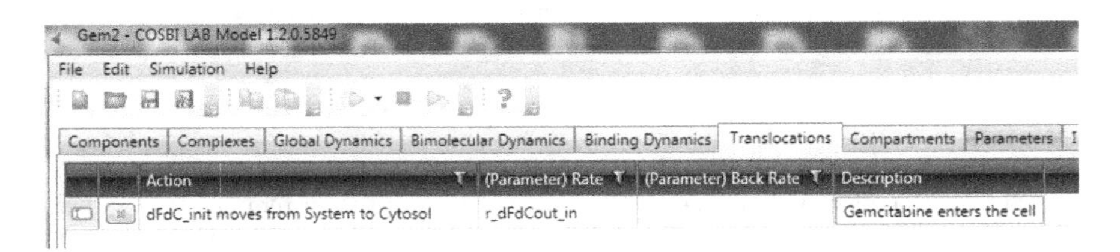

Figure 7.9 The translocation keyword `move` requires the source and target compartments as arguments. The forward rate and the backward rate describe bidirectional movements.

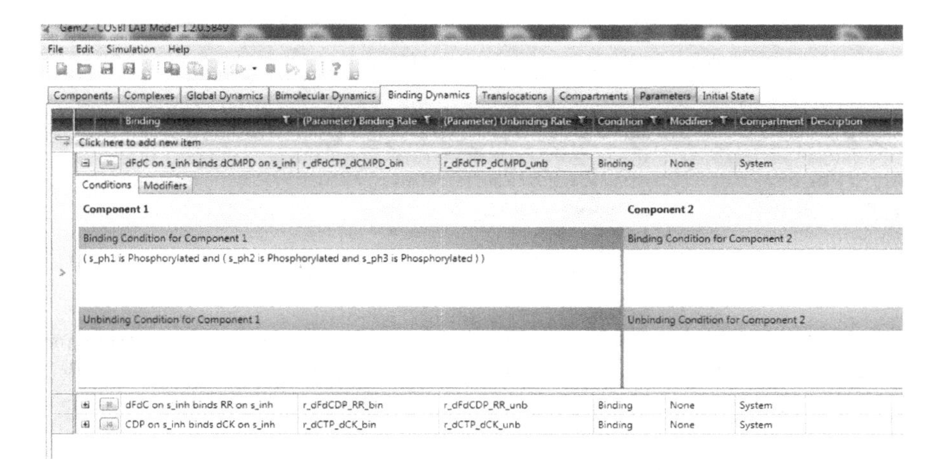

Figure 7.10 Binding dynamics are defined by the **binds to** keyword and two components and one site for each component that is the one creating the binding. Reversibility is expressed by proving binding/unbinding rates as well as binding and unbinding conditions on the two components. It is possible to specify modifiers: actions that are performed on the complex after the binding or on the components after the unbinding.

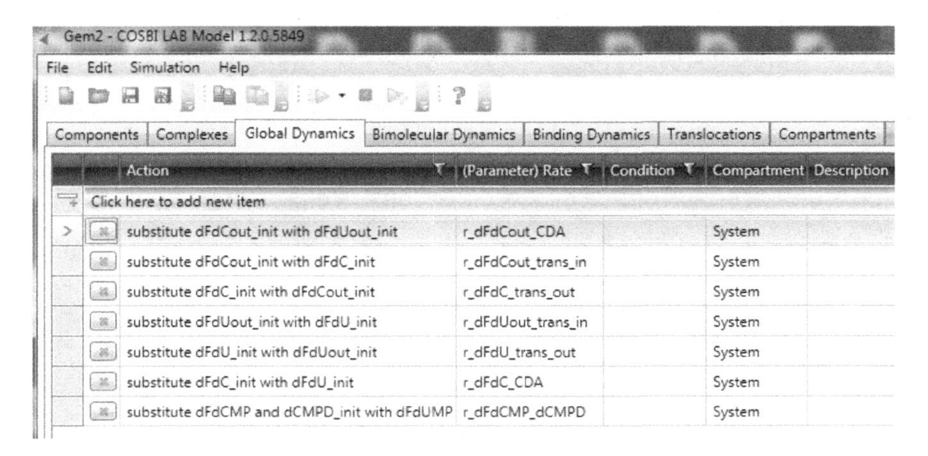

Figure 7.11 Global dynamics are useful to express perturbation experiments during the simulation (similar to BlenX events). The action has a rate and may have a condition. The typical statement is the substitution **substitute** of some components with others. The compartment in which the action occurs is also specified.

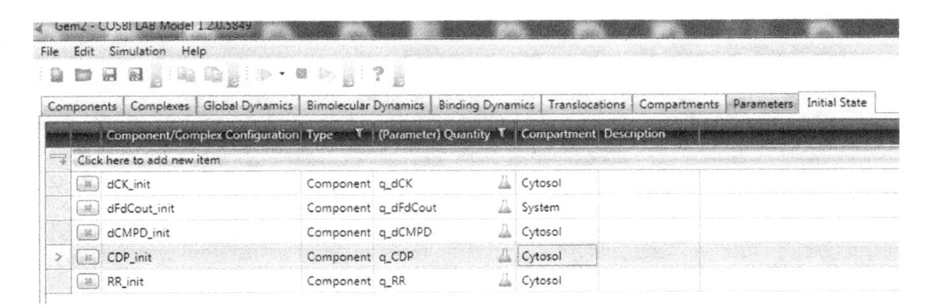

Figure 7.12 The initial state of the system is specified by listing the components and complexes available, their quantities and the compartment in which they are located.

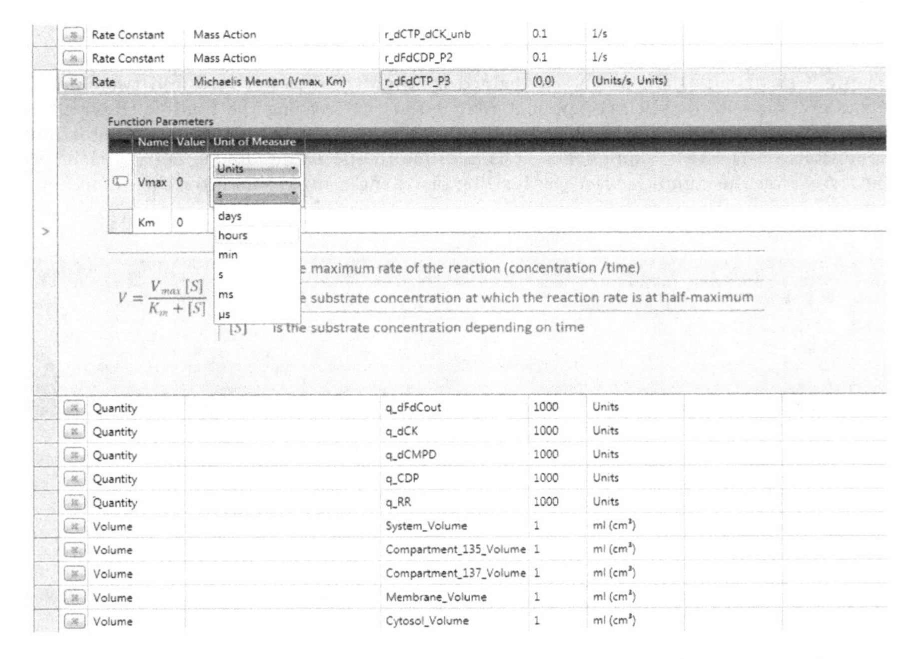

Figure 7.13 Parameters are instantiated in this table that can also contain references to literature or data sets that are used to infer them. The table describes the type of a parameter, the name used in the other tables for a parameter, its value and unit of measure. The rate constant specifies also the kinetic law, chosen from a pre-defined list, and it provides the characterizing values of the selected law (see the example of a Michaelis–Menten law).

Example 7.8. [Gemcitabine in tabular format]
Consider Example 7.5. There is an entry for each entity in the system and we map it to components or complexes. The attributes of the entities are described by the component or complex configurations (see Figs. 7.4 and 7.5), which also define the naming conventions. The attributes *modifications, bindings and translocations* in Example 7.5 are recorded in the tables for the dynamics of the system. For instance, Example 7.5(c)i is reported in Fig. 7.8, Example 7.5(d)i is reported in Fig. 7.10 and Example 7.5(e)i is reported in Fig. 7.9. The references as described in Example 7.7 can be included in the Description field of the tables that are associated with each component, complex, dynamic and parameter.

7.4 From the model schema to a concrete model

The translation of the model schema from the graphical representation of the system to a list of chemical reactions is quite easy (compare Fig. 7.3 and Fig. 7.14). The translation can also be automated according to the algorithm in Fig. 5.20.

r_1 : dFdCout ⟶ dFdC
r_2 : dFdC ⟶ dFdCout
r_3 : dFdUout ⟶ dFdU
r_4 : dFdU ⟶ dFdUout
r_5 : dFdC + dCK ⟶ dFdC-MP + dCK
r_6 : dFdC-MP ⟶ dFdC
r_7 : dFdC-MP ⟶ dFdC-DP
r_8 : dFdC-DP ⟶ dFdC-MP
r_9 : dFdC-DP ⟶ dFdC-TP
r_{10} : dFdC-TP ⟶ dFdC-DP
r_{11} : dFdU ⟶ dFdU-MP
r_{12} : dFdU-MP ⟶ dFdU
r_{13} : dFdU-MP ⟶ dFdU-DP
r_{14} : dFdU-DP ⟶ dFdU-MP
r_{15} : dFdU-DP ⟶ dFdU-TP
r_{16} : dFdU-TP ⟶ dFdU-DP
r_{17} : dFdC ⟶ dFdU
r_{18} : dFdC-MP + dCMPD ⟶ dFdU-MP + dCMPD
r_{19} : dFdC-TP ⟶ dFdC-TP-DNA
r_{20} : dFdU-TP ⟶ dFdU-TP-DNA
r_{21} : ⟶ CDP
r_{22} : CDP + RR ⟶ dCDP + RR
r_{23} : dCDP ⟶ dCTP
r_{24} : dCTP ⟶ CTP-DNA
r_{25} : dCTP + dCK ⟶ dCTP:dCK
r_{26} : dCTP:dCK ⟶ dCDP + dCK
r_{27} : dFdC-DP + RR ⟶ dFdC-DP:RR
r_{28} : dFdC-TP + dCMPD ⟶ dFdC-TP:dCMPD
r_{29} : dFdC-TP:dCMPD ⟶ dFdC-TP + dCMPD

Figure 7.14 Reaction-based model of the gemcitabine metabolism (see Example 7.1 and Fig. 7.3).

As already mentioned in the previous section, the translation from the tabular form to language-based models can be done automatically. We spot the intuition of the translation in Examples 7.9 and 7.10.

Example 7.9. [BlenX gemcitabine model]
Each component is mapped into a box. For instance, the gemcitabine in the extracellular domain is coded as

```
let dFdC: bproc = #(s1,c_dFdCout1)#(s2,c_dFdCout2)
                              [p_main | rep start_p_main?().p_main];.
```

The translocation of dFdC within the cytoplasm is modeled by changing the type of the sites (removing the out substring) in the code for the `p_main` process. The rate of the change primitives is defined in Fig. 7.9. After changing the types of the binders the process `p_main` is restarted with an output signal. The other actions that can affect `dFdC` have to be specified as alternatives in the body of `p_main`.

```
let p_main: pproc =
      if (s1,c_dFdCout1) then
           ch(r_dFdCout_in, s1,c_dFdC1).
           ch(r_dFdCout_in,s2,c_dFdC2).start_p_main!()
      endif
      + if (s1,c_dFdC1) then
           s1?().ch(s1,c_dFdCMP1).ch(s2,c_dFdCMP2).start_p_main!()
      endif
      + ....
```

The first phosphorylation of gemcitabine enabled by dCK (see Fig. 7.8) is coded as a bimolecular interaction between `dCK` and `dFdC`. It is managed by the second alternative in `p_main`. Note that no rate is expressed in the change primitive because it is assumed to be immediate after the synch signal on `s1`. The bimolecular interaction is enabled by the affinity between the binders `s` and `s1`.

```
Affinity: (c_dCK, c_dFdC1, r_dFdC_dCK)

let dCK: bproc = #(s,c_dCK)[s!().startdCK!() |
       rep startdCK?().s!().startdCK!()];.
```

The inhibition mechanism modeled in Fig. 7.10 is coded as a binding dynamics between `dFdC` in its `TP` configuration and `dCMPD` by including the affinity for complex/decomplex on the binders `c_dFdCTP1` and `c_dCMPD`.

```
Affinity: (c_dFdCTP1, c_dCMPD, r_dFdCTP_dCMPD_bin,
                          r_dFdCTP_dCMPD_unb, 0)

let dCMPD: bproc = #(s,c_dCMPD)[p_dCMPD];.
```

The global dynamics is coded with `when` events for each substitute in Fig. 7.11.

Example 7.10. [ℓ gemcitabine model]
The easiest way to code the gemcitabine model in ℓ is to have a dyn reaction for each reaction in Fig. 7.14 that corresponds to the actions described in the graphical and tabular schema of the model. For instance, the first phosphorylation of dFdC by dCK is

```
# r5 : dFdC + dCK -> dFdC_MP + dCK
dyn [dFdC] [dCK] rate r_dFdC_dCK react
       reagent_1.kill();
       spawn(1:dFdC_MP);.
```

Note that we can also apply Algorithm 6.4 starting from the reactions in Fig. 7.14. The algorithm produces 29 substitute declarations. For instance, reaction $\# \ r5$ would be translated into

```
substitute [dFdC] [dCK] with [dFdC_MP{}] [dCK{}] rate r_dFdC_dCK;.
```

7.5 Model calibration, evaluation and refinement

The concrete model has to be evaluated against the experimental observations and the acceptance criteria we set at the beginning of the modeling process. To tune the outcome of the simulation we must calibrate the values of the parameters by relying on sensitivity analysis or parameter scanning. These techniques generate many instances of the model with different parameter values and checks how the outcome of the simulation varies. The modeler then tunes the parameters to obtain an outcome that is the closest to experimental observations.

Sometimes, parameter tuning is not enough to match the expected threshold of accuracy. In this case a refinement of the model that affects its structure and hence an iteration that jumps back to the model schema are needed (see Fig. 7.1).

7.6 Summary

This chapter codifies a modeling process that starts from the definition of a question and a suitable knowledge base to form hypotheses. These steps allow us to derive a model schema that is then mapped to a concrete model for simulation and analysis. We used as an example the metabolism of the anti-cancer drug gemcitabine to illustrate the phases of the modeling process. We mapped the model schema of the running example into concrete models represented as chemical reactions, BlenX and ℓ programs. We finally briefly discussed model calibration, evaluation and refinement as an iterative cycle to reach the objectives of the modeling process.

7.7 Further reading

The narrative approach to represent the dynamics of biological systems has been introduced in [Guerriero *et al.* (2007a)] and applied to a biological example in [Guerriero *et al.* (2009a)]. An interesting requirement analysis for graphical representation of biological systems is [Saraiya *et al.* (2005)]. The tabular approach is also discussed in [Priami *et al.* (2009a)]. Model calibration is studied in [Bock *et al.* (2013); Kremling (2014)].

Details on the biological content of this chapter can be found in [Alberts *et al.* (2002)] and [Lodish *et al.* (2004)]. Deeper treatment of the metabolism of gemcitabine is available in [Heinemann *et al.* (1992); Veltkamp *et al.* (2008)]. The case study is developed in [Kahramanoğulları *et al.* (2012)].

Chapter 8

Simulation

Simulation is the process of model execution to reproduce the dynamics of a system. Digital simulation (hereafter just simulation because it is the kind we focus on) is the process of making a computer behave like a system of interest and therefore relates models and computers. Modeling is the process of creating a model and of connecting it to a real system, and simulation is the process of connecting a model to a digital computer.

Investigating the mechanistic properties of a system through simulation is faster (e.g., many cell cycles of mammalian cells can be simulated in seconds on a computer, while it takes days in a lab) and cheaper (e.g., simulating crash tests on a computer avoids destroying physical objects or simulating the effect of perturbation on biological systems may save animals).

The outcome of a simulation is the dynamic behavior of a model, i.e., the approximate reproduction of time behavior of a real system. If we model a disease and we find, through simulation, a drug that can cure the disease or even prevent it, we cannot be sure that the result would be the same in the real world. We need to follow the protocol for drug validation before delivering it to real patients. The advantage of simulation is discarding alternatives that will not work and helping experimenters to design focused experiments, thus saving time and money.

This chapter is organized as follows. The next section introduces the basic notions of model execution. Then, we discuss how to build random number generators that are the basis of each simulator. We then concentrate on discrete-event simulation and the stochastic algorithms that are the core of stochastic simulation engines.

8.1 Model execution

Simulation experiments on partially unknown phenomena to figure out how the real system works. Lack of knowledge of mechanistic behavior can be modeled through nondeterminism (inability to make any assumption on the choices that the system performs to select the next event). Nondeterminism arises in artificial systems by ignoring the factors that influence the selection of the next event. If a system can perform different actions, but we do not know its operational rules, we can say that the system chooses nondeterministically among all the enabled actions. Nondeterminism is a qualitative concept: no measure can be associated with the actions among which to select (see Sect. 3.1.1.2). Biological systems behave differently depending on quantities such as temperature, pH or concentrations. Therefore simulation must be driven by quantities and cannot be nondeterministic.

Stochastic processes

Let $X(t)$ be the state of a dynamical system at time t and assume that $X(t_0) = x_0$ for some fixed initial time t_0. $X(t), t > t_0$ is a random variable. $X(t)$ is a *stochastic process* if the joint density function of $X(t_1), .., X(t_n)$ is

$$f_n^1(x_n, t_n; ..; x_1, t_1 \,|\, x_0, t_0)dx_n..dx_1 = Prob\{X(t_i) \in [x_i, x_i + dx_i) \\ \text{given } X(t_0) = x_0 \,|\, i \in [1..n], t_0 \leq t_1 \leq .. \leq t_n\}$$

where the indexes of f_i^j denote the number of parameters before (i) and after (j) the vertical bar.

The Markov property states that the ability to predict a future state of the system given $X(t_j) = x_j$ is not enhanced by the knowledge of $X(t_i)$ with $t_i \leq t_j$. Formally,

$$f_1^{j+1}(x_{j+1}, t_{j+1} \,|\, x_j, t_j; ..; x_1, t_1; x_0, t_0) = f_1^1(x_{j+1}, t_{j+1} \,|\, x_j, t_j) = \\ f(x_j, t_j \,|\, x_{j-1}, t_{j-1}).$$

A stochastic process that has the Markov property is called a *Markov process*. For a Markov process the following holds:

$$f_n^1(x_n, t_n; ..; x_1, t_1 \,|\, x_0, t_0) = \prod_{i=1}^{n} f(x_i, t_i \,|\, x_{i-1}, t_{i-1}).$$

Simulation studies the time behavior of systems. Probabilistic choices connect to time in stochastic processes (e.g., the time that a chemical reaction needs to complete is a relevant parameter in a biochemical simulation).

Stochastic processes are usually an alternative to deterministic processes that provide the same results on each run (e.g., the solution of a system of ordinary differential equations always provides the same result, the execution of a sequential imperative program always produces the same output for the same input). The future behavior of stochastic processes has a degree of indeterminacy governed by probabilistic distributions. Many biological processes are stochastic in nature due to multiple levels of regulation that contribute to the noise and indeterminacy in the observable behavior of the cells (e.g., stochasticity is a main property of gene regulatory networks).

The three key choices to decide the kind of simulation to perform are the *time advance method*, the *execution type* and the *execution mode* (Fig. 8.1). Time advance methods define how time is advanced during simulation. *Time-stepped* methods advance time in fixed increments that determine the synchronous time points in which the system state is updated. *Discrete-event* methods let each component of the system evolve according to its own timescale. Events trigger time advances and determine the new simulation time. As a consequence, time does not advance by fixed increments. *Time parallel* methods partition time in multiple segments that are executed in parallel. It is a parallel simulation with respect to time.

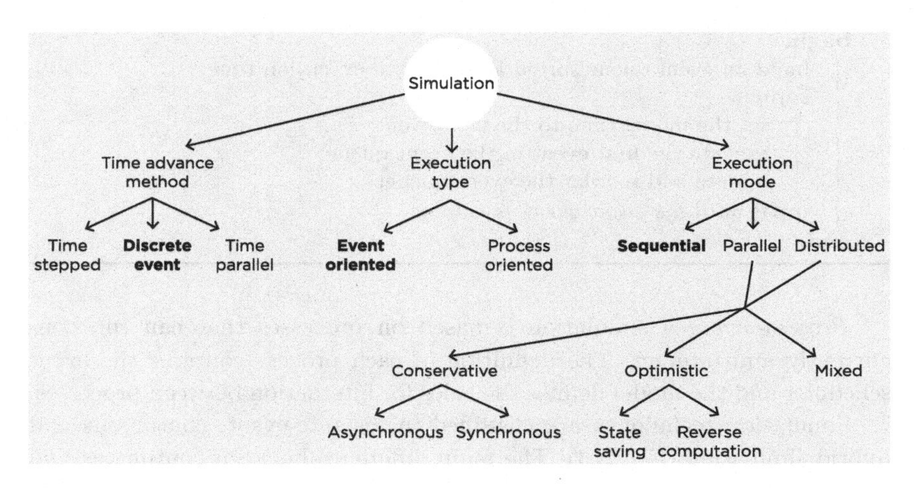

Figure 8.1 Major design elements of a simulation. The words in bold identify the simulation solutions dealt with in this book.

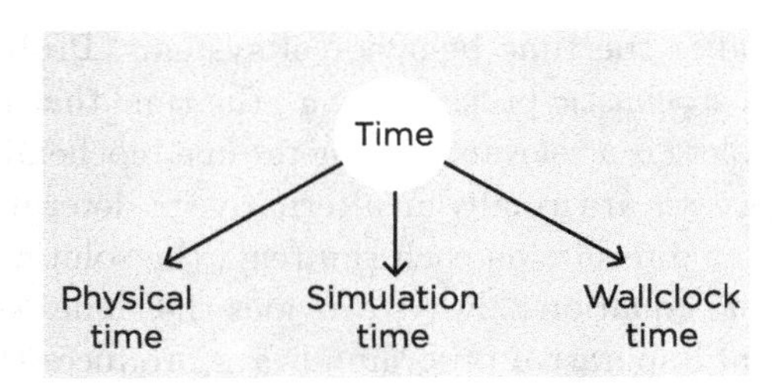

Figure 8.2 Timelines for simulation.

There are also three different time axes that are related to simulation (see Fig. 8.2). *Physical time* is the time of the physical system modeled. *Simulation time* is the representation of the physical time used in the simulation. *Wallclock time* or *runtime* is the duration of the model execution measured by the hardware clock.

We mainly focus on the event-oriented execution type. Algorithm 8.1 describes an *event scheduling* (ES) procedure. The user can only schedule events, but cannot manipulate the model time or the event queue.

Algorithm 8.1: Event scheduling

Input: A model to be simulated.
Output: The simulation of the model usually represented as a list of pairs (event, time).

begin
  build an event queue sorted according to execution time;
  repeat
    set the model time to the next event;
    execute the first event in the event queue;
    update and reorder the event queue;
  until *until the event queue is empty*;
end

Process-oriented simulation is based on processes that can run concurrently and interact. The definition of each process contains the event scheduler and the model defines the rules for interaction between processes.

Simulation techniques are classified in discrete-event, continuous and hybrid simulation (Fig. 8.3). The main difference between continuous and discrete-event simulation is on the time advance mechanism (fixed increments vs variable increments) and on the number of entities present in

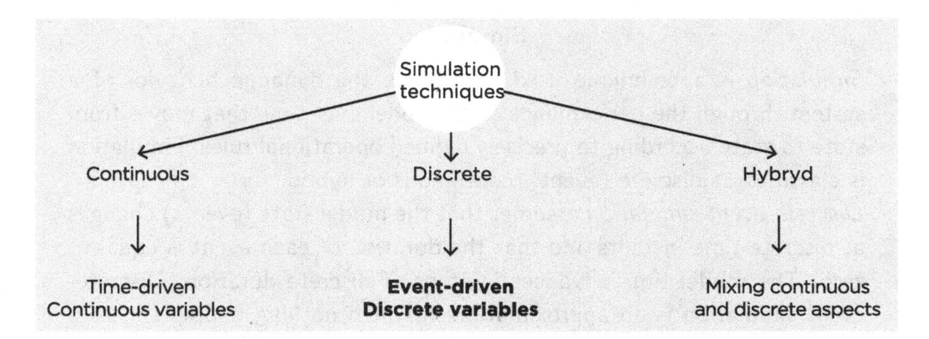

Figure 8.3 Simulation techniques. The words in bold identify the simulation solutions dealt with in this book.

the system (continuous variables versus discrete variables). The evolution of the system in the continuous simulation is time-driven (the increments determine when the variables have to be updated), while discrete-event simulation is event-driven (the events update the variables and determine the size of the time increment). Hybrid simulation contains entities that are managed as continuous and other that have handlers for events.

Consider now the execution mode. *Sequential simulation* uses a single processor, the activities/events are globally ordered and exactly one is consumed at a time. When more than one activity/event can occur simultaneously (e.g., multiple processing units are available), we have *parallel* or *distributed simulation* depending on whether the processing units have shared memory or high-speed interconnections. Parallel and distributed simulations are grouped in the category of *concurrent simulations.* Concurrency is used to speed up simulations (reducing their wallclock time).

Concurrent simulation must ensure that it produces the same results as the corresponding sequential simulation. Therefore a main issue is synchronizing events on different processors to ensure a global totally ordered timescale equal to that produced by sequential simulations. There are two main strategies to running concurrent simulations: *conservative* and *optimistic.* There also exists a *mixed* strategy in which one tries to consider optimistic strategies under specific conditions. Conservative simulation blocks the execution of the first scheduled event until there is a guarantee that no event with a smaller timestamp can be allocated to the processor. Conservative simulations never undo executed events. On the contrary, optimistic simulations let each processor run independently and synchronize when events are dispatched from another processor. The synchronization phase

Simulation

Simulation is a technique used to observe the dynamic behavior of a system through the performance of its model over time that moves from state to state according to precisely defined operational rules. Simulation is classified as discrete (event), continuous or hybrid.
Discrete-event simulation assumes that the model state (events) changes at discrete-time instants and that the duration of each event is equal to zero. The model time advances in steps of discrete duration. Discrete-event simulation is an approximation in which nothing happens in zero time.
Continuous simulation assumes that the model state changes continuously in time. The model time advances continuously. Continuous simulation is only approximated on digital computers where nothing is continuous.
Hybrid simulation mixes in the same model discrete-event components and continuous components.

may need roll-back of executed events to restore the previous status of the model and retract events to have been dispatched to other processors. Optimistic simulation must save the state of the system to restore it in case of synchronization conflicts or implement reverse handlers to perform the actions that must be undone in reverse order.

Hereafter we only concentrate on sequential, discrete-event simulation. The next section introduces random number generators that are the core of stochastic simulation engines.

8.2 Random number generation

Although many programming languages and simulation environments offer libraries and pre-defined macros to generate random numbers, it is useful to understand their algorithms. We discuss procedures to numerically generate sample values of a random variable with a given density function. All the computational processes that makes use of random number generators are called *Monte Carlo computations.* The main applications of Monte Carlo computations are numerical simulations of random processes (e.g., Markov processes) and numerical evaluation of definite integrals. We concentrate

How to evaluate a $U(0,1)$ random number generator

The *goodness test* (χ^2, see Appendix B.3.3) checks how close the random numbers are to $U(0,1)$ and the *independence test* checks to what extent the numbers are independent.

For the χ^2 test, we divide the unit interval into k subintervals. If the random sequence is $U(0,1)$, a random number would fall in a fixed subinterval with probability $1/k$. Over n observations, O_i (the number of random numbers in subinterval i) will be distributed as $Bin(0,k)$ and $\mu_{O_i} = n/k$. We set

$$\chi^2 = \sum_{i=1}^{k} \frac{(O_i - \mu_{O_i})^2}{\mu_{O_i}}.$$

The observations are $U(0,1)$ if $\chi^2 > \chi^2_{\alpha,k-1}$, where $\chi^2_{\alpha,k-1}$ is the $1-\alpha$ quantile from a χ^2 table (usually α is 0.05 or 0.1). The large number of observations from random number generators allows the safe approximation

$$\chi^2_{\alpha,k-1} \sim (k-1)\left[1 - \frac{2}{9(k-1)} + z_\alpha\sqrt{\frac{2}{9(k-1)}}\right]^3$$

where z_α is the appropriate normal quantile. Other goodness tests are Kolmogorov–Smirnov, Anderson–Darling or Crush randomness tests.

For the independence test, consider a *run* as a sequence of similar observations (e.g., a sequence of heads or a sequence of tails in coin tossing) and call A the number of runs in a set of n observations. If $n > 20$, it turns out that A is $N((2n-1)/3, (16n-29/90))$. We set $Z_0 = \frac{A-\mu_A}{\sqrt{\sigma^2(A)}}$ and the sequence of random numbers is independent if $|Z_0| \leq z_{\alpha/2}$. Other tests are the correlation test, the gap test or the birthday test.

here on simulation by assuming a basic knowledge of random variables (Appendix B.2) and statistics (Appendix B.3).

8.2.1 *Uniform random number generators*

Random number generators require the generation of a sequence of (pseudo)-random numbers that appear to be independent and identically distributed in $U(0,1)$. The goal is to have fast and repeatable procedures to generate these numbers (evaluation criteria for these generators are in the box on the previous page). Algorithm 8.2 is widely used to generate $U(0,1)$ random numbers.

Algorithm 8.2: Linear congruential generator $U(0,1)$

Input: Integer numbers N_0, A, C, M.
Output: A sequence of pseudo-random numbers that are independent and identically distributed in $U(0,1)$.

begin
  generate a sequence $N_i = (AN_{i-1} + C) \mod M$;
  the sequence of random numbers is $R_i = N_i/M$
end

N_0 in Algorithm 8.2 is called *seed.* If $C = 0$ the LCG is said to be *multiplicative.* A good choice of integers is fundamental for obtaining a long *period* of the generator (the number of outcomes before the generator starts repeating the sequence). For multiplicative generators, a good choice of integers is $M = 2^{31} - 1, A = 7^5, N_0 \in [1, M-1]$. These integers ensure that the generator has a period of $M - 1$, never produces $R = 0$ and $R = 1$ and has good scores for statistical randomness.

How to choose a $U(0,1)$ LCG random number generator

Here are some theoretical results that help in choosing a good generator. The generator $N_i = (AN_{i-1}) \mod 2^M$ has a maximal period of 2^{M-2} when N_0 is odd and $A = 8k + 3$ or $A = 8k + 5$ for some k.
The generator $N_i = (AN_{i-1} + C) \mod M$, $C > 0$, has a maximal period if C and M are relatively prime, $A - 1$ is multiple of any prime that divides M, $A - 1$ is multiple of 4 if $M = 4k$ for some k.
A corollary is that $N_i = (AN_{i-1} + C) \mod 2^M$, $C, M > 1$ has a maximal period if C is odd and $A = 4k + 1$ for some k.
The multiplicative generator $N_i = (AN_{i-1}) \mod M$, M prime, has a maximal period $M - 1$ if M divides $A^{M-1} - 1$ and $\forall i < M - 1, M$ does not divide $A^{i-1} - 1$.

An alternative is *Tausworthe's generator* (TG) in Algorithm 8.3. This generator produces sequences of random bits, which are then turned into random numbers. The period is $2^q - 1$. Due to the conversion of bits into

Algorithm 8.3: Tausworthe's generator $U(0,1)$

Input: Integer numbers r, q and a sequence of bits $b_1, .., b_q$.
Output: A sequence of pseudo-random numbers that are independent and identically distributed in $U(0,1)$.

begin
generate a sequence $b_i = (b_{i-r} + b_{i-q}) \mod 2, \quad 0 < r < q$;
the sequence of random numbers is $R_i = (\text{k-bit binary integer})/2^k$
end

numbers, this generator is usually too slow for practical purposes. It is, however, interesting from a conceptual point of view.

Mersenne prime and binary field $\mathbb{F}_2$

A *Mersenne prime* is a prime number of the form $M_n = 2^n - 1$.
A field is a set F that is a commutative group with respect to two operations (e.g., addition and multiplication), the two operations enjoy the distributivity property and the additive identity has no multiplicative inverse. The binary field $\mathbb{F}_2$ has two elements, 0 and 1, and the arithmetic operations over $\mathbb{F}_2$ are equivalent to arithmetic modulo 2.

The *Mersenne twister* generator has period usually chosen $2^{19937} - 1$. It is based on a matrix linear recurrence over a finite binary field $\mathbb{F}_2$. A recent improvement of the Mersenne generator is the *WELL generator* still based on the linear recurrence modulo 2 approach (Algorithm 8.4).

Algorithm 8.4: Linear recurrence modulo 2 generator $U(0,1)$

Input: Positive integer numbers s, w, $s \times s$ transition matrix $\mathbf{A}$, $w \times s$ output transformation matrix $\mathbf{B}$, seed state s-vector $\mathbf{x}_0$.
Output: A sequence of pseudo-random numbers independent and identically distributed in $U(0,1)$.

begin
The sequence of random numbers is $R_i = \sum_{k=1}^{w} y_{i,k-1} 2^{-k}$, $\mathbf{x}_i = \mathbf{A}\mathbf{x}_{i-1}$ and $\mathbf{y}_i = \mathbf{B}\mathbf{x}_i$
end

Special cases of this generator obtained with appropriate choice of **A** and **B** are Tausworthe's generator and the Mersenne twister generator. The WELL generator decomposes the size of the state vector as $s = rw - p$ with r and p the unique integers such that $r > 0$ and $0 \leq p < w$. The state vector is then decomposed in w-bit blocks and the last block is completed with p 0-bits. The matrix A is built out of $w \times w$ transformation matrices applied to the blocks of the state vector. All the transformations are applied at the state vector, i.e., **B** acts as the identity.

8.2.2 *General random number generators*

We show here two methods of transforming unit uniform random numbers into sample values of arbitrarily specified random variables. The first method is the *inversion generating method* (Algorithm 8.5) and the second one is the *rejection generating method* (Algorithm 8.6).

Algorithm 8.5: Inversion generating method

Input: A distribution function F of a continuous random variable X.
Output: A sample value of X.

```
begin
    generate R with a generator U(0, 1);
    generate the sample value of X as x = F^{-1}(R)
end
```

Algorithm 8.6: Rejection generating method

Input: A density $P = c \cdot h(x)$ of a continuous random variable X, with c a normalizing constant, $0 \leq h(x) \leq B, x \in [a, b]$ and $h(x) = 0, x \notin [a, b]$.
Output: A sample value of X.

```
begin
    generate R_1 and R_2 with a generator U(0, 1);
    done := false;
    while not done do
        x_t = a + (b - a)R_1;
        if h(x_t)/B >= R_2 then
            done := true;
            return(x_t)
        end
    end
end
```

The intuition of the inversion generating method is that mapping a unit uniform random number on the y-axis of the distribution function F allows us to recover a value of the x-axis. The x-axis value is a sample value of X because the range of F is always $[0, 1]$ (see Fig. B.2). Note that the assignment of the value to x comes from the equation $F(x) = R$.

Example 8.1 shows an application of the inversion generating method to sample values of real random variables in $U(a, b)$ and $E(a)$.

Example 8.1. [Samples for uniform and exponential random variables]
The distribution function of a random variable $U(a, b)$ is $F(x) = (x-a)/(b-a)$ if $a \leq x < b$. Generate a unit uniform random number R and solve $(x-a)/(b-a) = R$ for x, we obtain $x = a + (b-a)R$ and x is $U(a, b)$.
The distribution function of a random variable $E(a)$ is $F(x) = 1 - e^{-ax}$ if $0 \leq x$. Since $(1-R)$ is $U(0, 1)$ when R is $U(0, 1)$, we solve for x the equation $1 - e^{-ax} = 1 - R$, yielding $x = (1/a)\ln(1/R)$ and x is $E(a)$.

The efficiency of the rejection generating method (the average acceptance ratio of the generated sample values) is $(\int_a^b h(x)dx)/(B(b-a))$.

8.3 Stochastic simulation algorithms

Gillespie introduced the *stochastic simulation algorithm* (SSA — Algorithm 8.7) that, averaging over many runs, gives the same results as the solution of the *chemical master equation* (CME — see the box on page 314).

8.3.1 *Direct method*

Algorithm 8.7 describes the time evolution of a well-stirred chemical solution within a constant reacting volume. It has been recently applied to biochemical systems, because there are many biological processes that involve small numbers of reactants and hence discreteness and stochasticity play an important role.

The initialization phase before the main **while** loop computes the changes in the S_i molecular population, so that the new state vector after performing R_j will be $\mathbf{X}(t) + \mathbf{v}_j$ (last step of the loop).

To compute the probability of observing a reaction R_j, we must know the multiplicity (or redundancy) of the molecules acting as reactants and consider the number of ways the reaction can occur based on the reactant

Algorithm 8.7: SSA — Direct method (DM)

Assumptions: A chemical reaction occurs when two affine molecules randomly collide within a well-stirred mixture of molecules (thermal, but not chemical, equilibrium) within a constant volume V. The molecules are randomly and uniformly distributed at all times.
Input: The species $S_1, ..., S_N$ and the reactions $R_1, ..., R_M$ with

$$R_j = v_{1j}S_1 + \cdots + v_{Nj}S_N \xrightarrow{c_j} v'_{1j}S_1 + \cdots + v'_{Nj}S_N.$$

The *base rate* c_j is a constant depending on physical properties of the reactants; the state vector $\mathbf{X}(t_j) = \mathbf{x}_j$ with $X_i(t_j) = |S_i|$ at time t_j; simulation time t_{max}.
Output: The new state vector $\mathbf{X}(t+\tau)$ at each iteration.

begin
  $t := 0$;
  $a_0(\mathbf{x}) := 0$;
  for each R_j compute the vector $\mathbf{v}_j$ where its ith element is $v'_{ij} - v_{ij}$
  while $t < t_{max}$ **do**
    for $j \in \{1, .., M\}$ **do**
      case R_j **of**
        $S_i \xrightarrow{c_j} ..$ (monomolecular) $\Rightarrow$ $h_j(\mathbf{x}) = X_i(t)$
        $S_i + S_l \xrightarrow{c_j} ..$ (bimolecular) $\Rightarrow$ $h_j(\mathbf{x}) = X_i(t)X_l(t)$
        $2S_i \xrightarrow{c_j} ..$ (homodimerization) $\Rightarrow$ $h_j(\mathbf{x}) = \frac{X_i(t)(X_i(t)-1)}{2}$
        $.. \xrightarrow{c_j} ..$ (constant rate) $\Rightarrow$ $h_j(\mathbf{x}) = 1$
      endcs
      $a_j(\mathbf{x}) := h_j(\mathbf{x})c_j$;
      $a_0(\mathbf{x}) := a_0(\mathbf{x}) + a_j(\mathbf{x})$
    end
    generate two random numbers r_0 and r_1 in $U(0,1)$;
    $\tau := \frac{1}{a_0(\mathbf{x})} ln(\frac{1}{r_0})$;
    select j as the smallest value such that $\sum_{k=1}^{j} a_k(\mathbf{x}) > r_1 a_0(\mathbf{x})$;
    $\mathbf{X}(t+\tau) := \mathbf{X}(t) + \mathbf{v}_j$;
    $t := t + \tau$
  end
end

populations (independent combinations of reactants). The **case** statement consider the four alternatives: (1) monomolecular reactions, (2) bimolecular reactions, (3) bimolecular reactions between two molecules of the same species and (4) the special case of constant rates where the number of molecules does not affect the rate of the reaction. If R_j is a monomolecular

Propensity function and some useful probabilities

Under the assumptions and notation in Algorithm 8.7, given $\mathbf{X}(t) = \mathbf{x}$, the probability that in the next infinitesimal time interval $[t, t + dt)$

(1) *exactly one reaction* R_j will occur is equal to $h_j(\mathbf{x})c_j dt + o(dt)$, where $o(dt)$ goes to 0 faster than dt when $dt \to 0$. Then, $a_j(\mathbf{x}) = h_j(\mathbf{x})c_j$ is called the *propensity function*;
(2) *no reaction* will occur is equal to $1 - \sum_{j=1}^{M} h_j(\mathbf{x})c_j dt + o(dt)$;
(3) *more than one reaction* will occur is equal to $o(dt)$.

reaction, quantum mechanics states that there exists some constant c_j such that $c_j dt$ is the probability that some molecule will react in the next dt. The number $h_j(\mathbf{x})$ of combination of molecules that can react through R_j determines the probability $a_j(\mathbf{x})dt = h_j(\mathbf{x})c_j dt$ of observing R_j in the next dt. Similarly, we obtain the same result for bimolecular reactions. Recall that stochastic rates and deterministic kinetic constants are different (see Sect. 5.2.1 and Fig. 5.6).

The algorithm considers the probability that the next reaction will occur in the infinitesimal interval $[t+\tau, t+\tau+d\tau)$ and it is R_j given $\mathbf{X}(t) = \mathbf{x}$ and denoted $P(\tau, j \mid \mathbf{x}, t)d\tau$ (see the box on the next reaction density function on page 315).

Samples of τ can be obtained by applying the inverse generating method (Algorithm 8.5 and Example 8.1). We draw two random numbers r_0 and r_1 from $U(0,1)$ and we set τ as in Algorithm 8.7. To select the reaction randomly we consider an interval $(0,1)$ divided in the subintervals $a_1(\mathbf{x})/a_0(\mathbf{x}), a_2(\mathbf{x})/a_0(\mathbf{x}), .., a_M(\mathbf{x})/a_0(\mathbf{x})$ and we check in which subinterval the random number r_1 is falling, i.e.,

$$\sum_{k=1}^{j-1} \frac{a_k(\mathbf{x})}{a_0(\mathbf{x})} < r_1 \leq \sum_{k=1}^{j} \frac{a_k(\mathbf{x})}{a_0(\mathbf{x})} \Rightarrow \sum_{k=1}^{j-1} a_k(\mathbf{x}) < r_1 a_0(\mathbf{x}) \leq \sum_{k=1}^{j} a_k(\mathbf{x})$$

from which the formula is in Algorithm 8.5.

We finally update the simulation time and state vector of the system. Note that the reactions subdivide the interval $(0,1)$ according to their propensities and the random number r_1 selects one subinterval and its corresponding reaction.

The time advance mechanism here is exact and it is not an approximation as it happens in numerical solutions of deterministic ODE systems.

Chemical master equation (CME)

An estimate $P(\mathbf{x}, t \,|\, \mathbf{x}_0, t_0) = Prob\{\mathbf{X}(t) = \mathbf{x} \,|\, \mathbf{X}(t_0) = \mathbf{x}_0\}$, where t_0 is the initial time and $\mathbf{x}_0$ the initial state vector, starting from the propensity functions and the other useful probabilities reported in the box on the previous page, it is

$$P(\mathbf{x}, t + dt \,|\, \mathbf{x}_0, t_0) = P(\mathbf{x}, t \,|\, \mathbf{x}_0, t_0)[1 - \sum_{j=1}^{M} h_j(\mathbf{x})c_j dt + o(dt)] + \sum_{j=1}^{M} P(\mathbf{x} - \mathbf{v}_j, t \,|\, \mathbf{x}_0, t_0)[h_j(\mathbf{x} - \mathbf{v}_j)c_j dt + o(dt)] + o(dt).$$

The first term on the right side of the equation is the probability that no reaction occurs in $[t, t + dt)$ and hence the system must already be in state $\mathbf{x}$ at time t. An alternative is to reach the state $\mathbf{x}$ at time $t + dt$ through exactly one reaction R_j. Since the update of the reaction R_j to the system state is obtained by summing $\mathbf{v}_j$ (see Algorithm 8.7) to the current state, it means that the state at time t should equal $\mathbf{x} - \mathbf{v}_j$. Furthermore, the probability of firing exactly R_j is $h_j(\mathbf{x})c_j dt + o(dt)$. All the possible reactions could lead to the new state and hence we sum over all the reactions obtaining the second term of the right end of the equation. Finally, the probability that more than one reaction fires in $[t, t + dt)$ to lead to $\mathbf{x}$ at time $t + dt$ from the start at time t is $o(dt)$. The three terms are added, because they denote mutually exclusive events. By subtracting $P(\mathbf{x}, t \,|\, \mathbf{x}_0, t_0)$ from both sides of the previous equation, dividing by dt and letting $dt \to 0$, we obtain a time evolution equation called *chemical master equation* (CME):

$$\frac{\partial P(\mathbf{x}, t \,|\, \mathbf{x}_0, t_0)}{\partial t} = \sum_{j=1}^{M} [a_j(\mathbf{x} - \mathbf{v}_j) P(\mathbf{x} - \mathbf{v}_j, t \,|\, \mathbf{x}_0, t_0) - a_j(\mathbf{x}) P(\mathbf{x}, t \,|\, \mathbf{x}_0, t_0)].$$

The CME is usually an infinite set of ODEs and it is impractical to solve it numerically to obtain the probability density function of $\mathbf{X}(t)$. Therefore the algorithm generates simulated trajectories by building random samples from the probability density function of $\mathbf{X}(t)$.

As a final remark, the variable advance of simulation time reduces the possibility of introducing errors, typical of fixed advance time solutions when there are events that are faster than the fixed time step.

Next reaction density function

The *next reaction density function* is the joint probability density function of the random variables τ (time to the next reaction, $0 \leq \tau \leq \infty$) and j (index of the firing reaction, $0 \leq j \leq M$) given $\mathbf{X}(t) = \mathbf{x}$. If we consider the interval $[t+\tau, t+\tau+d\tau)$ divided into $k+1$ subintervals of length $\epsilon = \tau/k$, $P(\tau, j \mid \mathbf{x}, t)d\tau$ is the probability that no reaction occurs in the first k subintervals and R_j occurs in the last interval. This is

$$P(\tau, j \mid \mathbf{x}, t)d\tau = [1 - a_0(\mathbf{x})\epsilon + o(\epsilon)]^k [h_j(\mathbf{x})c_j d\tau + o(d\tau)]$$

where $a_0(\mathbf{x})$ is defined in Algorithm 8.7. The limit by $d\tau \to 0$ and then by $k \to \infty$ shows that

$$P(\tau, j \mid \mathbf{x}, t) = a_j(\mathbf{x})e^{-a_0(\mathbf{x})\tau}.$$

This implies that τ is an exponential random variable with mean (and standard deviation) $1/a_0(\mathbf{x})$ and j is an integer random variable with point probability $a_j(\mathbf{x})/a_0(\mathbf{x})$.

8.3.2 *Some SSA variants*

In this subsection we present some variants of the direct method that have all been proved to be equivalent. We start with the *first reaction method* (FRM) (Algorithm 8.8). The main difference from Algorithm 8.7 is that it generates M random numbers in $U(0,1)$ and computes the time τ_l of all the reactions. Then the algorithm selects the smallest time and fires the corresponding reaction, updating the vector state accordingly. Note that the computational cost of the first reaction method is easily larger than that of the direct method.

A generalization of the direct and the first reaction method is the *first family method* (FFM) (Algorithm 8.9). The reaction set M is partitioned into Q families $F_1, .., F_Q$ so that each family contains the reactions $R_1^q, .., R_{M_q}^q$. Intuitively, each family is considered an abstract reaction and the selection of the family to fire is done as in the first reaction method by selecting the smallest τ_q of the abstract reaction. Once a family is selected, the actual reaction to fire within the family is selected as in the direct method. Therefore, if we consider one family we have the direct method, while if each family contains one reaction we have the first reaction method. This method is, however, more flexible because it is possible to determine Q to minimize the computational cost, depending on the size of M.

Algorithm 8.8: First reaction method (FRM)

Assumptions, Input, Output: Same as Algorithm 8.7.

begin
  $t := t_0$;
  for each R_j compute the vector $\mathbf{v}_j$, where its ith element is $v'_{ij} - v_{ij}$;
  while $t < t_{max}$ **do**
    compute the propensities $a_j(\mathbf{x})$ and $a_0(\mathbf{x})$ as in Algorithm 8.7;
    generate M random numbers $r_1, .., r_M$ in $U(0, 1)$;
    for $l \in \{1, .., M\}$ **do**
      $\tau_l := \frac{1}{a_l(\mathbf{x})} ln(\frac{1}{r_l})$
    end
    $\tau_j := min\{\tau_1, .., \tau_M\}$;
    $\mathbf{X}(t + \tau_j) := \mathbf{X}(t) + \mathbf{v}_j$;
    $t := t + \tau_j$
  end
end

Algorithm 8.9: First family method (FFM)

Assumptions, Output: Same as Algorithm 8.7.
Input: Same as Algorithm 8.7 except for the reaction set that is partitioned in $F_1, .., F_Q$ with $F_q = \{R_1^q, .., R_{M_q}^q\}, q \in \{1, .., Q\}$.

begin
  $t := t_0$;
  for each R_j compute the vector $\mathbf{v}_j$ where its ith element is $v'_{ij} - v_{ij}$;
  while $t < t_{max}$ **do**
    generate Q+1 random numbers $r_1, .., r_{Q+1}$ in $U(0, 1)$;
    for $q \in \{1, .., Q\}$ **do**
      $a_0^q(\mathbf{x}) = \sum_{k=1}^{M_q} a_k(\mathbf{x})$;
      $\tau_q := \frac{1}{a_0^q(\mathbf{x})} ln(\frac{1}{r_q})$
    end
    $\tau_q := min\{\tau_1, .., \tau_Q\}$;
    select j as the smallest value such that $\sum_{k=1}^{j} a_k^q(\mathbf{x}) > r_{Q+1} a_0^q(\mathbf{x})$;
    $\mathbf{X}(t + \tau_j) := \mathbf{X}(t) + \mathbf{v}_j$;
    $t := t + \tau_j$
  end
end

Algorithm 8.10: Next reaction method (NRM)

Assumptions, Input, Output: Same as Algorithm 8.7.

```
begin
    t := 0;
    generate M random numbers r_1, .., r_M in U(0, 1);
    for j ∈ {1, .., M} do
        compute the vector v_j where its ith element is v'_ij − v_ij;
        compute h_j as in Algorithm 8.7;
        a_j(x) := h_j(x)c_j;
        τ_j := (1/a_j(x)) ln(1/r_j);
        update(τ_j, heap)
    end
    generate dependency graph;
    while t < t_max do
        idx := min_heap;
        X(τ_idx) := X(t) + v_idx;
        t := τ_idx;
        for j ∈ {k ≠ idx | R_k depends on R_idx} do
            a_j^old(x) := a_j(x);
            compute h_j as in Algorithm 8.7;
            a_j(x) := h_j(x)c_j;
            τ_j := t + (a_j^old(x)/a_j(x)) (τ_j − t);
            update(τ_j, heap)
        end
        compute h_idx as in Algorithm 8.7;
        a_idx(x) := h_idx(x)c_idx;
        generate a random numbers r in U(0, 1);
        τ_idx := t + (1/a_idx(x)) ln(1/r);
    end
end
```

A more computationally efficient variant of the first reaction method is the *next reaction method* (NRM) (Algorithm 8.10). This method avoids recomputing the propensities for all the reactions at each iteration. It uses a dependency graph to store the relationships between reactions. A reaction depends on the amount of its reactants and affects the amount of its reactants and products. The intuition is that only the reactions whose reactants are modified by the fired reactions change their propensity. Furthermore, an indexed priority queue is used to store the putative times of the reactions so that the lowest is always on the head of the heap and so that searching

Dependency graph

Let $V = \{R_1, .., R_M\}$ be a set of reactions. Let $R(R_j)$ and $P(R_j)$ denote the reactants and the products of R_j. The *dependency graph* generated by V is the direct graph (V, E) with

$$E = \{(R_i, R_j) \mid R(R_i) \cap [R(R_j) \cup P(R_j)] \neq \emptyset, i, j = 1, .., M\}.$$

Note that $\forall i = 1, .., M$, $(R_i, R_i) \in E$. An edge (R_i, R_j) means that reaction R_i depends on R_j.

for the reaction to fire is much faster (the computational complexity of the previous methods is linear in M, while it is logarithmic in M for NRM). Finally, the random number generation is reduced because only one random number is needed at each iteration instead of the minimum two of the direct method. Since the generation of a random number costs about ten operations, reducing the number of calls to the generator reduces the overall cost of the algorithm. The idea is to modify the unused reaction times (corresponding to not-selected reactions) without the need to generate random numbers and resample the firing times (second **for** loop of Algorithm 8.10; see the box on reusing τs in NRM on the next page). The new time τ_j is obtained by a random variable transformation from relative times as in the first reaction method to absolute times (see also the summation of the current time t to the modification of the unused firing times). Note that no summation of the current time is needed in the initialization phase because the initial time is assumed to be 0. The firing time of the selected reaction is newly generated from an exponential distribution (statements following the second **for** loop).

A further computational improvement is provided by the *optimized direct method* (ODM) (Algorithm 8.11) that reduces the cost of updating the heap data structure of the next reaction method. The ODM indexes the reactions to minimize the steps in the reaction selection (second **for** loop in Algorithm 8.11). The reactions that are executed more frequently are at the beginning of the search list. To estimate the most common reactions, this algorithm performs a short pre-run with some other SSAs (the first three statements in the algorithm). The algorithm also uses the dependency graph to optimize the computation of the total propensity.

The *sorting direct method* (SDM) (Algorithm 8.12) reorders reactions dynamically so that no pre-run is needed. When a reaction is fired its index

Reusing τs in NRM

Let t be the current absolute time, $\mathbf{x}_t$ the state vector at time t, $a_j(\mathbf{x}_t)$ the propensity functions and $T_j(t)$ the internal time of each reaction. The internal time of a reaction is a clock that determines when the reaction would fire if no other reaction fires before the expiration of its internal time. Furthermore, let Δt_j be the amount of absolute time remaining before the firing of reaction R_j if $a_j(\mathbf{x}_t)$ is not changing in $[t, t + \Delta t_j)$, from which the absolute firing time τ_j of R_j is $t + \Delta t_j$.
The internal firing time is $T_j(t) + a_j(\mathbf{x}_t)\Delta t_j$ and the next reaction will fire after $\Delta = \min_j\{\Delta t_j\}$. The new current time will be $t' = t + \Delta$ and the updated internal times for all the reactions will be $T_j(t') = T_j(t) + a_j(\mathbf{x}_t)\Delta$. Let us denote the new propensities as $a'_j(\mathbf{x}_{t'})$.
The internal time of each reaction that has not been selected is still $T_j(t) + a_j(\mathbf{x}_t)\Delta t_j$. We also have the updated internal time deriving from the changed propensity and the absolute time Δ consumed for firing the selected reaction, i.e., $T_j(t') = T_j(t) + a_j(\mathbf{x}_t)\Delta$. The remaining absolute time before firing R_j is obtained by subtracting the consumed internal time from the initial internal time

$$[T_j(t) + a_j(\mathbf{x}_t)\Delta t_j] - [T_j(t) + a_j(\mathbf{x}_t)\Delta] = a_j(\mathbf{x}_t)(\Delta t_j - \Delta).$$

From the new current time and the new propensities we compute the remaining absolute time before the next reaction R_j fires as $a'_j(\mathbf{x}_{t'})\Delta t'_j = a_j(\mathbf{x}_t)(\Delta_{t_j} - \Delta)$ yielding $\Delta_{t'_j} = (a_j(\mathbf{x}_t)/a'_j(\mathbf{x}_{t'}))(\Delta t_j - \Delta)$.
Finally we compute the absolute firing time τ'_j by summing the current time to the remaining absolute time yielding

$$(a_j(\mathbf{x}_t)/a'_j(\mathbf{x}_{t'}))(\Delta t_j - \Delta) + t' = (a_j(\mathbf{x}_t)/a'_j(\mathbf{x}_{t'}))((t + \Delta t_j) - (t + \Delta)) + t' =$$

$$(a_j(\mathbf{x}_t)/a'_j(\mathbf{x}_{t'}))(\tau_j - t') + t'.$$

is exchanged with the next lower index. Along the simulation run, this exchange step tends to establish the right ordering of the reactions.

The *rejection-based SSA* (RSSA) (Algorithm 8.13) reduces the number of propensity computations: in most simulation loops RSSA does not need to compute propensities at all. The exact population $X_i(t)$ of each S_i is abstracted with an interval $[\underline{X_i}, \overline{X_i}]$ (note that we do not highlight time dependence because the interval applies to more than one step). The

Algorithm 8.11: Optimized direct method (ODM)

Assumptions, Input, Output: Same as Algorithm 8.7.

```
begin
    Simulate the system for some t ≪ t_max;
    Associate reactions with the number of their firing occurrences;
    Reorder reactions so that R_i is the i^th most executed reaction;
    t := 0;
    a_0(x) := 0;
    for j ∈ {1, .., M} do
        compute the vector v_j, where its ith element is v'_ij − v_ij;
        compute h_j as in Algorithm 8.7;
        a_j(x) := h_j(x)c_j;
        a_0(x) := a_0(x) + a_j(x);
    end
    generate dependency graph;
    while t < t_max do
        generate r_0 and r_1 in U(0, 1);
        τ := (1/a_0(x)) ln(1/r_0);
        sel := a_0(x)r_1;
        for i ∈ {1..N} do
            sel := sel − a_i(x);
            if sel <= 0 then
                j := i;
                Break
            end
        end
        X(τ_j) := X(t) + v_j;
        t := t + τ;
        for j ∈ {k | R_k depends on R_idx} do
            a_0(x) := a_0(x) − a_j(x);
            compute h_j as in Algorithm 8.7;
            a_j(x) := h_j(x)c_j;
            a_0(x) := a_0(x) + a_j(x)
        end
    end
end
```

propensity of each R_j is abstracted with the interval $[\underline{a_j}, \overline{a_j}]$ encompassing all the values that the actual propensity function can assume over $[\underline{\mathbf{X}}, \overline{\mathbf{X}}]$. Mass action (and Michaelis–Menten) kinetics generate monotonic propensity functions of the state $\mathbf{X}(t)$, hence the extremes of the interval are the minimum and maximum actual propensity values.

Algorithm 8.12: Sorting direct method (SDM)

Assumptions, Input, Output: Same as Algorithm 8.7.

begin
  Same as Algorithm 8.11 without the first three statements (pre-run) and before the end of the **while** loop we add

  if $j \neq 1$ **then**
    swap R_j with R_{j-1} in the ordered list of reactions
  end
end

The selection of the reaction is performed according to the abstract propensities (see Fig. 8.4). Note that the white zones in the figure represent the cases in which the abstract propensities correctly predict the evolution of the actual model. Experiments show that this is the common case. The non-white zones represent the cases in which the actual propensity value has to be computed to test whether the arrow points within the slanted (still accept) zone, or dark (reject) zone.

When a reaction is accepted and fired, the state vector is updated and a new loop is initiated.

The algorithms presented so far need to recompute the propensities of at least those reactions that are dependent on the one selected in the current iteration. Instead RSSA checks whether $\underline{X}_j \leq X_j(t) \leq \overline{X}_j$ still holds for all the species S_j affected by the reaction fired. Quite often, this is the case, because a reaction only affects a few molecules and RSSA does not have to recompute the abstract propensities $\underline{a_j}, \overline{a_j}$. When this is not the case a new interval and its abstract propensities have to be recomputed.

To keep the algorithm exact, the simulation time has to be advanced at every loop of RSSA by a quantity exponentially distributed as $\overline{a_0}$, even when reactions are rejected. If we have $k > 1$ loops with $k - 1$ rejections and 1 acceptance, the simulation time is then advanced by the quantity

$$\frac{1}{\overline{a_0}} ln(1/r_1) + .. + \frac{1}{\overline{a_0}} ln(1/r_k) = \frac{1}{\overline{a_0}} \ln(1/\prod_{i=1}^{k} r_i), r_i \in U(0,1)$$

that is an Erlang distribution with parameters k and $\overline{a_0}$.

Algorithm 8.13: Rejection-based SSA (RSSA)

Assumptions, Input, Output: Same as DM Algorithm 8.7;
Note: Propensities are computed as in Algorithm 8.7.

begin
- $t := 0$;
- **for** $j \in \{1, .., M\}$ **do**
  - compute the vector $\mathbf{v}_j$ where its ith element is $v'_{ij} - v_{ij}$;
- **end**
- **while** $t < t_{max}$ **do**
  - choose an abstract state $[\underline{\mathbf{X}}, \overline{\mathbf{X}}]$ including $\mathbf{X}(t)$;
  - derive the abstract propensities $\underline{a_j}, \overline{a_j}$ for each reaction R_j;
  - compute the sum $\overline{a_0}$ of propensities upper-bounds;
  - **repeat**
    - $u := 1$;
    - $accepted :=$ false;
    - **repeat**
      - generate r_1 and r_2 in $U(0, 1)$;
      - select a reaction R_j according to probability $\overline{a_j}/\overline{a_0}$;
      - **if** $r_1 \leq (\underline{a_j}/\overline{a_j})$ **then**
        - $accepted :=$ true;
      - **else**
        - compute $a_j(\mathbf{x})$ using the concrete state $\mathbf{X}(t)$;
        - **if** $r_1 \leq (a_j(\mathbf{x})/\overline{a_j})$ **then**
          - $accepted :=$ true;
        - **end**
      - **end**
      - $u := u \cdot r_2$;
    - **until** $accepted$;
    - $\tau := \frac{1}{\overline{a_0}} ln(\frac{1}{u})$;
    - $\mathbf{X}(t + \tau) := \mathbf{X}(t) + \mathbf{v}_j$;
    - $t := t + \tau$;
  - **until** $\mathbf{X}(t) \notin [\underline{\mathbf{X}}, \overline{\mathbf{X}}]$ **or** $t \geq t_{max}$;
- **end**

end

8.3.3 *SSA-based reaction-diffusion*

All the algorithms presented so far assume that particle diffusion does not affect the dynamics of the system, i.e., the kinetics of the system is described by reaction-limited rates. Space is becoming more and more important in the simulation of the dynamics of biological systems that can be diffusion-influenced. A simple modification of the stochastic simulation algorithms

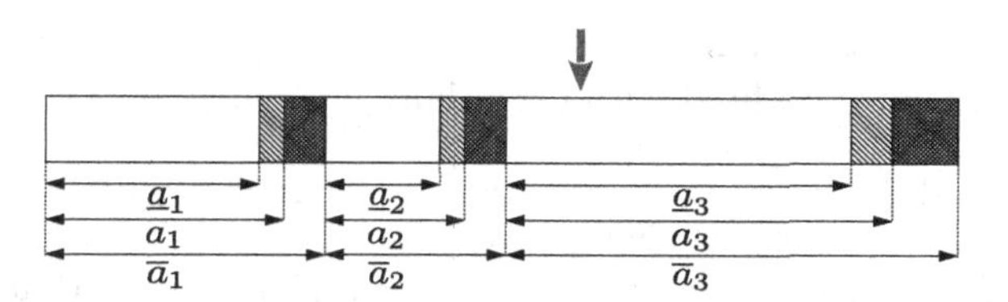

Figure 8.4 RSSA reaction selection. The arrow points to a random uniform number in $[0, \overline{a_0}]$. After a reaction is selected according to the abstract propensities, we check the arrow. If the arrow points within a white zone, we can accept it immediately. Note that this can be checked by exploiting $\underline{a_j}$ alone, and without having to compute the concrete value $a_j(\mathbf{x})$. Otherwise, we do compute $a_j(\mathbf{x})$ to identify the dark rejection zone $\overline{a_j} - a_j(\mathbf{x})$. On a rejection, the selection procedure is repeated.

introduced in the previous subsection is to discretize the volume V in small homogeneous subvolumes $V_1, .., V_s$ in which the actual positions of particles do not affect the kinetics. Diffusion of a particle is a monomolecular reaction (translocation) of the form $A_i \xrightarrow{d} A_j$ that moves a particle A from a subvolume i to some of its neighboring subvolumes j. The rate of translocation reactions is computed from a diffusion coefficient D and the size of the subvolumes h as $d = D/h$, if we consider a two-dimensional volume with squared subvolumes. Three-dimensional volumes with cubical subvolumes originate $d = D/h^2$. The formulas for computing d for volumes partitioned by generic shapes must be multiplied by an *ad hoc* correction factor.

We have considered so far diffusive processes that describe the movement of entities such as proteins within an aqueous environment such as a cell. The speed and direction is temperature dependent and the usual representation is a random walk of the molecules. There are, however, active transport mechanisms in cells that are worth representing (see Fig. 2.10). We can still use the same approach as long as the rate of the corresponding translocation does not depend on the diffusion coefficient, but reflects the underlying biological mechanisms of transmembrane or vesicular trafficking.

8.3.4 *The τ-leaping approximation*

The main computational shortcomings of the exact algorithms presented in the previous subsections are that:

(1) the population of molecules is very large ($h_j(\mathbf{x})$ and $a_j(\mathbf{x})$ are large) and hence reactions fire very frequently;

(2) the reaction rate of reversible reactions is large and many back and forth reactions are fired with essentially no change to $\mathbf{X}(t)$.

The assumption (*leap condition*) of the τ-leaping approximation (Algorithm 8.14) is that no $a_j(\mathbf{x})$ changes its value by a significant amount in the interval $[t, t+\tau)$. A consequence is that the number of times that R_j fires in $[t, t+\tau)$ is a Poisson random variable K_j with mean and variance $a_j(\mathbf{x})\tau$. Under this assumption, given $\mathbf{X}(t) = \mathbf{x}$, the basic τ-leaping formula is

$$\mathbf{X}(t+\tau) \approx \mathbf{x} + \sum_{j=1}^{M} K_j(a_j(\mathbf{x})\tau)\mathbf{v}_j.$$

If $\mathbf{v}$ are the changes of the reactions fired in $[t, t+\tau)$, τ must be such that $|a_j(\mathbf{x}+\mathbf{v}) - a_j(\mathbf{x})| < \epsilon a_0(\mathbf{x})$. The error control parameter satisfies $\epsilon \ll 1$. Note that if the resulting $\tau < tm/a_0(\mathbf{x})$ (with tm being a control parameter) there is no considerable advantage in the approximated algorithm. The solution is then to run some steps of an exact algorithm to move into another configuration that could be more favorable for approximation.

The critical step in the algorithm is the selection of τ so that the leap condition is satisfied. The theoretical foundation of the calculations performed in Algorithm 8.14 are explained in the box on page 326.

The species with few copies in system may become negative during the leap. A new integer parameter n_c of the method distinguishes critical reactions that consume the species with few copies from the others (computation of R^c in the algorithm — the other reactions are non-critical and are denoted by R^{nc}). All the critical reactions are restricted to single firings during the leap to avoid negative populations ($K_j = 1$ and control on the negative components of the state vector in the algorithm).

8.3.5 *Language-based simulation*

The semantics of a stochastic program is a continuous time Markov chain whose states are all the possible configurations the program can reach and the arcs are the transitions that make a program move from one state to the next (see Appendix C). Since the transitions from a state are selected stochastically, they represent the reactions considered in the algorithms.

The major difference is that programs written in languages such as BlenX or ℓ may generate new species and new reactions during their execution. As a consequence the initialization phase of the stochastic algorithm could be re-executed (at least for the species and the reactions created

Algorithm 8.14: Explicit τ-leaping method (eTLM)

Assumptions: Same as Algorithm 8.7 plus leap condition.
Input: Same as Algorithm 8.7 and the error control parameter ϵ, the critical reaction parameter n_c, the τ control parameter tm and s the number of SSA steps. Typical values are $\epsilon = 0.03$, $n_c = tm = 10$, $s = 100$.
Output: Same as Algorithm 8.7.

begin
- $t := 0$; for each reaction compute $\mathbf{v}_j$, where its ith element is $v'_{ij} - v_{ij}$;
- **while** $t < t_{max}$ **do**
  - $R^c := \left\{ j \mid \min_{\{1 \le i \le n; \mathbf{v}_j[i] < 0\}} \left\lfloor \frac{X_i(t)}{|\mathbf{v}_j[i]|} \right\rfloor < n_c \right\}$; $R^{nc} = \{1, .., M\} - R^c$;
  - I_r reactant indexes; compute $h_j(\mathbf{x}), a_j(\mathbf{x})$ as in Algorithm 8.7;
  - $\mu_i := \sum_{j \in R^{nc}} \mathbf{v}_j[i] a_j(\mathbf{x}), i \in I_r$; $\sigma_i^2 := \sum_{j \in R^{nc}} (\mathbf{v}_j[i])^2 a_j(\mathbf{x}), i \in I_r$;
  - g_i defined as in the box and $\epsilon_i := \epsilon / g_i, i \in I_r$;
  - $\tau_1 := \min_{I_r} \left\{ \frac{\max\{\epsilon_i X_i(t), 1\}}{|\mu_i(\mathbf{x})|}, \frac{\max\{\epsilon_i X_i(t), 1\}^2}{\sigma_i^2(\mathbf{x})} \right\}$;
  - done:=false;
  - **while** *not done* **do**
    - **if** $\tau_1 < tm / a_0(\mathbf{x})$ **then**
      - run s SSA steps from $\mathbf{x}$; done:= true
    - **else**
      - $a_0^c(\mathbf{x}) := \sum_{j \in R^c} a_j(\mathbf{x})$;
      - generate a random number r_0 in $U(0, 1)$; $\tau_2 := \frac{1}{a_0^c(\mathbf{x})} ln(\frac{1}{r_0})$;
      - $\tau := \min\{\tau_1, \tau_2\}$;
      - for each $j \in R^{nc}$ generate K_j Poisson with mean $a_j(\mathbf{x})\tau$;
      - **if** $\tau = \tau_1$ **then**
        - for each $j \in R^c, K_j := 0$
      - **else**
        - select $j \in R^c$ according to $a_j(\mathbf{x}) / a_0^c(\mathbf{x})$; $K_j := 1$;
        - $\forall i \neq j \in R^c, K_i := 0$;
      - **end**
      - **if** $\exists$ *a negative component of* $\mathbf{X}(t) + \sum_{j=1}^{M} K_j \mathbf{v}_j$ **then**
        - $\tau_1 := \tau_1 / 2$
      - **else**
        - $\mathbf{X}(t + \tau) := \mathbf{X}(t) + \sum_{j=1}^{M} K_j \mathbf{v}_j$; $t := t + \tau$; done:= true
      - **end**
    - **end**
  - **end**
- **end**

end

Selection of the τ

Let $I_r \subseteq \{1,..,N\}$ be the set of indexes of the reactant species and $\Delta_\tau X_i(t)$ be the change of $X_i(t)$ over $[t, t+\tau)$. Impose $\forall i \in I_r, |\Delta_\tau X_i(t)| \leq \max\{\epsilon_i X_i(t), 1\}$. The parameters ϵ_i are defined to make each propensity change bound by the error parameter ϵ

$$\epsilon_i = \frac{\epsilon}{g_i}, \; g_i = \begin{cases} 1 & hor(i) = 1 \\ 2 & hor(i) = 2 \text{ with a single copy of } S_i \\ 2 + 1/(X_i(t) - 1) & hor(i) = 2 \text{ with two copies of } S_i \end{cases}$$

where $hor(i)$ is the order of the higher-order reaction in which S_i appears as a reactant. Recall that reactions of order greater than 2 have almost 0 probability to occur and we do not consider them. According to the τ-leaping equation, it is

$$\Delta_\tau X_i(t) = \sum_{j \in R^{nc}, i \in I_r} K_j(a_j(\mathbf{x})\tau)\mathbf{v}_j[i]$$

where R^{nc} are the non-critical reactions that cannot generate negative abundances of molecules (see the text).
The Poisson random variable K_j are statistically independent and the mean and variance of the sum are

$$\mu_{\Delta_\tau X_i(t)} = \sum_{j \in R^{nc}, i \in I_r} K_j(a_j(\mathbf{x})\tau)\mathbf{v}_j[i]$$

$$\sigma^2_{\Delta_\tau X_i(t)} = \sum_{j \in R^{nc}, i \in I_r} K_j(a_j(\mathbf{x})\tau)\mathbf{v}_j[i]^2.$$

The bound condition on the propensities is satisfied if it holds for the absolute mean and the standard deviation. Imposing this condition we get

$$\tau \leq \frac{\max\{\epsilon_i X_i(t), 1\}}{|\mu_i(\mathbf{x})|}, \qquad \tau \leq \frac{\max\{\epsilon_i X_i(t), 1\}^2}{\sigma_i^2(\mathbf{x})}.$$

dynamically) during the simulation loop. This is, however, just a computational concern that does not affect the soundness and feasibility of the approach.

8.4 Summary

We introduced the notion of simulation and discussed its main approaches. We reported some classification attributes to introduce the connections between the different solutions. We reported the main algorithms for random number generation and stochastic simulation of discrete-event models, and connected them with language-based modeling.

8.5 Further reading

Extensive material on the general concept of simulation is in [Raczynski (2006); Fishwick (2007a); Schwartz (2008); Sokolowski and Banks (2009)] and on discrete-event simulation in [Zeigler *et al.* (2000)]. Parallel discrete-event simulation is studied in [Fujimoto (2000); Fujimoto *et al.* (2007)]. Stochastic processes, random variables, Monte Carlo simulation and random number generators are studied in [Gillespie (1992a)].

A survey of stochastic simulation algorithms is in [Gillespie (2007)] and references for the variants of the SSA algorithms are [Gillespie (1977); Lok and Brent (2005); Cao *et al.* (2004); McCollum *et al.* (2006)]. A rigorous derivation of the chemical master equation is in [Gillespie (1992b)]. The details of the next reaction method and the proof of the correctness of the new recomputed unused times at each iteration are in [Gibson and Bruck (2000); Anderson (2007a)]. There are some approaches to accelerate stochastic simulation algorithms either by reaction grouping [Mauch and Stalzer (2011); Schulze (2008); Slepoy *et al.* (2008); Blue *et al.* (1995); Thanh and Zunino (2012, 2014); Vo *et al.* (2014)] or reactant grouping [Ramaswamy *et al.* (1995); Indurkhya and Beal (2010)]. Reaction-diffusion solutions are extensively dealt with in [Rodrìguez (2009)]. Approximate algorithms are in [Cao *et al.* (2005, 2006, 2007)]. Some variants of the τ-leaping approach are in [Tian and Burrage (2004); Pettigrew and Resat (2007); Auger *et al.* (2006); Cai and Xu (2007)].

A seminal paper on simulation languages is [Buxton and Laski (1962)]. Details on the connection between language-based models and Monte Carlo simulations are in [Priami (2002); Hillston (1996)]. A discussion of discrete-event formalisms with respect to the stochastic π-calculus is in [Uhrmacher and Priami (2005)] and of beta-binders in [Himmelspach *et al.* (2006); Leye *et al.* (2008)]. Language-based reaction-diffusion simulation is considered in

[Lecca *et al.* (2010a,b)]. A relevant aspect of model simulation is quantitative parameter acquisition. For the technologies presented in this book we refer the interested reader to [Lecca *et al.* (2010c, 2011, 2012); Lecca and Priami (2012)].

Deterministic simulation algorithms mainly used for ODE systems are surveyed in [Jones and Sleeman (2003); Butcher (2008)]. Deterministic and stochastic approaches are mixed in hybrid solutions [Pahle (2009)]. It is worth mentioning that there are also variants of stochastic simulation algorithms that take delays into account [Cai (2007); Bratsun *et al.* (2005); Anderson (2007b); Leier *et al.* (2008); Yi *et al.* (2012); Barbuti *et al.* (2011)].

Chapter 9

Perspectives and conclusions

This book has introduced techniques (both static and dynamic) to produce models of biological systems with the aim of understanding, predicting and controlling their behavior.

There are many approaches (mainly based on statistics and network theory) that operate with very large systems described by qualitative structures and observed through omics technologies (see Sect. 2.6 and Chapter 4). The studies here (i.e., omic studies) tend to identify components of the system (e.g., genes) that exhibit strong correlation with the perturbation and/or phenotype of interest. This is the *top-down* approach to systems biology, in which the modeler observes as much as possible and tries to extract point-wise knowledge.

The complementary approach is *bottom-up*. Starting from very specific knowledge of some mechanisms (e.g., reaction rates), the modeler tries to enlarge the context of the mechanisms by adding new aspects of the system dynamics validated by experimental observations. The systems considered in bottom-up studies are small and very detailed, in terms of quantitative parameters as well.

Moving forward, our proposal is to mix the best of static and dynamic modeling technologies to identify modules of systems that are responsible for phenotypes of interest. A typical working process may entail starting from a complete interaction network of an organism (either built from experimental data or retrieved from public databases) and pruning it according to omic data analysis. The resulting subset of genes identified by the analysis as being correlated with a perturbation and/or phenotype also identifies one or more subnetworks of the complete interaction map. These subnetworks or modules may then serve as a starting point for dynamic modeling (Fig. 9.1).

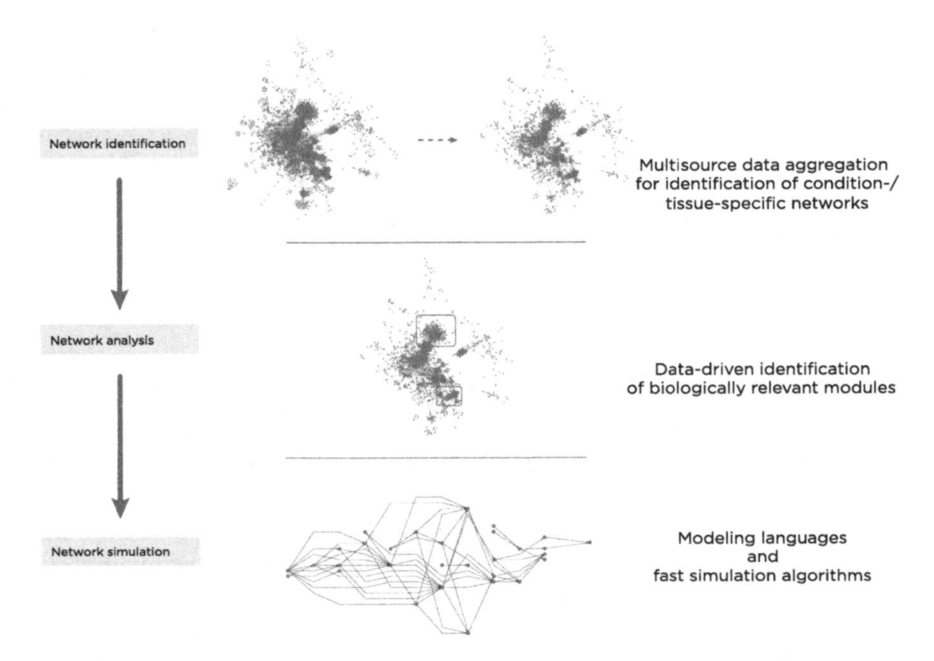

Figure 9.1 From large-scale, qualitative networks to small, quantitative modules. The pruning process of a large network through network analysis methods driven by omics data. The intermediate result of the pruning is a subnetwork that can be used as a starting point for dynamic modeling.

The advantage of our proposal is that the selection of the context of the system relevant for a phenotype is unbiased because it is determined by experimental data (rather than by topological properties of networks or by canonical pathways). A main consequence of this approach is that classical pathways are replaced by subnetworks identified in a data-driven manner (typically, a network module is the intersection of multiple canonical pathways).

As a final remark, although modeling is becoming fundamental in biology, it is worth recalling that no modeling approach is good for everything. Furthermore, we will never be able to prove the correctness of a model we can only possibly prove that it is wrong. In spite of these philosophical and practical limitations, modeling is the only tool we have to better understand complex systems without experimenting directly on their physical realizations. There are systems on which it is highly impractical or even impossible to experiment with, such as ecosystems or the live human brain.

Modeling is a process similar to computer programming or algorithm design (which, according to Knuth [Knuth (2012)], is an art) where the experience of the modeler is fundamental to making the right choices. The complexity of the modeling process and of the applicative domain makes the analysis of biological systems an highly interdisciplinary field in which each problem must be approached by teams of experts with different skills. This is the reason why we included in this book biological (Chapter 2), mathematical (Appendices A and B) and computational background material (Appendix C). The set of technologies described in the book should provide a toolbox for approaching a wide range of problems currently studied in systems biology, and the examples should provide guidelines for solving these problems.

Appendix A

Basic math

A.1 Sets, relations and functions

Intuitively, a *set* is an unordered collection of *elements* or *members*. The expression $x \in A$ means that x is an element of set A. The negation of being a member is written $x \notin A$. We equate two sets if they have the same elements

$$(\forall x \,.\, (x \in A \;\Leftrightarrow\; x \in B)) \;\Leftrightarrow\; A = B.$$

A set A is a *subset* of a set B, if any element of A is also element of B

$$\forall x \,.\, x \in A \;\Rightarrow\; x \in B$$

and we write $A \subseteq B$. To express that $A \subseteq B$, but $A \neq B$, we write $A \subset B$.

To define a set we only need to enumerate all its elements, if possible. The special set which has no element is called *empty set* and is denoted by $\emptyset$. The empty set is unique because we distinguish sets according to their elements, and the empty set has no element. Furthermore, for any set A we have $\emptyset \subseteq A$. A *singleton* is a set with a single element.

The definition by enumeration is not applicable for sets such as $\mathbb{N}$ or $\mathbb{R}$ which have infinite elements. Alternatively, the elements of a set can be identified by giving a property which they all satisfy. Let A be a set and $P(x)$ be a property which makes sense for the elements of A, then

$$\{x \in A \mid P(x)\}$$

is the subset of A whose elements satisfy $P(x)$. Sometimes we write $\{x \mid P(x)\}$ when the domain of P is clear from the context. For the sake of notation we often omit braces in the definition of singletons.

Note that we can define a set through a property P if we already have a set and we specify one of its subsets. Without this assumption, we have *Russel's paradox.* In fact, assume that the expression $\{x \mid x \notin x\}$ is a set s. It would be the set of all sets that do not contain themselves as elements. If $s \in s$ it must be $s \notin s$ because this is the defining property of s and must be satisfied by all its elements. On the other hand, if $s \notin s$ it must be $s \in s$ by definition of s. Hence, we have the paradox. We can conclude that the set of all sets does not exist.

We now report some fundamental operations on sets. The *union* of two sets $A \cup B$ is a set whose elements belong to A or B, $A \cup B = \{x \mid x \in A \vee x \in B\}$. The *intersection* of two sets $A \cap B$ is a set whose elements belong to both A and B, $A \cap B = \{x \mid x \in A \wedge x \in B\}$. If A and B have no element in common, then $A \cap B = \emptyset$. In this case we say that the two sets are *disjoint.* The *difference* of two sets $A - B$ is the set of the elements of A which are not in B, $A - B = \{x \mid x \in A \wedge x \notin B\}$. The *powerset* of a set A is the set 2^A whose elements are all subsets of A, $2^A = \{X \mid X \subseteq A\}$. We write 2^A_f for the powerset of A consisting of the finite subsets of A only.

The *product* of two sets $A \times B$ is the set of the ordered pairs whose first component is an element of A and whose second component is an element of B, $A \times B = \{(x, y) \mid x \in A \;\wedge\; y \in B\}$. Note that we can similarly define the product of n sets. The *disjoint union* of two sets $A_1 \uplus A_2$ is the union of the two sets $\{1\} \times A_1$ and $\{2\} \times A_2$. Their elements are pairs whose first component is an index that uniquely identifies one of the two sets, $A_1 \uplus A_2 = (\{1\} \times A_1) \cup (\{2\} \times A_2)$. Unlike union, the operation above keeps distinct two equal elements coming from distinct sets by duplicating them.

Union and intersection can be extended to cope with possibly infinite families of sets. Let F be a family of sets, then the *big union* of F is the set whose elements belong to some set of F

$$\bigcup_{X \in F} X = \{x \mid \exists X \,.\, X \in F \;\wedge\; x \in X\}.$$

The *big intersection* of F is the set whose elements belong to all sets of F

$$\bigcap_{X \in F} X = \{x \mid \forall X \,.\, X \in F \;\Rightarrow\; x \in X\}.$$

Hereafter, we write $\forall x \in A$ $(\forall x \notin A)$ for $\forall x.\, x \in A$ $(\forall x.\, x \notin A)$. A similar notation is adopted for the existential quantifier.

A set $A \subseteq B$ is *closed* for an operation $\star : B^n \to B$ that applied to n elements of B returns an element of B, if whenever $b_1, \ldots, b_n \in A$ we have $\star(b_1, \ldots, b_n) \in A$.

A *binary relation* between two sets A and B is an element of $2^{A \times B}$, or equivalently a subset of $A \times B$. Given a relation $R \subseteq A \times B$, we often write xRy for $(x, y) \in R$.

A relation $f \subseteq A \times B$ is called *function* or *mapping* (written $f : A \to B$) if for any $x \in A$ there is *at most a single* $y \in B$ such that xfy. In this case we write $f(x) = y$ and we say that $f(x)$ is *defined.* If there is no $y \in B$ such that $f(x) = y$, $f(x)$ is *undefined.* If for any $x \in A$ there exists $y \in B$ such that $f(x) = y$, f is *total*, otherwise it is *partial.* If we want to stress that function f is partial, we write $f : A \rightharpoonup B$. Set A is the *domain* of the function, and set B is its *codomain.* Given $X \subseteq A$, we write $f(X)$ for the set whose elements are related to some elements of X through f

$$f(X) = \{y \mid y \in B \ \wedge \ (\exists x \in X \, . \, f(x) = y)\}$$

or simply $f(X) = \{f(x) \mid x \in X\}$. A function $f : A \to B$ is *surjective* if $f(A) = B$. Function f is *injective* if

$$\forall x_1 \in A, \forall x_2 \in A \, . \, f(x_1) = f(x_2) \Rightarrow x_1 = x_2.$$

Function f is *bijective* if it is both surjective and injective.

The basic operation between relations, and thus between functions, is *composition.* The composition of two relations $R \subseteq A \times B$ and $S \subseteq B \times C$ is a relation between A and C

$$S \circ R = \{(x, z) \in A \times C \mid \exists y \in B \, . \, (x, y) \in R \ \wedge \ (y, z) \in S\}.$$

The composition of two functions $f : A \to B$ and $g : B \to C$ is then function $g \circ f : A \to C$ defined as $g \circ f = \{(x, g(f(x))) \mid x \in A\}$.

Any set A is equipped with a special function I_A called *identity* on A defined as $I_A = \{(x, x) \mid x \in A\}$. Function $f : A \to B$ has an *inverse* $g : B \to A$ iff $\forall x \in A \, . \, g(f(x)) = x \ \wedge \ \forall y \in B \, . \, f(g(y)) = y$. In this case sets A and B are in 1-1 *correspondence.* Sets which are in 1-1 correspondence with $\mathbb{N}$ are said to be *countable.*

An *equivalence relation* on a set A is a reflexive, symmetric and transitive relation, i.e., $\forall x \in A, \forall y \in A, \forall z \in A$

$$xRx \ \wedge \ xRy \Rightarrow yRx \ \wedge \ (xRy \wedge yRz) \Rightarrow xRz.$$

The set of all elements equivalent to x is denoted $[x]$ and is called the *equivalence class* of x. More formally, $[x] = \{y \mid y \in A \wedge xRy\}$. Two equivalence classes are disjoint or coincide. Thus, the set F of all equivalence classes is a subset of 2^A such that

- $\forall S \in F \,.\, S \neq \emptyset$
- $\forall S \in F, \forall T \in F \,.\, S \neq T \Rightarrow S \cap T = \emptyset$
- $\bigcup_{S \in F} S = A$.

A family F of subsets of A with the properties above is said to be a *partition* of A. A family with only the last property is said to be a *covering* of A. An equivalence relation R on A singles out a partition of A that is called *quotient* of A with respect to R and it is written $A_{/R}$.

A binary relation on a set A is an *ordering* if it is reflexive, transitive and antisymmetric relation. If the antisymmetric property does not hold, the relation is a *preordering*.

We end this section by introducing *multisets*. A multiset over a set A is a total function $f : A \to \mathbb{N}$, associating natural numbers, possibly 0, with each element of A. We denote finite multisets by listing their elements between the symbols $\{|$ and $|\}$. Note that we list an element $a \in A$ as many times as $f(a)$. The operations on multisets are the same of those on sets, except that replication of elements is allowed.

A.2 Logics

We start with the definition of an arbitrary *first-order language* L. The symbols of L are:

- a countable infinite sequence $v_1, \ldots, v_n, \ldots$ of *variables*;
- for any $n \in \mathbb{N}$ a set of *n-ary function symbols* (the 0-ary function symbols, if any, are called *constants*);
- for any positive $n \in \mathbb{N}$ a set of *n-ary predicate symbols* (for at least one n this set must be non empty);
- the connectives *negation* $\neg$ and *implication* $\Rightarrow$;
- the *universal quantifier* $\forall$;
- the parentheses (and).

Variables, connectives and the universal quantifier are called *logical* symbols. Note that variables are ordered according to an *alphabetic* ordering.

A *string* of L is a finite (possibly empty) sequence of symbols of L. Hereafter, we are only interested in particular kind of strings: *terms* and *formulae.* Terms are strings originated according to the rules:

- any string made up of a single occurrence of a variable is a term;
- if f is an n-ary function symbol and $t_1, \ldots, t_n$ are terms, then $f(t_1, \ldots, t_n)$ is a term and $t_1, \ldots, t_n$ are its *arguments.*

Formulae are originated according to the rules:

- if P is an n-ary predicate symbol and $t_1, \ldots, t_n$ are terms, then $P(t_1, \ldots, t_n)$ is a formula;
- if A is a formula, then so is $\neg A$;
- if A and B are formulae, then so is $A \Rightarrow B$;
- if A is a formula and v is a variable, then $\forall v.\, A$ is a formula.

The definitions above are based on the economy of symbols. In practice other connectives (*conjunction* $\wedge$, *disjunction* $\vee$ and *bi-implication* $\Leftrightarrow$) and another quantifier (*existential* $\exists$) are used quite often. Let A and B be formulae and v be a variable. The new logical symbols are defined as:

$$A \wedge B \;=\; \neg(A \Rightarrow \neg B) \qquad\qquad A \vee B \;=\; \neg A \Rightarrow B$$

$$A \Leftrightarrow B \;=\; ((A \Rightarrow B) \wedge (B \Rightarrow A)) \qquad \exists v.\, A \;=\; \neg\forall v.\, \neg A.$$

To use parentheses with parsimony, we assume that $\Rightarrow$ has precedence over $\wedge$ and $\vee$.

We need the notion of *interpretation* I to specify the intended meaning of formulae. An interpretation is a structure consisting of:

- a non-empty class U called *universe* whose elements are called *individuals*;
- a mapping that assigns to each function symbol f of L an operation f^i on U with the same arity;
- a mapping that assigns to each predicate symbol P of L a relation P^i on U with the same arity.

We then define a *valuation* σ as an interpretation I together with an assignment of value $v^\sigma \in U$ to each variable v. We extend this notation to functions and predicates by writing f^σ and P^σ to denote f^i and P^i once variables have been instantiated. We write $\sigma(v/u)$ to mean the valuation that coincides with σ on every variable other than v, while $v^{\sigma(v/u)} = u$. Finally, we also write t^σ to indicate the value of t under σ. The meaning of terms and formulae according to a valuation σ with universe U is as follows:

$$(f(t_1, \ldots, t_n))^\sigma = f^\sigma(t_1^\sigma, \ldots, t_n^\sigma)$$

$$(P(t_1, \ldots, t_n))^\sigma = \begin{cases} \text{true if } \langle t_1^\sigma, \ldots, t_n^\sigma \rangle \in P^\sigma \\ \text{false otherwise} \end{cases}$$

$$(\neg A)^\sigma = \begin{cases} \text{true if } A^\sigma = \text{false} \\ \text{false otherwise} \end{cases}$$

$$(A \Rightarrow B)^\sigma = \begin{cases} \text{true if } A^\sigma = \text{false or } B^\sigma = \text{true} \\ \text{false otherwise} \end{cases}$$

$$(\forall v.\, A)^\sigma = \begin{cases} \text{true if } A^{\sigma(v/u)} = \text{true for every } u \in U \\ \text{false otherwise.} \end{cases}$$

A valuation σ *satisfies* a formula A (a set of formulae S), written $\sigma \models A$ ($\sigma \models S$) if and only if the truth valuation induced by σ maps A (any A in S) to true. In symbols we have

$$\sigma \models A \text{ if and only if } A^\sigma = \text{ true}$$

$$\sigma \models S \text{ if and only if for any } A \text{ in } S,\ A^\sigma = \text{ true.}$$

If every valuation that satisfies a set of formulae S also satisfies a formula A, we say that A is a *logical consequence* of S and we still write $S \models A$. If A is satisfied by every valuation, we say that it is *valid*. Finally, A is *satisfiable* if there is a valuation σ such that $\sigma \models A$.

A formula whose predicate symbol is $=$ is called an *equation* and its first and second arguments are called *left-hand side* and *right-hand side*. In particular, we have

$$(t = s)^\sigma = \begin{cases} \text{true if } t^\sigma = s^\sigma \\ \text{false otherwise.} \end{cases}$$

We now designate certain formulae as *axioms*. These will be used to build deductions starting from a set of formulae (the hypotheses). Some notation could help. A *generalization* of a formula A is any formula of the form $\forall v_1 \ldots \forall v_k\, A$, where $k \geq 1$ and $v_1, \ldots, v_k$ are any variables, not necessarily distinct. An occurrence of a variable v in a formula A is *bound* if it is within a subformula of A having the form $\forall v.\, B$. All other occurrences of v in A are *free*. We say that the variable v is free in A if v has at least one free occurrence in A. Given a term t, a formula A and a variable v, we define $A\{t/v\}$ as the formula obtained from A when all free occurrences of

v in A are replaced by occurrences of t. We say that t is *free for v in A* if no free occurrence of v is within a subformula of A having the form $\forall v'.\, B$, where v' occurs in t. Finally, the axiom scheme of first-order logic is the following:

(1) $A \Rightarrow B \Rightarrow A$;
(2) $(A \Rightarrow B \Rightarrow C) \Rightarrow (A \Rightarrow B) \Rightarrow A \Rightarrow C$;
(3) $(\neg A \Rightarrow B) \Rightarrow (\neg A \Rightarrow \neg B) \Rightarrow A$;
(4) $\forall v.\, (A \Rightarrow B) \Rightarrow \forall v.\, A \Rightarrow \forall v.\, B$,
where A and B are formulae and v is any variable;
(5) $A \Rightarrow \forall v.\, A$,
where A is a formula and v is a variable not free in A;
(6) $\forall v.\, A \Rightarrow A\{t/v\}$,
where A is any formula and t is any term free for v in A;
(7) $t = t$,
where t is any term;
(8) $t_1 = t_{n+1} \Rightarrow \ldots t_n = t_{2n} \Rightarrow f(t_1, \ldots, t_n) \Rightarrow f(t_{n+1}, \ldots, t_{2n})$,
where f is any function symbol and $t_1, \ldots, t_{2n}$ are any terms;
(9) $t_1 = t_{n+1} \Rightarrow \ldots t_n = t_{2n} \Rightarrow P(t_1, \ldots, t_n) \Rightarrow P(t_{n+1}, \ldots, t_{2n})$,
where P is any predicate symbol and $t_1, \ldots, t_{2n}$ are any terms;
(10) all generalizations of axioms of the preceding groups.

Let S be a set of formulae. A *deduction from* S is a finite, non-empty sequence of formulae $A_1, \ldots, A_n$ such that, for all k $(1 \leq k \leq n)$, either

- A_k is an axiom, or
- $A_k \in S$, or
- $\exists i, j < k \,.\, A_j \equiv A_i \Rightarrow A_k$.

The set S is called a set of *hypotheses.* A *proof* is a deduction from the empty set of hypotheses. We use $\vdash$ to denote *deducibility.* More precisely, we write $S \vdash A$ to assert that A is deducible from S. If S is empty, we say that A is *provable* (or it is a *theorem*) and we write $\vdash A$. Recall that

$$S \vdash A \;\Rightarrow\; S \models A.$$

In the above description, we have assumed as a unique *inference rule* the *modus ponens*, i.e. the operation of passing from two formulas A and $A \Rightarrow B$ to the formula B. In other words, we have

$$\{A, A \Rightarrow B\} \models B.$$

An inference rule has a set of *premises* and a *conclusion*. Notation

$$\frac{P_1 \dots P_n}{c}$$

means that from the premises $P_1, \dots, P_n$ we can infer the conclusion c. Note that axioms are inference rules with empty premises. This allows us to arrange proofs in a tree-structure called a *deduction tree*. The root of the tree is the theorem and its leaves are the hypotheses. Any internal node is a premise of a rule. If we view axioms as inference rules we can say that a proof is a finite, non-empty sequence of formulae in which any formula is obtained from the preceding ones by applying any inference rule. Sometimes, we need a larger set E of inference rules. In these cases, we annotate it in the deduction symbol, i.e. we write $S \vdash_E A$.

A.3 Algebra

An algebra is a mathematical structure which is widely used in computer science to model programming features.

A *signature* Σ is a set of *function symbols*. In other words, Σ is the set of symbols of a first-order language (see Sect. A.2) made up of function symbols only. The arity of a signature is a mapping $ar : \Sigma \to \mathbb{N}$ that associates each symbol with its arity. We write Σ_n to denote the set of function symbols in Σ with arity n.

Given a signature Σ, a Σ-*algebra* is a pair $\langle A, \Sigma_A \rangle$ where A is a set called *carrier* and Σ_A is a set of functions $\{f_A : A^n \to A \,|\, f \in \Sigma \wedge ar(f) = n\}$. Essentially a Σ-algebra is an interpretation of a signature Σ. A signature may have many different interpretation even over the same carrier.

A special interpretation for a signature Σ is its *term* or *free algebra*. This algebra is a purely syntactic object: the carrier is the set of strings (*terms*) built with the symbols in Σ and the functions only syntactically manipulate them. More formally, the set T_Σ of terms over Σ is the least set of strings that satisfy

- $f \in \Sigma \wedge ar(f) = 0 \Rightarrow f \in T_\Sigma$, and
- $f \in \Sigma \wedge ar(f) = k > 0 \wedge t_1, \dots, t_k \in T_\Sigma \Rightarrow f(t_1, \dots, t_k) \in T_\Sigma$.

Note that if no constant is in the signature, the set of terms is empty. The functions of the term algebra construct new terms. For any $f \in \Sigma$ with $ar(f) = k$, we let $f_{T_\Sigma} : T_\Sigma^k \to T_\Sigma$ be the function which maps tuples of terms $\langle t_1, \ldots, t_k \rangle$ into the term $f(t_1, \ldots, t_k)$. By abuse of notation we will denote the term algebra as T_Σ.

We now introduce the concept of *structural induction* which we frequently use throughout this work. The carrier of T_Σ is defined inductively. It is the least set of strings that contains constants and is closed under the operations f_Σ. The method of structural induction says that to prove a property P of all terms in T_Σ, we only need to prove that P holds of

- all constant symbols in Σ, and
- the term $f(t_1, \ldots, t_k)$ for every $f \in \Sigma$ with $ar(f) = k > 0$, assuming that the property holds of the terms $t_1, \ldots, t_k$.

Structural induction may be also used to define a function g on T_Σ:

- define g over constants, and
- define g over $f(t_1, \ldots, t_k)$ in terms of $g(t_1), \ldots, g(t_k)$, for every $f \in \Sigma$ with $ar(f) = k > 0$.

In logical notation the principle of structural induction can be expressed as

$$\begin{aligned}
&\forall f \in \Sigma \,.\, [ar(f) = 0 \Rightarrow P(f) \wedge ar(f) = k \Rightarrow \\
&(\forall t_1, \ldots, t_n \in T_\Sigma \,.\, P(t_1) \wedge \ldots \wedge P(t_n) \Rightarrow P(f(t_1, \ldots, t_n))] \\
&\Rightarrow \forall t \in T_\Sigma \,.\, P(t)
\end{aligned}$$

where P is the property investigated.

Some terminology: the first conjunct before the last implication is the *basis* of the induction, the second conjunct is the *induction step*, and the P-part of the left-hand side of the induction step is the *induction hypothesis*.

It is possible to define functions (called Σ-*homomorphisms*) between carriers of Σ-algebras which preserve the structure of their domain. Let $\langle A, \Sigma_A \rangle$ and $\langle B, \Sigma_B \rangle$ be two Σ-algebras, and $h : A \to B$ a function. Then h is a Σ-*homomorphism* if

$$\forall f \in \Sigma,\ ar(f) = k,\ h(f_A(a_1, \ldots, a_k)) = f_B(h(a_1), \ldots, h(a_k)).$$

The *fundamental property* of term algebras says that for every Σ-algebra $\langle A, \Sigma_A \rangle$, there exists a *unique* Σ-homomorphism $i_A : T_\Sigma \to A$. Let $\mathcal{C}$ be a class of Σ-algebras. A Σ-algebra I is *initial* in $\mathcal{C}$ if for every $J \in \mathcal{C}$ there is a unique Σ-homomorphism from I to J. Thus, we can say that T_Σ is initial in the class of all Σ-algebras.

A Σ-*congruence* is an equivalence relation on Σ-algebras which preserves the structure induced by Σ. Given a Σ-algebra $\langle A, \Sigma_A \rangle$, a relation C over A is a Σ-congruence if it is an equivalence relation and

$$\forall f \in \Sigma,\ ar(f) = k, \forall i \,.\, 0 \le i \le k,\ \langle a_i, a_i' \rangle \in C \ \Rightarrow$$
$$\langle f_A(a_1, \ldots, a_k), f_A(a_1', \ldots, a_k') \rangle \in C.$$

Let $[a]_C = \{a' \mid \langle a, a' \rangle \in C\}$ be the equivalence class of a induced by C. Then, the set of equivalence classes is $A_{/C} = \{[a]_C \mid a \in A\}$.

A Σ-algebra A *satisfies* a Σ-congruence C if $i_A(t) = i_A(t')$ whenever $\langle t, t' \rangle \in C$. We denote with $\mathcal{C}(C)$ the class of all Σ-algebras that satisfy C. The extension of the fundamental property of term algebras to congruences says that the Σ-algebra $T_{\Sigma/C}$ is initial in the class $\mathcal{C}(C)$.

Particular classes of Σ-algebras can be defined through equations. An equation is determined by two terms possibly with variables. The valuation of these terms is modulo an assignment of values to variables.

Let V be a set of variables ranged over by $v, v_1, v_i, \ldots$. We extend a signature Σ to $\Sigma(V)$ to include variables. For any $f \in \Sigma$ we also have $f \in \Sigma(V)$. Furthermore, any $v \in V$ is a function symbol of arity 0 in $\Sigma(V)$. The term algebra of $\Sigma(V)$ is denoted by $T_\Sigma(V)$ and its *closed* or *ground* terms are those which contain no variables.

Given a Σ-algebra A, an A-*assignment* is a mapping $\rho_A : V \to A$ that associates to every variable in V an element in A. The *fundamental property* of $T_\Sigma(V)$ is that there is a unique Σ-homomorphism $h_A : T_\Sigma(V) \to A$ such that $\forall v \in V,\ h_A(v) = \rho_A(v)$. By abuse of notation, we denote the above Σ-homomorphism as ρ_A. For $t, t' \in T_\Sigma(V)$, we let $t =_A t'$ if for every A-assignment ρ_A, $\rho_A(t) = \rho_A(t')$. We write $t\rho$ for the application of the $T_\Sigma(V)$-assignment ρ to term t.

A relation R over $T_\Sigma(V)$ satisfies a set of equations E if $E \subseteq R$. A Σ-algebra A satisfies a set of equations E if $E \subseteq =_A$. Let $\mathcal{C}(\mathcal{E})$ be the class of Σ-algebras which satisfy equations E. Then, $\mathcal{C}(\mathcal{E})$ has an initial Σ-algebra. Let $=_E$ be a relation on $T_\Sigma(V)$ closed under substitutions defined as

$$t =_E t' \ \Leftrightarrow \ \vdash_E t = t'.$$

It turns out that $=_E$ is a Σ-congruence. We also have that $T_\Sigma(V)_{/=_E}$ is initial in $\mathcal{C}(\mathcal{E})$. Thus, $T_\Sigma(V)_{/=_E}$ is a particular representation (up to Σ-isomorphism) of the unique initial Σ-algebra of $\mathcal{C}(\mathcal{E})$.

We end this section with the definition of some useful algebras. We start with labeled trees.

Let RT be the set of finite rooted labeled trees and let A be the set of labels. Then, RT can be viewed as a Σ-algebra whose signature is

$$\Sigma = \{\mathbf{0}\} \cup \{a\cdot \mid a \in A\} \cup \{+\}.$$

We then interpret

$$Ref: t = t \qquad Sym: \frac{t = t'}{t' = t}$$

$$Tra: \frac{t = t',\ t' = t''}{t = t''} \qquad Ins: \frac{t = t'}{t\rho = t'\rho}$$

$$Sub: \frac{t_1 = t'_1, \ldots, t_k = t'_k}{f(t_1, \ldots, t_k) = f(t'_1, \ldots, t'_k)} \qquad Eq: \frac{\langle t, t' \rangle \in E}{t = t'}$$

- $\mathbf{0}$ as the empty tree that acts as neutral element for $+$,
- $a\cdot$ as prefixing an arc labeled a to a tree, and
- $+$ as (associative, commutative and idempotent) sum of trees that glues the roots of its operands.

The above algebra allows us to write $T = \sum_{i\in I} a_i \cdot T_i$ for a tree with edges a_i exiting from its root and subtrees T_i. With this notation we assume that $\mathbf{0} = \sum_{i\in\emptyset} a_i \cdot T_i$.

Appendix B

Probability and statistics

B.1 Probability

We rely on the *frequency interpretation* of probability that assumes the possibility of an experiment being repeated under identical conditions to yield a series of outcomes within the sample space of the experiment $\{o_1, o_2, ...\}$ on different trials. We write $\#_n(o)$ to denote the number of observations of o out of n trials of an experiment. We then define the *probability* of o with respect to an experiment as

$$p(o) \equiv \lim_{n \to \infty} \frac{\#_n(o)}{n}.$$

Logical combinations of two or more outcomes are called *events*. For instance, $ev = \ not\ o$ is the event interpreted as the nonobservation of o as outcome of an experiment, $ev = o_1 \wedge o_2$ as the simultaneous observation of o_1 and o_2, $ev = o_1 \vee o_2$ as the observation of either o_1 or o_2. The cardinality of an event is the number of outcomes used to define it. Probability can be extended to events by replacing the outcome o in the definition of probability with the logical characterization of an event.

The *conditional probability* describes the observation of an outcome o_j given the observation of o_i

$$p(o_j|o_i) \equiv \lim_{n \to \infty} \frac{\#_n(o_j \wedge o_i)}{\#_n(o_i)}.$$

Note that if o_i never occurs, $\#_n(o_i) = 0$ and $p(o_j|o_i)$ is undefined.

It is possible to derive three *laws of probability*:

(1) (*Range*) The probability of an outcome o is a real number such that

$$0 \leq p(o) \leq 1$$

with $p(o) = 0$ meaning that o never occurs and $p(o) = 1$ meaning that o always occurs.

(2) (*Addition*) For m mutually exclusive outcomes, it is

$$p(o_1 \vee .. \vee o_m) = p(o_1) + .. + p(o_m).$$

(3) (*Multiplication*) For any two outcomes o_i and o_j it is

$$p(o_i \wedge o_j) = p(o_i)p(o_j|o_i) = p(o_j)p(o_i|o_j).$$

A set S of N outcomes of an experiment is *statistically independent* if and only if for any outcome $o \in S$, $p(o) = p(o|ev)$ for all the events ev obtained as $\wedge$-combination of outcomes in $S/\{o\}$ with cardinalities in $1..|S/\{o\}|$. We then have

(4) (*Independence*) The outcomes $o_1, .., o_N$ are statistically independent iff

$$p(o_1, .., o_N) = p(o_1)p(o_2)..p(o_N).$$

B.2 Random variables

A variable can be considered as an entity with a value that can be measured in a given sampling context. Intuitively, if the sampling context does not determine a unique value, the variable is called random. Formally, X is a (real) *random variable* iff $\exists f : \mathbb{R} \rightarrow \mathbb{R}$ such that

$$f(x)dx = Prob\{X \in [x, x + dx)\}.$$

f is the *density function* of X and completely defines the random variable X. From the previous equation it follows that

$$Prob\{X \in [a, b)\} = \int_a^b f(x)dx.$$

Since dx is positive and $f(x)dx$ is a probability, it must be

$$\forall x, f(x) \geq 0 \wedge \int_{-\infty}^{+\infty} f(x)dx = 1. \tag{B.1}$$

The integral in (B.1) is called the *normalization condition*. All the functions satisfying (B.1) are density functions defining some random variable X. The intuition of f is that a normalized histogram of the sample values of X approaches the curve $f(x)$ when the samplings approaches ∞.

The *distribution function* F of a random variable X is defined as $F(x) \equiv Prob\{X < x\}$. The relationship with the density function is given by

$$F(x) = \int_{-\infty}^{x} f(z)dz \quad \text{and} \quad f(x) = F'(x)$$

where $F'(x)$ is the derivative of F with respect to x. A random variable X can also be completely determined by its distribution function (we write $X \sim F$ for X has distribution F). Note also that F is a monotone increasing function on $\mathbb{R}$ assuming values in $[0.1]$ (each function with this property is a distribution function of some random variable X).

The nth *moment* of a random variable X is

$$\mu_X^n = \int_{-\infty}^{\infty} x^n f(x)dx.$$

The zeroth moment always exists and it is $\mu_X^0 = 1$. Higher moments may not exist. Three important notions are the *mean*, the *variance* and the *standard deviation* of X defined as

$$\mu_X = \int_{-\infty}^{\infty} xf(x)dx, \quad \sigma_X^2 = \mu_X^2 - (\mu_x)^2 \quad \text{and} \quad \sigma_X = \sqrt{\sigma_X^2}.$$

It is $\mu_X^2 \geq (\mu_x)^2$ with equality obtained when $\sigma_X^2 = 0$ (i.e., the randomness of X vanishes). Intuitively, the mean is the best approximation value of X and the standard deviation σ_X measures how adequate the approximation is by measuring the difference between a sample X and its mean. Usually a sample of X will get a value in $[\mu_X - \sigma_X..\mu_X + \sigma_X]$.

Other two important parameters are the *skewness* (that measures the asymmetry of a distribution with respect to its mean) and the *kurtosis* or *excess* (that measures the tails of a distribution). The skewness is denoted by γ_1 and its square by β_1, while the kurtosis is denoted by β_2. The general definition is

$$\gamma_1 = E[(x - \mu)^3]/\sigma^3 \quad \text{and} \quad \beta_2 = E[(x - \mu)^4]/\sigma^4$$

where $E[-]$ is the expected value.

B.2.1 *Useful random variables*

We introduce some important random variables starting from the *uniform random variable* (see Fig. B.1) that is uniformly distributed over the interval

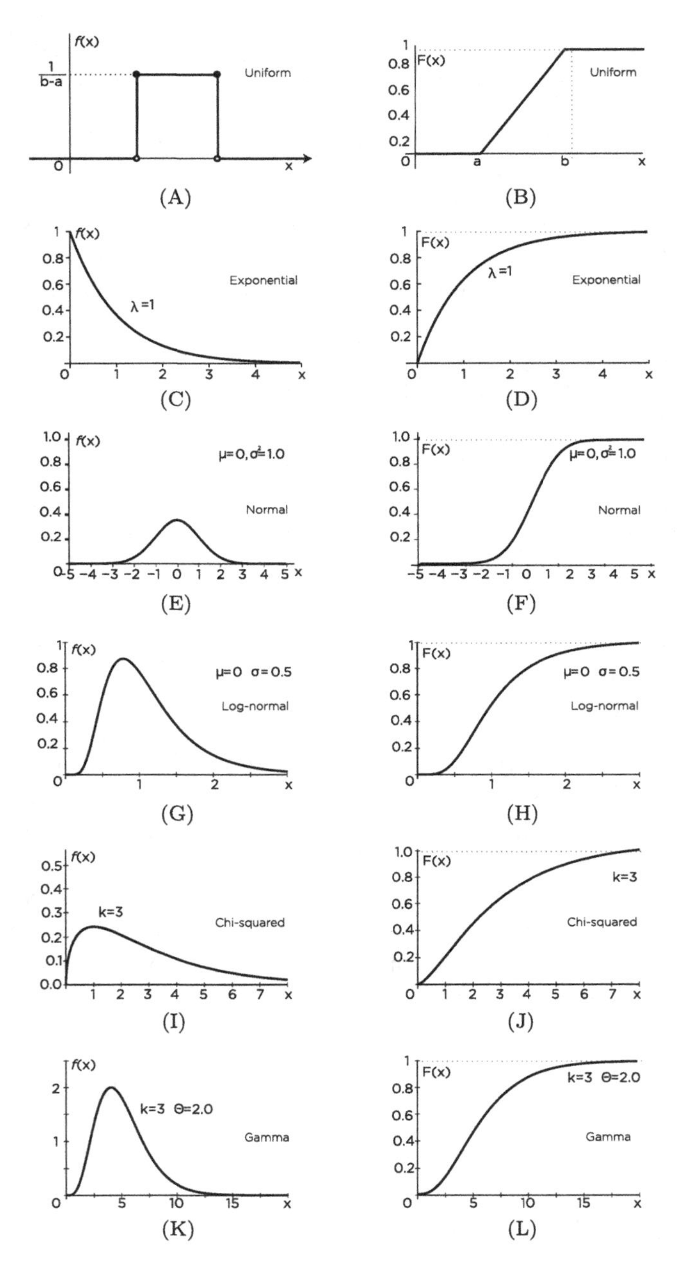

Figure B.1 Density (left column) and the corresponding distribution functions (right column) of important random variables.

$[a, b)$ (usually denoted by $U(a, b)$ or simply by U when the reference interval is $[0, 1)$) and is defined by the density function

$$f(x) = \begin{cases} 1/(b-a) & \text{if } a \leq x < b \\ 0 & \text{otherwise} \end{cases}$$

with $-\infty < a < b < \infty$. It is $\mu_X^n = (1/(n+1))\Sigma_{j=0}^n a^j b^{n-j}$ from which we derive

$$\mu_X = \frac{a+b}{2}, \quad \sigma_X^2 = \frac{(b-a)^2}{12} \quad \text{and} \quad \sigma_X = \frac{b-a}{2\sqrt{3}}.$$

Note that when b approaches a, the mean of X approaches a as well and the variance approaches 0. The skewness of X is 0 and its kurtosis is $-6/5$. The distribution function of X is

$$F(x) = \begin{cases} 0 & \text{if } x < a \\ (x-a)/(b-a) & \text{if } a \leq x < b \\ 1 & \text{if } x \geq b \end{cases}.$$

The *exponential random variable* (see Fig. B.1) that is exponentially distributed with decay constant λ (usually denoted by $Exp(\lambda)$) is defined by the density function

$$f(x) = \begin{cases} \lambda e^{-\lambda x} & \text{if } x \geq 0 \\ 0 & \text{otherwise} \end{cases}$$

with $\lambda \in \mathbb{R}^+$. It is $\mu_X^n = n!/\lambda^n$ from which we derive

$$\mu_X = 1/\lambda, \quad \sigma_X^2 = 1/\lambda^2 \quad \text{and} \quad \sigma_X = 1/\lambda.$$

Note that when λ approaches ∞, the mean and the variance of X approach 0. The skewness of X is 2 and its kurtosis is 6. The distribution function of X is

$$F(x) = \begin{cases} 1 - e^{-\lambda x} & \text{if } x \geq 0 \\ 0 & \text{otherwise} \end{cases}.$$

The *normal random variable* (see Fig. B.1) that is normally distributed with mean μ and variance σ^2 (usually denoted by $N(b, a^2)$) is defined by the density function

$$f(x) = \frac{1}{\sqrt{2\pi a^2}} e^{-\frac{(x-b)^2}{2\sigma^2}}$$

with $0 < a < \infty$ and $-\infty < b < \infty$. It is

$$\mu_X^n = n! \Sigma_{k=0}^n \frac{b^{n-k}(a^2)^{k/2}}{(n-k)!(k/2)!2^{k/2}}, \quad k \text{ even}$$

from which we derive $\mu_X = b$, $\sigma_X^2 = a^2$ and $\sigma_X = a$. The skewness of X is 0 and its kurtosis is 0. There is no simple formula for $F(x)$, so it is usually tabulated.

A *log-normal random variable* X with parameters m and s (see Fig. B.1) is a random variable whose logarithm is normally distributed. The parameters of the log-normal distribution are the mean and standard deviation of the natural logarithm of X. For instance, $X = e^{m+sZ}$ where Z is a normally distributed random variable $N(0,1)$. The density function is

$$f(x) = \frac{1}{xs\sqrt{2\pi}} e^{-\frac{(\ln x - m)^2}{2s^2}}, \ x > 0.$$

It is $\mu_X^n = e^{nm+n^2s^2/2}$ from which we derive $\mu_X = e^{m+s^2/2}, \sigma_X^2 = (e^{s^2} - 1)e^{2m+s^2}$. The skewness of X is $(e^{\sigma^2}+2)\sqrt{e^{\sigma^2}-1}$ and its kurtosis is $e^{4\sigma^2} + 2e^{3\sigma^2} + 3e^{2\sigma^2} - 6$.

The *chi-squared random variable* (see Fig. B.1) with v degrees of freedom (usually denoted by $\chi^2(v)$) is defined by the density function

$$f(x) = \begin{cases} \frac{1}{2^{v/2}\Gamma(v/2)} x^{(v/2)-1} e^{-(x/2)} & \text{if } x > 0 \\ 0 & \text{otherwise} \end{cases}$$

where the Γ function is

$$\Gamma(r) = \int_0^{\infty} u^{r-1} e^{-u} du, \ r > 0$$

and its upper and lower incomplete versions are

$$\Gamma(r,x) = \int_x^{\infty} u^{r-1} e^{-u} du \quad \text{and} \quad \gamma(r,x) = \int_0^{x} u^{r-1} e^{-u} du.$$

Note that if $r \in \mathbb{Z}^+$, $\Gamma(r) = (r-1)!$ and that $\Gamma(r,x) + \gamma(x,r) = \Gamma(r)$. It is

$$\mu_X^n = 2^m \frac{\Gamma(m + (v/2))}{\Gamma(v/2}$$

from which we derive $\mu_X = v$, $\sigma_X^2 = 2v$ and $\sigma_X = \sqrt{2v}$. The skewness of X is $\sqrt{8/v}$ and its kurtosis is $12/v$. There is no simple formula for $F(x)$, so it is usually tabulated.

The *gamma random variable* (see Fig. B.1) with shape parameter k and scale parameter θ, $k, \theta \in \mathbb{R}^+$ (usually denoted by $Gamma(k, \theta)$) is defined by the density function

$$f(x) = \frac{x^{k-1} e^{-x/\theta}}{\theta^k \Gamma(k)}, \ x > 0, \ k > 0 \text{ and } \theta > 0.$$

It is $\mu_X = k\theta$, $\sigma_X^2 = k\theta^2$ and $\sigma_X = \sqrt{k}\theta$. The skewness of X is $2/\sqrt{k}$ and its kurtosis is $6/k$. The distribution function of X is

$$F(x) = \frac{\gamma(k, x/\theta)}{\Gamma(k)}.$$

B.2.2 *Joint random variables*

The random variables $X_1, .., X_n$ are jointly distributed if the n random variables can be sampled simultaneously and there exists an n-variate function f (called the *joint density function*) such that

$$f(x_1, .., x_n)dx_1..dx_n = Prob\{X_i \in [x_i, x_i + dx_i) \,|\, i = 1, .., n\}$$

from which we derive

$$Prob\{X_i \in [a_i, b_i) \,|\, i = 1, .., n\} = \int_{a_2}^{b_2} dx_1 \int_{a_1}^{b_1} dx_2 .. \int_{a_n}^{b_n} dx_n f(x_1, .., x_n)$$

and

$$\int_{-\infty}^{\infty} dx_1 \int_{-\infty}^{\infty} dx_2 .. \int_{-\infty}^{\infty} dx_n f(x_1, .., x_n) = 1$$

is the normalization condition and $f(x_1, .., x_n) \geq 0$.

The *joint distribution function* is defined as

$$\begin{aligned} F(x_1, .., x_n) &= Prob\{X_i < x_i \,|\, i = 1, .., n\} \\ &= \int_{-\infty}^{x_1} dz_1 \int_{-\infty}^{x_2} dz_2 .. \int_{-\infty}^{x_n} dz_n f(z_1, .., z_n) \end{aligned}$$

and it is

$$f(x_1, .., x_n) = \frac{\partial^n}{\partial_1 \partial_2 .. \partial_n} F(x_1, .., x_n).$$

Let $I = \{1, .., n\}$ and $J, K \subseteq I$. Then, the *marginal density functions* are defined $\forall J \subseteq I$ by the family of functions

$$f_J(\mathbf{x})d\mathbf{x} = Prob\{X_j \in [x_j, x_j + dx_j) \,|\, j \in J\}$$

where $\mathbf{x}$ is a subset of $\{x_1, .., x_n\}$ determined by the indexes in J. The random variables $X_1, .., X_n$ are statistically independent iff

$$f(x_1, .., x_n) = \prod_{i=1}^{n} f_{\{i\}}(x_i).$$

The *conditional density functions* are defined $\forall J, K \subseteq I, J \cap K = \emptyset$, by

$$f_J^K(\mathbf{x}|\mathbf{x}')d\mathbf{x} = Prob\{X_j \in [x_j, x_j + dx_j),\ \text{given}\ X_k = x_k \,|\, j \in J, k \in K\}$$

where $\mathbf{x}$ and $\mathbf{x}'$ are disjoint subsets of $\{x_1, .., x_n\}$ determined by the indexes in J and K, respectively.

An important measure of two jointly distributed random variables is their *covariance* defined as $cov\{X_i, X_j\} = \mu_{X_i X_j} - \mu_{X_i}\mu_{X_j}$. Note that the covariance of a single random variable coincides with its variance, i.e., $cov\{X, X\} = \sigma_X^2$. It is also $|cov\{X_i, X_j\}| \leq \sigma_{X_i}\sigma_{X_j}$. It is now possible to define the *correlation coefficient* of the two variables as

$$corr\{X_i, X_j\} = \frac{cov\{X_i, X_j\}}{\sigma_{X_i}\sigma_{X_j}}$$

and it holds $-1 \leq corr\{X_i, X_j\} \leq 1$. The variables $X_1, .., X_n$ are *uncorrelated* iff $\forall i, j \,.\, 1 \leq i < j \leq n, cov\{X_i, X_j\} = 0$. Uncorrelation is a sufficient (but not necessary) condition for statistical independence.

B.2.3 *Some important results*

It is possible to define a set of random variables in terms of an operation applied to known random variables. Assume to have $X_1, .., X_n$ with joint density function $f(x_1, .., x_n)$. Define $Y_i = g_i(X_1, .., X_n), i \in [1..m]$ with g_i functions. Then the joint density function is

$$f_Y(y_1, .., y_m) = \int_{-\infty}^{\infty} dx_1 .. \int_{-\infty}^{\infty} dx_n f(x_1, .., x_n) \prod_{i=1}^{m} \delta(y_i - g_i(x_1, .., x_n))$$

where δ is the Dirac function defined by the equations

$$\delta(x - x_0) = 0, \text{ if } x \neq x_0 \quad \text{and} \quad \int_{-\infty}^{\infty} \delta(x - x_0)dx = 1.$$

A simple application is the *linear transformation* of a random variable X with density function $f(x)$. If λ and β are two constants, the random variable $Y = \beta X + \lambda$ has density function $f_Y(y) = f(\beta^{-1}(y - \lambda))|\beta|^{-1}$. Furthermore, if $X \sim N(b, a^2)$ it is $\beta X + \lambda \sim N(\beta b + \lambda, \beta^2 a^2)$. It is also $N(m, \sigma^2) = m + \sigma N(0, 1)$.

The sum of two normal random variables $X_1 \sim N(b_1, a_1^2)$ and $X_2 \sim N(b_2, a_2^2)$ is a normal random variable $X_1 + X_2 \sim N(b_1 + b_2, a_1^2 + a_2^2)$. An important result is also that if $X \sim N(\mu, \sigma^2)$ then $(X - \mu)/\sigma \sim N(0, 1)$. If $X_1, .., X_n$ are n random samples $N(\mu, \sigma^2)$, then $\sqrt{n}\frac{\overline{X} - \mu}{\sigma} \sim N(0, 1)$. Furthermore, if $X_1, .., X_n$ are $N(0, 1)$, then $X_1^2 + .. + X_n^2 \sim \chi^2(n)$.

The *distribution function theorem* states that if the random variable X has distribution $F(x)$, then $F(X) = U(0, 1)$. This means that if F is a differentiable function of x that increases monotonically from 0 to 1, then $F^{-1}(U)$ is the random variable with density function F', with U the unit uniform random variable. This result is the basis for the Monte Carlo simulation method.

The *linear combination theorem* allows us to state that given $X_1, .., X_n$ and constants $a_1, .., a_n$, it is

$$\mu_{\sum_{i=1}^{n} a_i X_i} = \sum_{i=1}^{n} a_i \mu_{X_i}$$

$$\sigma^2_{\sum_{i=1}^{n} a_i X_i} = \sum_{i=1}^{n} a_i^2 \sigma^2_{X_i} + 2 \sum_{i=1}^{n-1} \sum_{j=i+1}^{n} a_i a_j cov\{X_i, X_j\}$$

with special cases $\mu_{aX} = a\mu_X$, $\mu_{X_1+X_2} = \mu_{X_1} + \mu_{X_2}$, $\sigma^2_{aX} = a^2\sigma^2_X$ and $\sigma^2_{X_1+X_2} = \sigma^2_{X_1} + \sigma^2_{X_2} + 2cov\{X_1, X_2\}$. Note that for uncorrelated variables the *cov* terms vanish and hence the variance of the sum coincides with the sum of the variances.

A main result is the *central limit theorem*. Let $X_1, .., X_n$ be n statistically independent random variables, each defined by a common density function f with finite mean μ and finite variance σ^2. Define

$$S_n = \sum_{j=1}^{n} X_j \qquad \text{and} \qquad A_n = \frac{1}{n} \sum_{j=1}^{n} X_j.$$

Independently of f, it is $\lim_{n\to\infty} S_n = N(n\mu, n\sigma^2)$ and $\lim_{n\to\infty} A_n = N(\mu, \sigma^2/n)$. If we interpret X_j as the outcome of the jth independent sampling of a random variable X with density function f, mean μ and variance σ^2, then S_n and A_n can be interpreted as the n-sample sum of X and the n-sample average of X, respectively. When n is large enough, both S_n and A_n will be normally distributed.

An important corollary of the central limit theorem can be used to assign numerical *confidence limits* to the estimate of a measured physical quantity. In fact, the probability of A_n being within $[\mu - \epsilon, \mu + \epsilon]$ can be expressed in terms of the error function erf and a variable γ as follows

$$Prob\{|A_n - \mu| < \gamma\sigma n^{-1/2}\} \sim \text{erf}(\gamma/\sqrt{2}), \qquad \epsilon = \gamma\sigma n$$

whose approximations for $\gamma = 1, 2, 3$ are respectively $0.683, 0.954$ and 0.997. Note also that the confidence interval size is directly proportional to σ and inversely proportional to n, i.e., the larger the n, the closer A_n to μ.

B.2.4 *Some useful integer random variables*

A variable X is an *integer random variable* iff $\exists f : \mathbb{Z} \rightarrow \mathbb{R}$ such that

$$f(n) = Prob\{X = n\}$$

and f is still the *density function* of X. We then have

$$Prob\{X \in [n_1, n_2]\} = \sum_{n=n_1}^{n_2} f(n).$$

The normalizing condition for f is $\sum_{n=-\infty}^{+\infty} f(n) = 1$ and $0 \leq f(n) \leq 1$.
The *distribution function* F for an integer random variable X is

$$F(n) = Prob\{X \leq n\}, \text{ and hence } F(n) = \sum_{m=-\infty}^{n} f(m)$$

from which we also derive $f(n) = F(n) - F(n-1)$.
The *discrete uniform* random variable X is defined by the density function

$$f(n) = \begin{cases} (n_2 - n_1 + 1)^{-1} & \text{if } n_1 \leq n \leq n_2. \\ 0 & \text{otherwise} \end{cases}$$

It is

$$\mu_X = \frac{n_1 + n_2}{2}, \quad \sigma_X^2 = \frac{(n_2 - n_1)(n_2 - n_1 + 2)}{12}.$$

The *binomial* random variable X is defined by the density function

$$f(n) = \begin{cases} \frac{N!}{n!(N-n)!)} p^n (1-p)^{N-n} & \text{if } 0 \leq n \leq N \\ 0 & \text{otherwise} \end{cases}$$

where $N \in \mathbb{N}$ and $p \in \mathbb{R}$ such that $0 \leq p \leq 1$. It is $\mu_X = Np$ and $\sigma_X^2 = Np(1-p)$.
The *Poisson* random variable X is defined by the density function

$$f(n) = \begin{cases} \frac{e^{-a} a^n}{n!} & \text{if } 0 \leq n \\ 0 & \text{otherwise} \end{cases}$$

where $a \in \mathbb{R}^+$. It is $\mu_X = \sigma_X^2 = a$. Note that Poisson random variables with $\mu_X \gg 1$ can be approximated by normal random variables $N(\mu_X, \mu_X)$.

B.3 Statistics

Statistics deals with the collection and interpretation of data (a set of observations of random variables that can be either qualitative — non numeric — or discrete/continuous quantitative). All possible observations are collectively called the *population*. Observing the population is not feasible in practice and we restrict ourselves to a representative random subset called *sample*, in which all the individuals of the population have the same probability to be included.

The first step in statistical analysis is related to *summary statistics* for understanding the main features and quality of the data. It is important to identify *outliers* (extreme observations that are not consistent with the others and could compromise any analysis), the shape of the distribution of the random variables (to identify symmetries or asymmetries with respect to the central value) and *multi-modal* data (there are gaps or multiple peaks in the distribution).

B.3.1 *Sample measures*

We provide two classes of measures that quantify specific data properties (*measures of location*) and data variability (*measures of spread*).

Assume we have n observations resulting in measurements $x_1, .., x_n$. We denote by $x'_1, .., x'_n$ a permutation of the measurements arranged in increasing order. Furthermore, assume that $y_1, .., y_k$ are the possible outcomes of the observations with frequencies $f_1, .., f_k$ (note that $f_1 + .. + f_k = 1$).

The measures of location are different kind of averages of data and they are the *sample mean*, the *sample median* and the *sample mode*:

$$\text{mean} = \overline{x} = \frac{1}{n}\sum_{i=1}^{n} x_i = \frac{1}{n}\sum_{i=1}^{k} f_i y_i, \quad \text{median} = \begin{cases} x'_{\frac{n+1}{2}} & \text{n odd} \\ \frac{1}{2}x'_{\frac{n}{2}} + \frac{1}{2}x'_{\frac{n}{2}+1} & \text{n even.} \end{cases}$$

The median is usually preferred to the mean when the data has asymmetric distribution or there are outliers. The mode is the value that has greatest frequency in the observations. It works with discrete data or continuous data after grouping in small intervals. Then the mode is $\{y_k \mid f_k = \max_i\{f_i\}\}$.

If the data is symmetric the three measure are very close. If the data has many asymmetries some transformation can be applied. A long left tail (small values), i.e., *left* or *negatively skewed* data, we can square or exponentiate the data. A long right tail (large values), i.e., *right* or *positively skewed* data, we can square root or log the data.

The measures of spread are *range* and *mean absolute deviation*, *variance* and *standard deviation*, *quartiles* and *interquartile range* and *coefficient of variation*. The range is the difference between the largest and smallest observations and usually depends on the sample size range $= x'_n - x'_1$.

The mean absolute deviation is

$$\text{m.a.d.} = \frac{1}{n}\sum_{i=1}^{n} |x_i - \overline{x}|.$$

Sample variance (σ^2) and standard deviation (σ) are better than m.a.d.

$$s^2 = \frac{1}{n-1}\sum_{i=1}^{n}(x_i - \overline{x})^2 = \frac{1}{n-1}\sum_{i=1}^{k} f_i(y_i - \overline{x})^2, \qquad s = \sqrt{s^2}.$$

The quartiles are intended to partition the data into four groups: the lower quartile and the upper quartile border the lower quarter of the data and the upper quarter of the data. The second and third quarters are discriminated by the interquartile range defined as the difference between the upper and lower quartiles. Similar to quartiles are *percentiles*. The nth percentile is the value below which $n\%$ of the observations are found. Note that 25th percentile is the lower quartile, 50th percentile is the median and 75th percentile is the upper quartile. The last measure of spread is the coefficient of variation defined as the ratio of the standard deviation to the mean.

B.3.2 *Basic data visualization*

Frequency tables are useful for summary statistics because they categorize the possible observations and list for each category the number of actual observations. From frequency tables, it is then easy to build histograms (for continuous variables) that provide an approximate pictorial view of the shape of the distribution function of the random variable (see also Appendix B.2) if all the intervals on the x-axis have the same width (recall that the area of a rectangle in the histogram graph is the proportion of observations — density — falling in the rectangle). If the intervals are not homogeneous, the y-axis should be labeled by relative frequency rather than absolute frequency computed for each rectangle as $h = \frac{freq}{nb}$ where h (y-axis) and b (x-axis) are the height and the width of the rectangle and n is the number of observations. When observations are discrete values, we can use bar charts or frequency polygons (useful to compare different data sets) as well as pie charts (for qualitative data) in place of histograms (Fig. B.2).

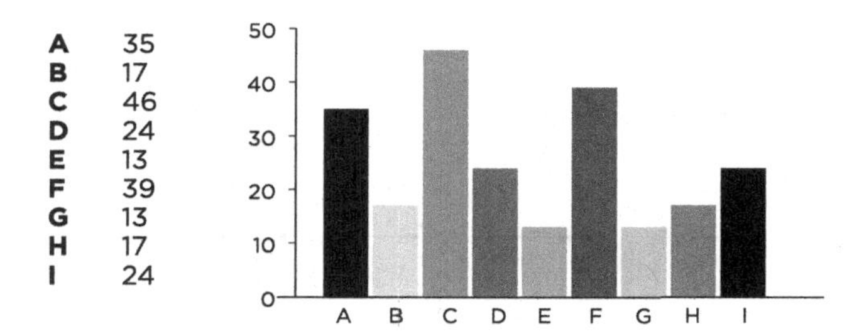

Figure B.2 A frequency table and its histogram.

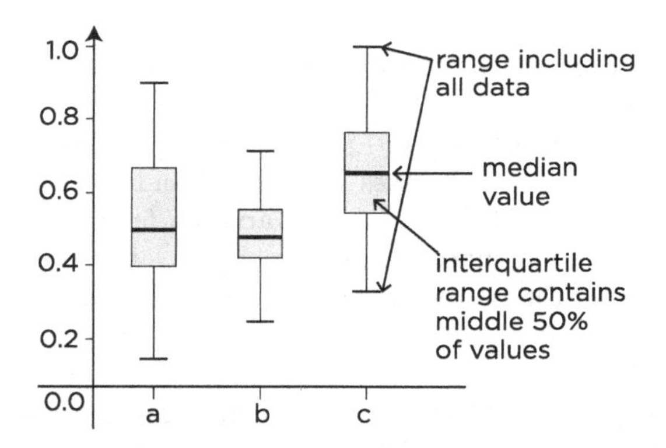

Figure B.3 A box plot.

A useful representation for the variability of data is the box plot (Fig. B.3). The lower and upper extreme of the rectangle are the lower and the upper quartiles, the line in between is the median and the lines exiting the rectangle connect with the maximum and minimum observations. At least ten observations per group are needed.

Scatter plot is a useful representation for determining the relation between two variables. The control/independent variable changed by the experimenter is plotted on the x-axis and the measured/dependent variable is plotted on the y-axis. The resulting plot shows *correlation* between the variables. The correlation is *positive* if the pattern of the values is rising from the origin of the axis towards the upper-right corner; *negative* if the pattern of the values falls from the upper-left corner towards the right part of the x-axis. The visualization of the pattern is also shown by a best fit line. Scatter plot can also be used to compare the relation between two data sets; in this case a reference line ($y = x$) is drawn because the more

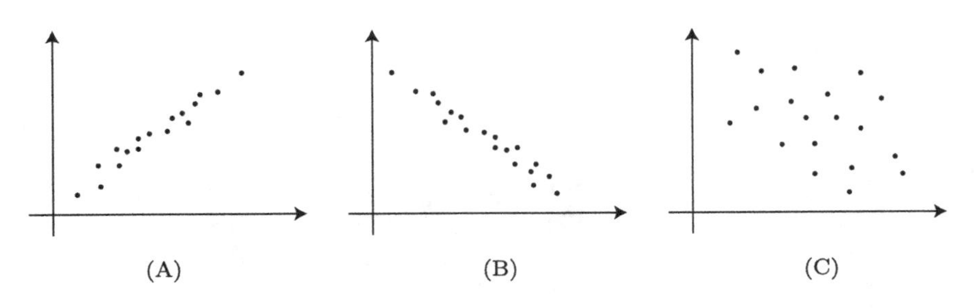

Figure B.4 A scatter plot.

similar the data sets are, the more the values fall close to the $y = x$ line. If no pattern is observed between the variables (or data sets) plotted, no correlation exists.

Correlation circle (see Fig. 4.9) plots the values of the variables in the factor space according to two axes (horizontal and vertical) representing relevant characteristics of the data set. Variables far from the center and close to each other are significantly positively correlated. If the variables are on opposite sides of the center they are significantly negatively correlated. If they are are orthogonal, they are not correlated. It is difficult to interpret variables that are close to the center according to the chosen axes.

Dendogram (see Fig. 4.11) is a tree diagram mainly used to represent hierarchical clustering of data. The leaves of the tree are the data points (observations) and the intermediate nodes represent clusters of observations. The arcs denote dissimilarity between clusters and the difference between the depths of connected nodes is proportional to the dissimilarity of the clusters.

B.3.3 *Hypothesis testing*

Statistical hypothesis testing is a method based on measures performed in a sample to test whether the hypothesis about a parameter in a population is likely to be true. This approach quantifies the compatibility of a sample with a hypothesis. It deals with the acceptance of one out of two mutually exclusive hypotheses (the *null hypothesis* — H_0 — assumed to be true and the *alternative hypothesis* H_a) on unknown population parameters. Usually H_a is vague and corresponds to H_0 is false. Let $z \in \zeta$ be a parameter for which we define the hypothesis. Then the null hypothesis and the alternative hypothesis define a partition of ζ, i.e., $H_0 : z \in Z \subset \zeta$ and $H_a : z \in (\zeta - Z)$. Therefore, H_0 is accepted iff H_a is rejected, and similarly H_a is accepted iff

H_0 is rejected. Assuming real-valued parameters, an alternative hypothesis for which the considered parameter lies completely on an interval in which one extreme is determined by a value in H_0 is said *one-tailed* (e.g., $H_0 : z = n, H_2 : z > n$). An alternative hypothesis for which the parameter lies within two intervals is said *two-tailed* (e.g., $H_0 : z = n, H_2 : z \neq n$).

The hypothesis testing problem is to choose among H_0 and H_a based on the evidence provided by our incomplete observations. Probability theory drives the choice and therefore there can be two types of errors that can be quantified:

Type I error is the probability α of rejecting H_0 by mistake

$$\alpha = Prob\{H_0 \text{ rejected} \mid H_0 \text{ true}\};$$

Type II error is the probability β of accepting H_0 by mistake

$$\beta = Prob\{H_0 \text{ accepted} \mid H_0 \text{ false}\}.$$

The probability α is also called *level of significance* because we fix it to decide whether to accept the null hypothesis. Usually $\alpha \in \{0.01, 0.05, 0.1\}$. The probability $1 - \alpha$ is the *confidence level* of acceptance of H_0. The probability $1 - \beta$ (sometimes denoted π) is called *power* and it is the probability of rejecting a false null hypothesis. The probabilities α and β are related: as one increases, the other decreases. Note that enlarging the sample size makes both probabilities decrease as the sampling error decreases. Small sample size usually leads to frequent Type II errors.

Hypothesis testing is done in three steps:

(1) define a criterion to make a decision on whether either H_0 or H_a is true. The criterion is based on the difference between the sample mean and H_0 and is determined by the level of significance chosen;
(2) select a random sample and compute the sample mean for the parameter; and
(3) compare the sample with the expected observation if H_0 is true.

The central limit theorem ensures that for large samples the possible sample means of the values of the parameter in H_0 are normally distributed in the population with a center on the population mean. We can then compute the mean of the parameter in the population according to H_0. If the mean of the measured parameter in the sample falls in the tails of the normal distribution, it is unlikely for H_0 to be true. The selection of the level of significance coincides with the selection of how much of the tails of the normal distribution for the parameter in the population falsifies H_0.

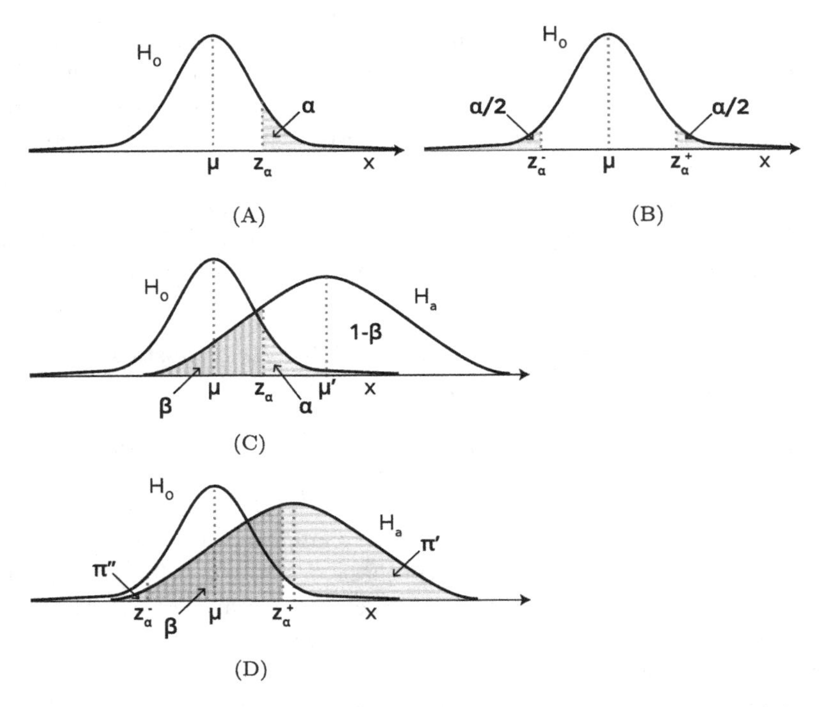

Figure B.5 Hypothesis testing with one-tailed H_0 (A) and two-tailed H_0 (B). The grayed area under the normal distributions is the probability of rejecting H_0 by mistake (also known as the level of significance α).

The area under the part of tails selected is the probability of making a Type I error and coincides with the chosen α. The x-axis values corresponding to the cut-off points for the tails (called z_α^+ for the right tail and z_α^- for the left tail) determine the *rejection region* under the tails and the *acceptance region* as the remaining area under the curve (Fig. B.5). In the case of H_0 one-tailed, there is only one cut-off point called z_α. The cutoff points are computed from the equations below by solving for z_α.

$$\begin{aligned}
\alpha &= Prob\{X > z_\alpha\} = Prob\{Z > \tfrac{z_\alpha - \mu}{\sigma}\}, \quad \text{one-tailed} \\
\alpha &= Prob\{X < z_\alpha^-\} + Prob\{X > z_\alpha^+\} = \\
&\quad 1 - Prob\{z_\alpha^- < X < z_\alpha^+\} = \\
&\quad 1 - Prob\{\tfrac{z_\alpha^- - \mu}{\sigma} < Z < \tfrac{z_\alpha^+ - \mu}{\sigma}\}, \qquad \text{two-tailed}
\end{aligned}$$

Another interesting measure is the *p value*, also called *observed significance level.* Given a value x' of X, the p value is the probability of rejection of H_0 for all significance levels $0 < \alpha < p$.

$$p = Prob\{X > x'\} = Prob\{Z > \tfrac{x'-\mu}{\sigma}\}, \qquad \text{one-tailed}$$
$$p = 2\min(Prob\{X > x'\}, Prob\{X < x'\}) =$$
$$2min(Prob\{Z > \tfrac{x'-\mu}{\sigma}\}, Prob\{Z < \tfrac{x'-\mu}{\sigma}\}), \quad \text{two-tailed}$$

It is the lowest level of significance that allows us to reject H_0 for the value x' from the sample. In other words, $p = \alpha$ when x' is chosen as the cut-off point. Indeed $p = \alpha$ if $x' = z_\alpha$ in the case of H_0 one-tailed. The null hypothesis is rejected if $p < \alpha$, i.e., the observed value is greater than z_α and falls in the rejection region (case of one-tailed null hypothesis). A small p value is evidence against H_0. The graphical interpretation of p value for the one-tailed alternative hypothesis is in Fig. B.6.

The probability of not rejecting (accepting) H_0 by mistake is the area under the curve of H_a delimited on the right by z_α (Fig. B.5(C))

$$\beta = Prob^{H_a}\{X < z_\alpha\} = Prob^{H_a}\{Z < \tfrac{z_\alpha - \mu'}{\sigma'}\}, \quad \text{one-tailed}$$
$$\beta = Prob^{H_a}\{z_\alpha^- < X < z_\alpha^+\} =$$
$$1 - Prob^{H_a}\{\tfrac{z_\alpha^- - \mu'}{\sigma'} < Z < \tfrac{z_\alpha^+ - \mu'}{\sigma'}\}, \qquad \text{two-tailed}$$

by solving for z_α. The probability of correctly rejecting H_0 when false is then $1 - \beta$ (the power of the test).

We further classify hypotheses. A hypothesis is *simple* if it completely fixes a population (e.g., H_0: the sample is $N(2, 5)$) and it is *composite* if it has free parameters (e.g., H_0: the sample is drawn from a normal distribution). According to this classification of hypotheses, we consider tests to discriminate between two simple hypotheses (*likelihood ratio test*),

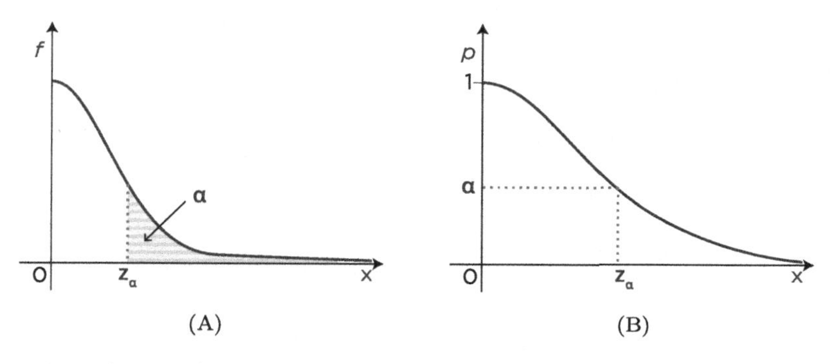

Figure B.6 Graphical interpretation of p value and its relationship with α for one-tailed alternative hypotheses.

to test whether H_0 holds or not when H_a is vaguely defined (*goodness of fit tests* — GOF), to verify if two samples belong to the same population (*two-sample tests*) and to check whether a small signal is significant (*signal tests*).

Given the sample values $x_1, .., x_n$, a *test statistic* is a real random variable $T(x_1, .., x_n)$. T should determine how close a sample result is to the population result stated by H_0 to discriminate between the null and the alternative hypothesis. Therefore, T must be defined to make the difference between the distribution of T given H_0 with the distributions of H_a as large as possible.

A test is *biased* if $\alpha \geq 1 - \beta$ and it is *consistent* if $\lim_{n \to \infty} \beta = 1$, where n is the sample size. These conditions must hold for all the distributions represented by H_a.

To apply the test we need a *rejection region* or *critical region* K that allows the rejection of H_0 when $T \in K$. Note that $Prob\{T \in K \mid H_0\} = \alpha$. Also, note that the p value is a function of the test statistic, i.e., for an observed value *obs* of the test statistic, $p = Prob\{t \geq obs \mid H_0\}$ is the probability of obtaining a value $t \geq obs$ if H_0 is true.

Appendix C

Semantics of modeling languages

Calculi and their derived programming languages are defined by a syntax that determines how sentences can be written by composing the basic lexical elements according to pre-defined rules encoded in the productions of context-free grammars. The meaning associated with each sentence (or program) written in a language is determined by the *semantics* of the language. Since each interesting language is made of infinite strings (i.e., we can write infinite different programs), we cannot assign a meaning to each string by enumeration. We need a finite description of the meaning of the infinite strings. A logic approach that defines the semantics of the language by structural induction on the syntax of the language can be applied. We start by introducing the basics of grammars, languages and semantics. Then we define the semantics of π-calculus and its stochastic variant, of β-binders (a subset of BlenX) and of ℓ.

C.1 Languages and grammars

An *alphabet* is a finite, non-empty set of symbols. The concatenation of two symbols a and b of the alphabet is $a \cdot b$ or simply ab. A *string* is a finite concatenation of symbols from an alphabet. The length of a string x is the number of its symbols (written $|x|$). A special string is the empty string ϵ that has length 0. The *concatenation* of two strings x and y is the string $x \cdot y$, or simply xy obtained by following the symbols of x by the symbols y. *Exponentiation* of strings represent iterated concatenation (e.g., x^3 denotes xxx). By convention, we let $x^0 = \epsilon$. A *prefix* of x is a string obtained by discarding 0 or more trailing symbols of x. A *suffix* of x is a string obtained by deleting 0 or more leading symbols of x. A *substring* of x is a string obtained by erasing a prefix and a suffix of x. A string y is a *proper* prefix, suffix or substring of x if $y \neq x$.

A *language* is a set of strings formed from a specific alphabet. Two particular languages are $\emptyset$ (the empty language) and $\{\epsilon\}$ (the language made up of the empty string alone). Note that the definition of language does not assign any meaning to its strings. Therefore, languages are purely syntactic objects. We extend concatenation to languages as follows:

$$LM = \{xy \,|\, x \in L \ \wedge \ y \in M\}.$$

Similarly to strings, we define exponentiation of languages, assuming that $\{\epsilon\}L = L\{\epsilon\} = L$. A set of languages with concatenation and neutral element $\{\epsilon\}$ is a *monoid.* We write L^* for the concatenation of L with itself any number of times. More formally,

$$L^* = \sum_{i=0}^{\infty} L^i$$

where $L^0 = \{\epsilon\}$ by convention. Hereafter, we use the shorthand L^+ for $L(L^*)$. We call the postfix operator $*$ *closure* and $+$ *positive closure* of a language. The definition of closure allows us to characterize the languages over an alphabet A as the subsets of A^*. Since languages are sets, they are equipped with the fundamental operations on set, as well.

All interesting languages are made up of an infinite number of strings. Therefore, it is impossible to define these languages by enumerating their strings. There are three ways to finitely define infinite languages. A language is the set of strings generated by a finite structure called *grammar.* Alternatively, a language is the set of strings recognized or accepted by a finite structure called *automaton.* Finally, a language is the solution of a system of *algebraic relations.*

A *grammar* is a quadruple $G = (N, \Sigma, P, S)$ where N is a finite set of *non-terminal* symbols, Σ is an alphabet of *terminal* symbols, $P \subseteq (N \cup \Sigma)^* \times (N \cup \Sigma)^*$ is the finite set of productions and $S \in N$ is the start symbol. We adopt the following conventions: $A, B, .. \in N$, $a, b, .. \in \Sigma$, $X, Y, .. \in (N \cup \Sigma)$, $x, y, .. \in \Sigma^*$, and $\alpha, \beta, .. \in (N \cup \Sigma)^*$. We often write $\alpha \longrightarrow \beta$ for a production $\langle \alpha, \beta \rangle \in P$.

To simplify the presentation of a grammar, we sometimes only give its productions (from which one can easily recognize the terminal and non-terminal symbols). Furthermore, the productions with the same left part are grouped by writing on the right of the arrow all the possible alternatives separated by $|$. If we replace the arrow $\longrightarrow$ with the symbol ::= in the above productions, we obtain the Backus–Naur form (BNF) usually used to define the syntax of programming languages.

The language defined by G is $L(G) = \{w \mid w \in \Sigma^*,\ S \longrightarrow^* w\}$ where $\longrightarrow^*$ is the reflexive and transitive closure of $\longrightarrow$.

C.2 Structural operational semantics

In this section we present transition systems and their intentional definition through inference rules.

C.2.1 *Transition systems*

A transition system is used to represent the states that the execution of a program can generate and to determine how these states are connected. A transition system is characterized at time t by its internal *state* (*configuration*) that is made of a *control program* and some *data*. The set of configurations that a system may pass through and their relations characterize the dynamic behavior.

A *transition system* is a structure $(\Gamma, \longrightarrow)$, where Γ is a set of elements γ called *configurations* and the binary relation $\longrightarrow \subseteq \Gamma \times \Gamma$ is called *transition relation*. Hereafter, we write $\gamma \longrightarrow \gamma'$ for $\langle \gamma, \gamma' \rangle \in \longrightarrow$.

Transition systems can specify many concepts from formal language theory. For instance, finite state automata are obtained by imposing Γ to be a finite set and by picking out a set of *terminal* states. The resulting transition system is a *terminal transition system*.

A *terminal transition system* is a structure $(\Gamma, \longrightarrow, T)$ where $(\Gamma, \longrightarrow)$ is a transition system, and $T \subseteq \Gamma$ is the set of *final* configurations such that $\forall \gamma \in T, \gamma' \in \Gamma \,.\, \gamma \not\longrightarrow \gamma'$.

We now formalize the representation of a finite state automaton

$$M = (\Sigma, S, \delta, s_0, F)$$

as a terminal transition system. We define $\Gamma = S \times \Sigma^*$ and $\langle s, aw \rangle \vdash \langle s', w \rangle$ whenever $q' \in \delta(s, a)$. The set of terminal configurations is

$$T = \{\langle s, \epsilon \rangle \mid s \in F\},$$

where ϵ denotes the empty string.

The behavior of a finite state automaton is determined by the set of strings that it accepts $L(M) = \{w \in \Sigma^* \mid \delta^*(\langle s_0, w \rangle) \in F\}$, where δ^* is the reflexive and transitive closure of δ. The same behavior can be obtained by a terminal transition system with initial state.

An *initial terminal transition system* is a structure $(\Gamma, \longrightarrow, T, I)$ where $(\Gamma, \longrightarrow, T)$ is a terminal transition system, and $I \subseteq \Gamma$ is the set of *initial* configurations such that whenever $\gamma' \longrightarrow \gamma''$ there exists $\gamma \in I$ and $\gamma \longrightarrow^* \gamma'$.

Finally, automaton M is completely described by the initial terminal transition system $(S \times \Sigma^*, \vdash, F \times \{\epsilon\}, \{S_0\})$.

Similarly, we can express context-free grammars through transition systems. The initial terminal transition system corresponding to $G = (N, \Sigma, P, S)$ is $((N \cup \Sigma)^*, \longrightarrow, T, \{S\})$, where $wXv \longrightarrow wxv$ if $(X, x) \in P$ and T is the set of configurations which do not contain symbols from N.

Transition systems allows us to study properties which are independent of the kind of individual transition. For instance, if one is interested in all configurations which are reachable from a given one, the kind of transition is irrelevant. Other properties need more information on transitions to be investigated. For example, to study the frequency of a given action's occurrence, we must assign to each transition a name that identifies the corresponding action. This leads to the definition of labeled transition systems.

A *labeled transition system* is a structure $(\Gamma, A, \longrightarrow)$ where Γ is a set of *configurations*, A is a set of *labels* (sometimes called *actions* or *operations*) and $\longrightarrow \subseteq \Gamma \times A \times \Gamma$ is the *transition relation*. Hereafter, we write $\gamma \xrightarrow{a} \gamma'$ for $\langle \gamma, a, \gamma' \rangle \in \longrightarrow$, $\gamma \not\xrightarrow{a} \gamma'$ for $\langle \gamma, a, \gamma' \rangle \notin \longrightarrow$, $\gamma \longrightarrow \gamma'$ for $\exists a \in A.\langle \gamma, a, \gamma' \rangle \in \longrightarrow$, and $\gamma \not\longrightarrow$ for $\forall a \in A, \forall \gamma' \in \Gamma . \gamma \not\xrightarrow{a} \gamma'$.

Labeled transition systems can be considered in their initial or terminal versions, as well. We now generalize the reflexive and transitive closure of the transition relation to labeled transition systems as

$$\gamma \xrightarrow{w}{}^{+} \gamma' \Leftrightarrow \exists \gamma_1, .., \gamma_n . \gamma = \gamma_0 \xrightarrow{a_1} \gamma_1 .. \xrightarrow{a_n} \gamma_n = \gamma', \ n > 0$$

$$\gamma \xrightarrow{w}{}^{*} \gamma' \Leftrightarrow \exists \gamma_1, .., \gamma_n . \gamma = \gamma_0 \xrightarrow{a_1} \gamma_1 .. \xrightarrow{a_n} \gamma_n = \gamma', \ n \geq 0$$

where $w = a_1..a_n$.

Labeled initial terminal transition systems may be used to simplify the specification of finite state automata. In particular, they allow us to drop the component Σ^* from configurations. The automaton M of the example above can now be defined as

$$(S, \Sigma, \longrightarrow, F, \{s_0\})$$

with $s_i \xrightarrow{a} s_j$ whenever $s_j \in \delta(s_i, a)$. Hereafter, we omit the adjectives labeled initial and terminal when the kind of transition system at hand is clear from the context.

We now show that a transition system can be conveniently represented through a graph in which the configurations coincide with the nodes and the arcs represent the possible transitions between them. The dynamic behavior of a system define the form of this graph, that we call *transition graph.* Some auxiliary definitions are needed. Hereafter, we assume as given a labeled transition system $(\Gamma, A, \longrightarrow)$.

Configuration γ' is an *immediate derivative* of γ, if $\gamma \xrightarrow{a} \gamma'$. It is simply a *derivative*, if $\gamma \xrightarrow{w}{}^{*}\gamma'$. Configuration γ is *acyclic* if there is no w such that $\gamma \xrightarrow{w}{}^{+}\gamma$. Sometimes is useful to identify the set of derivatives of a given configuration.

The *derivative set* of a configuration γ is

$$ds(\gamma) = \{\gamma' \mid \gamma \xrightarrow{w}{}^{*}\gamma'\}.$$

Given a configuration γ and its set of derivatives $ds(\gamma)$, the *derivation graph* of γ is

$$dg(\gamma) = \langle ds(\gamma), A, \{(\gamma_i, a, \gamma_j) \mid \gamma_i \in ds(\gamma) \ \wedge \ \gamma_i \xrightarrow{a} \gamma_j\}\rangle$$

where $ds(\gamma)$ is the set of nodes, A is the labeling alphabet and the third set defines the arcs.

The possible patterns of behavior of a system are obtained by visiting its derivation graph. The sequences of consecutive transitions that describes the behavior are called *computations.*

Let $\gamma_i \xrightarrow{a} \gamma_j$ be a transition. Then, γ_i is the *source* of the transition and γ_j is its *target.* A computation of γ is a sequence of transitions $\gamma = \gamma_0 \xrightarrow{a_0} \gamma_1 \xrightarrow{a_1} ..$ starting from γ, and such that the target of any transition coincides with the source of the next one. We let $\xi, \xi', \xi_1, ..$ range over computations, and we write ϵ for the empty computation. The notions of source and target are extended in the obvious way to computations. We let $C(\gamma)$ be the set of computations with source γ, and $C(\gamma, \gamma')$ be the set of computations with source γ and target γ'. Note that whenever Γ is a finite set, $C(\gamma, \gamma')$ is the language accepted by the automaton $(\Gamma, A, \longrightarrow, \{\gamma'\}, \{\gamma\})$. If we call *branching-free* regular languages those that may be expressed through regular expressions built without the operator $+$, then $C(\gamma, \gamma')$ is a branching-free language.

Sometimes is useful to have a linearization of all computations that a system may engage in. A possibility is to get the unfolding of the derivation graph, thus yielding a tree of computations.

Let γ be a configuration and $dg(\gamma)$ its derivation graph. Then, the *derivation tree* of γ is

$$dt(\gamma) = \sum_{\{(\gamma,a_i,\gamma_i)\in dg(\gamma)\}} a_i \cdot dt(\gamma_i).$$

Hereafter, isomorphic derivation trees will be identified. It is clear from the above definition that any path in the derivation tree of a configuration γ represents a computation starting at γ. As a consequence, the unfolding of a transition system with more than one initial state originates a forest.

C.2.2 *Structural operational semantic definitions*

A transition system is conveniently defined intentionally by a formal system of axioms and inference rules. The structure of the nodes of the transition system for describing the behavior of the execution of a program is usually a triple $(\mathcal{P}, Env, Sto)$ where $\mathcal{P}$ is the set of possible programs that can be written in the language, Env (*environments*) is a family of functions mapping identifiers into locations or constant values, and Sto (*stores*) is a family of functions mapping locations into values. Formally, given the set Id of possible identifiers of the language,

$$Env = \cup_{I\subseteq_f Id} Env_I$$

with $\subseteq_f$ meaning a *finite subset of* and $Env_I : I \to DVal \cup Loc \cup \{\bot\}$ has metavariable ρ. $DVal$ is the set of *denotable values* of the language (the values that can be recorded in the environment) and the symbol $\bot$ denotes undefined bindings. Given $I, I' \subseteq_f Id$, $\sigma \in Env_I$ and $\sigma' \in Env_{I'}$, the update of an environment ρ with another one ρ' is

$$\rho[\rho'](x) = \begin{cases} \rho'(x) & x \in I' \\ \rho(x) & x \in (I - I'). \end{cases}$$

Assuming an infinite set of locations Loc,

$$Sto = \cup_{L\subseteq_f Loc} Sto_L$$

with $Sto_L : L \to SVal \cup \{?, \bot\}$ has metavariable σ. $SVal$ is the set of *storable values* of the language (the values that can be recorded in memory locations) and the symbols ? and $\bot$ denote unused and undefined locations, respectively. Given $L, L' \subseteq_f Loc$, $\sigma \in Sto_L$ and $\sigma' \in Sto_{L'}$, the update of a store σ with another one σ' is

$$\sigma[\sigma'](l) = \begin{cases} \sigma'(l) & l \in L' \\ \sigma(l) & l \in (L - L'). \end{cases}$$

Intuitively, *Env* represents the set of memory addresses where variables stores their values and *Sto* is the memory of the abstract machine on which the programs run.

Structural operational semantic definitions are typical of the so-called *structural operational semantics* introduced by Plotkin. This approach permits compact semantic definitions as well as simple and powerful proof methods. For instance, structural induction and induction on the depth of the proof of transitions are among the most widely used proof methods in this setting. Furthermore, the set of inference rules that generates the transition systems constitutes an abstract machine for the language that is under specification.

We show some examples starting from the declaration of variables in imperative languages and the sequential composition of declarations. Declarations are *elaborated* to build the environments by associating names of denotable values like variables with locations and initializing the allocated memory location with values when available. Configurations are $\langle D, \sigma \rangle$ where D is a metavariable for declarations. The axiom defining the semantics of variable declaration is

$$\rho \vdash \langle var\ x : int = 2, \sigma \rangle \longrightarrow \langle [x : l], \sigma[l : 2] \rangle.$$

The wording corresponding to the axiom is that the declaration of the variable x creates a new binding in the environment between the name x and the new location l. Furthermore, the memory location l is written with the initialization value 2 expressed in the declaration.

The rule for the sequential composition of declarations are

$$1 : \frac{\rho \vdash \langle D_0, \sigma \rangle \longrightarrow \langle D'_0, \sigma' \rangle}{\rho \vdash \langle D_0; D_1, \sigma \rangle \longrightarrow \langle D'_0; D_1, \sigma' \rangle} \qquad 2 : \frac{\rho[\rho_0] \vdash \langle D_1, \sigma \rangle \longrightarrow \langle D'_1, \sigma' \rangle}{\rho \vdash \langle \rho_0; D_1, \sigma \rangle \longrightarrow \langle \rho_0, D'_1, \sigma' \rangle}$$

$$3 : \rho \vdash \langle \rho_0; \rho_1, \sigma \rangle \longrightarrow \langle \rho_0[\rho_1], \sigma \rangle.$$

The wording of rule 1 is that a declaration D_0, starting with an environment ρ and a store σ, can perform a step of its elaboration leading to a configuration in which the rest of the declaration to be elaborated is D'_0 from the store σ'. If the premise is satisfied then we can infer that the sequential composition of declarations does not affect the elaboration of the left argument and hence from the configuration $\langle D_0; D_1, \sigma \rangle$ the same step of the premise can be performed by D_0 leading to $\langle D'_0; D_1, \sigma' \rangle$ where the left argument of the sequential composition and the store are the same as in the right part of the premise. Rule 2 states that D_1 can be elaborated from the current environment ρ extended with the bindings ρ_0 generated by the elaboration of D_0. The result of one step of elaboration in the premise can also be obtained if D_1 sequentially follows ρ_0. Axiom 3 states that the sequential composition of two environments ρ_0 and ρ_1 is mapped into the first environment extended with the second one.

We define assignment and the sequential composition of commands. Commands are the basic computational step of imperative languages. Commands are *executed* to modify the store that will contain the result of the execution. Configurations are $\langle C, \sigma\rangle$, where C is a metavariable for commands (or programs). The axiom defining the semantics of the assignment is

$$\rho \vdash \langle x := 2, \sigma\rangle \longrightarrow \langle \epsilon, \sigma[l : 2]\rangle, \text{ if } \rho(x) = l.$$

The wording corresponding to the axiom is that the assignment looks for the location associated to its left operand x in the current environment ($\rho(x) = l$) and updates the corresponding location in the store with the right part of the assignment 2.

The inference rules for sequentialization are

$$1: \frac{\rho \vdash \langle C, \sigma\rangle \longrightarrow \langle C', \sigma'\rangle}{\rho \vdash \langle C; C'', \sigma\rangle \longrightarrow \langle C'; C'', \sigma'\rangle} \qquad 2: \frac{\rho \vdash \langle C, \sigma\rangle \longrightarrow \langle \epsilon, \sigma'\rangle}{\rho \vdash \langle C; C'', \sigma\rangle \longrightarrow \langle C'', \sigma'\rangle}.$$

The premise of rule 1 says that C transforms into C' in a single step and changes the store σ into σ'. Then, its conclusion introduces the sequentialization operator ;. The wording corresponding to the premise of rule 1 above is that a command C, starting with an environment ρ and a store σ, can perform a step of its execution leading to a configuration in which the rest of the command to be executed is C' from the store σ'. If the premise is satisfied then we can infer that the sequential composition of commands does not affect the execution of the left argument and hence from the configuration $\langle C; C'', \sigma\rangle$ the same step of the premise can be performed by C leading to $\langle C'; C'', \sigma'\rangle$ where the left argument of the sequential composition and the store are the same as in the right part of the premise. Rule 2 states that the command C is completed (to the empty one ϵ) and thus it is possible to derive that in a step the configuration $\langle C; C'', \sigma\rangle$ of the abstract machine moves to $\langle C'', \sigma'\rangle$. The wording corresponding to the premise of rule 1 above is that a command C, starting with an environment ρ and a store σ, can complete its execution leading to a configuration in which no other command is to be executed by generating a new store σ'. If the premise is satisfied then we can infer that the sequential composition of commands is reduced to the execution of the right argument of the sequential composition starting from the store σ' generated by the left argument execution.

We consider here the *evaluation* of expressions that produce *expressible values* without affecting the environment and the store that are only needed to retrieve values associated with identifiers. Configurations are $\langle E, \sigma \rangle$, where E is a metavariable for expressions. The inference rules for evaluating a binary expression are

$$1: \frac{\rho \vdash \langle E_0, \sigma \rangle \longrightarrow \langle E_0', \sigma \rangle}{\rho \vdash \langle E_0 \; bop \; E_1, \sigma \rangle \longrightarrow \langle E_0' \; bop \; E_1, \sigma \rangle}$$

$$2: \frac{\rho \vdash \langle E_1, \sigma \rangle \longrightarrow \langle E_1', \sigma \rangle}{\rho \vdash \langle k_0 \; bop \; E_1, \sigma \rangle \longrightarrow \langle k_0 \; bop \; E_1', \sigma \rangle}$$

$$3: \rho \vdash \langle k_0 \; bop \; k_1, \sigma \rangle \longrightarrow k \text{ if } k = k_0 \text{ bop } k_1.$$

The wording of rule 1 is that the left operand can be evaluated independently of the right operand until it is reduced to a value k_0. The we start applying rule 2 until the second operand E_1 is reduced to a value k_1. Finally the axiom is applied to compute the value k associated with the expression $E_0 \; bop \; E_1$. Note that the evaluation defined by the rules is left to right (first the left operand is completely evaluated and then the right operand is completely evaluated. Note also that bop is the semantic operator corresponding to the syntactic one *bop*.

Structural operational semantic definitions of primitives for handling concurrency such as parallel composition of processes or non-deterministic choices, as well as the stochastic extensions of these semantic definitions, are dealt with in the next sections where we formally define the calculi and languages introduced in Chapter 6.

C.3 The π-calculus and its stochastic extension

The syntax of the π-calculus is defined in BNF as

$$P ::= nil \mid \sum_{i \in I} A_i.P_i \mid P_1 | P_2 \mid (new \; x)P \mid !P$$
$$A ::= x?y \mid x!y$$

assuming an infinite, countable set of names $\mathcal{N}$ with metavariables $x, y, \ldots$ We assume here to have processes in normal forms with multiple, guarded choices as sequential components.

Different programs may compute the same function, i.e., exhibit the same behavior and may not be distinguishable by an observer that plays with them. For instance $P_1 \mid P_2$ must have the same semantics of $P_2 \mid P_1$ because the syntactic ordering in which processes are listed in a parallel composition is irrelevant from a semantic perspective. Similar syntactic differences with other operators should be neglected as well when defining the semantics of the language. To overcome this issue, we work with equivalence classes of syntactic programs that share the same semantics. The equivalence classes are determined by the structural congruence $\equiv$ defined as the minimal relation satisfying the following equations:

$$
\begin{aligned}
x?y.P &= x?z.(P\{z/y\}), \text{ if } z \notin fn(P), \text{ change bound names}\\
(new\ y)P &= (new\ z)(P\{z/y\}), \text{ if } z \notin fn(P), \text{ change bound names}\\
P \mid nil &= nil \mid P = P, \text{ neutral element}\\
P \mid Q &= Q \mid P, \text{ commutativity}\\
(P \mid Q) \mid R &= P \mid (Q \mid R), \text{ associativity}\\
P + nil &= nil + P = P, \text{ neutral element}\\
P + Q &= Q + P, \text{ commutativity}\\
(P + Q) + R &= P + (Q + R), \text{ associativity}\\
(new\ x)P &= P, \text{ if } x \notin fn(P), \text{ not used definition}\\
(new\ x)(new\ y)P &= (new\ y)(new\ x)P, \text{ commutativity}\\
(new\ x)P \mid Q &= (new\ x)(P \mid Q), \text{ if } x \notin fn(Q), \text{ scope extrusion}\\
!P &= P \mid !P, \text{ replication.}
\end{aligned}
$$

The first two equations manage the change of bound names without captures. The three rules on parallel composition and the corresponding three on choice states that the set of all processes with $\mid$ or $+$ and *nil* is a commutative monoid. The rules for definitions of new names state that the order of consecutive definitions is not relevant and that the definition of a name that is not used in its scope can be removed. The last rule of definitions states that we can include in the scope of a definition all the processes that do not have the very same name free and hence no capture is possible. The last rule defines replication of processes.

We now define the reaction rules that define the semantics of the language (transition systems with nodes corresponding to processes still to be executed) in SOS style:

$$\left(\sum_{j\in J} A_j.Q_j + x!z.Q\right) \mid \left(\sum_{i\in I} A_i.P_i + x?y.P\right) \longrightarrow Q \mid P\{z/y\}$$

$$\frac{P \longrightarrow P'}{P \mid Q \longrightarrow P' \mid Q} \quad \frac{P \longrightarrow P'}{(new\ x)P \longrightarrow (new\ x)P'}$$

$$\frac{Q \equiv P, P \longrightarrow P', P' \equiv Q'}{Q \longrightarrow Q'}.$$

The transitions that a system specified in π-calculus can perform are those that can be proved as theorems in the formal system defined by the semantic axiom and inference rules. The proof is a deduction tree that has the transition $P_i \to P_j$ as conclusion (P_i is the whole system specification from which we move and P_j is the target state that we reach through the transition) and axioms as leaves (see also Appendix A.2). Consider Example 6.9 and the transition $S_1 \to S_2$. The formal derivation of the transition is (assuming that $S_A = Gene\_A \mid RNA\_A \mid A$, $S_{TF} = Gene\_TF$ and $S'_{TF} = Gene\_TF \mid RNA\_TF$)

$$\frac{S_A|S_{TF}|S \equiv S_{TF}|S_A|S,\ \dfrac{S_{TF} \to S'_{TF}}{S_{TF}|S_A|S \to S'_{TF}|S_A|S},\ S'_{TF}|S_A|S \equiv S_A|S'_{TF}|S}{S_A|S_{TF}|S \to S_A|S'_{TF}|S}.$$

Note that building all the possible deduction trees from a state produces all the arcs exiting that state. Iterating the procedure, we can build the whole transition system $\langle \mathcal{P}, \to \rangle$ (with $\to \subseteq \mathcal{P} \times \mathcal{P}$) that is implicitly defined by the reaction rules and the set $\mathcal{P}$ of all possible processes.

To rule out nondeterminism, we associate a rate with name (hereafter we write x_r to mean that $rate(x) = r$). To associate the right distribution with each transition we count how many interactions on the same channel are available and we update the rate of transitions inductively while building their deduction trees. We use two auxiliary functions $In_x(P)$ and $Out_x(P)$ to count the receive and send enabled actions on a channel x in each configuration.

$$In_x(nil) = \emptyset$$

$$In_x\left(\sum_{i\in I} A_i.P_i\right) = |\{A_i \mid sbj(A_i) = x? \wedge i \in I\}$$

$$In_x(P_1 \mid P_2) = In_x(P_1) + In_x(P_2)$$

$$In_x(new\ z)P = \begin{cases} In_x(P) & \text{if } z \neq x \\ 0 & \text{otherwise} \end{cases}$$

The function $Out_x(P)$ is defined in the same way by replacing all occurrences of In_x with Out_x and $sbj(A_i) = x?$ with $sbj(A_i) = x!$ in the above definition. We also assume having a set of names $\mathcal{H}$ that are used to represent homodimerization reactions so that we can correctly compute the stochastic kinetics.

$$(\textstyle\sum_{j\in J} A_j.Q_j + x!z.Q) \mid (\sum_{i\in I} A_i.P_i + x?y.P) \xrightarrow{x_r \cdot 1 \cdot 1} Q \mid P\{z/y\},\ x \in \mathcal{H}$$

$$(\textstyle\sum_{j\in J} A_j.Q_j + x!z.Q) \mid (\sum_{i\in I} A_i.P_i + x?y.P) \xrightarrow{x_r/2 \cdot 2 \cdot 1} Q \mid P\{z/y\},\ x \notin \mathcal{H}$$

$$\frac{P \xrightarrow{x_r \cdot n_0 \cdot n_1} P'}{P \mid Q \xrightarrow{x_r (n_0 + In_x(Q))(n_1 + Out_x(Q))} P' \mid Q}$$

$$\frac{P \xrightarrow{x_r \cdot n_0 \cdot n_1} P'}{(new\ x)P \xrightarrow{x_r (n_0 + In_x(Q))(n_1 + Out_x(Q))} (new\ x)P'}$$

$$\frac{Q \equiv P, P \xrightarrow{x_r \cdot n_0 \cdot n_1} P', P' \equiv Q'}{Q \xrightarrow{x_r (n_0 + In_x(Q))(n_1 + Out_x(Q))} Q'}$$

The derivation of transitions now account for rates and multiplicities of actions.

C.4 β-binders

We assume a countably infinite set N of names (ranged over by lowercase letters). β-binders allow the description of the behavior of π-calculus processes wrapped into boxes with interaction capabilities (hereafter called *β-processes* or simply *boxes*). Processes encapsulated into boxes (ranged over by capital letters distinct from B) are given by the following syntax:

$$P ::= \mathsf{nil} \mid x?w.\,P \mid x!y.\,P \mid P|P \mid (new\ y)\,P \mid !\,P \mid$$

$$\mathsf{expose}(x,\ \Gamma)\,.\,P \mid \mathsf{hide}(x)\,.\,P \mid \mathsf{unhide}(x)\,.\,P.$$

The deadlocked process nil, input and output prefixes ($x?w.\,P$ and $x!y.\,P$, respectively), parallel composition ($P \mid P$), restriction ($(new\ y)\,P$) and the bang operator ($!$) have exactly the same meaning as in the π-calculus.

The expose, hide and unhide prefixes are intended for changing the external interface of boxes by adding a new site, hiding a site and unhiding a site which has been previously hidden, respectively.

The π-calculus definitions of *name substitution* and of *free* and *bound names* (denoted by $\mathsf{fn}(\text{-})$ and $\mathsf{bn}(\text{-})$, respectively) are extended to the processes generated by the above syntax in the obvious way. It is sufficient to state that neither $\mathsf{hide}(x)$ nor $\mathsf{unhide}(x)$ act as binders for x, while the prefix $\mathsf{expose}(x, \Gamma)$ in $\mathsf{expose}(x, \Gamma) \,.\, P$ is a binder for x in P.

β-processes are defined as processes prefixed by specialized binders. An *elementary β-binder* has either the form $\beta(x : \Gamma)$ or $\beta^h(x : \Gamma)$, where

(1) the name x is the $\mathsf{subject}$ of the β-binder, and
(2) Γ is the type of x. It is a non-empty set of names such that $x \notin \Gamma$.

The elementary β-binder $\beta(x : \Gamma)$ is used to denote an active (potentially interacting) site of the box. A β-binder $\beta^h(x : \Gamma)$ denotes a site that is hidden to forbid further interactions.

Composite β-binders are generated by the following grammar:

$$\vec{B} ::= \beta(x : \Gamma) \;\big|\; \beta^h(x : \Gamma) \;\big|\; \beta(x : \Gamma)\,\vec{B} \;\big|\; \beta^h(x : \Gamma)\,\vec{B}.$$

A composite β-binder is said to be $\mathsf{well\text{-}formed}$ when the subjects of its elementary components are all distinct. We let well-formed β-binders be ranged over by $\vec{B}, \vec{B}_1, \vec{B}_2, .., \vec{B}', ...$

The set of the subjects of all the elementary β-binders in $\vec{B}$ is denoted by $\mathsf{sub}(\vec{B})$, and we write $\vec{B} = \vec{B}_1\vec{B}_2$ to mean that $\vec{B}$ is the β-binder given by the juxtaposition of $\vec{B}_1$ and $\vec{B}_2$.

Also, the metavariables $\vec{B}^*, \vec{B}_1^*, \vec{B}_2^*, ..$ stay for either a well-formed β-binder or the empty string. The above notation for the subject function and for juxtaposition is extended to these metavariables in the natural way.

β-processes (ranged over by $B, B_1, .., B', ..$) are generated by the following grammar:

$$B ::= \mathsf{Nil} \;\big|\; \vec{B}[P] \;\big|\; B \parallel B.$$

Nil denotes the deadlocked box and is the neutral element of the parallel composition of beta-processes, written $B \parallel B$. But for Nil, the simplest form of beta-process is given by a process encapsulated into a β-binder ($\vec{B}[P]$). Notice that nesting of boxes is not allowed.

Each β-process consisting of n parallel components has a simple graphical notation, given by n distinct boxes, one per parallel component. Each box contains a process and has as many sites (hidden or not) as the number of elementary beta binders in the composite binder. The relative position of sites along the perimeter of the box is irrelevant, just as the relative positions of parallel boxes in the two-dimensional space.

β-processes are given an operational reduction semantics that makes use of both a structural congruence over β-processes and a structural congruence over processes. We overload the same symbol to denote both congruences, and let the context disambiguate the intended relation. The *structural congruence* over processes coincides with the one defined in Sect. C.3. *Structural congruence* over β-processes, denoted by $\equiv$, is the smallest relation which satisfies the laws listed below, where $\hat{\beta}$ is intended to range over $\{\beta, \beta^h\}$.

- $\vec{B}[P_1] \equiv \vec{B}[P_2]$ provided $P_1 \equiv P_2$
- $B_1 \parallel (B_2 \parallel B_3) \equiv (B_1 \parallel B_2) \parallel B_3$, $B_1 \parallel B_2 \equiv B_2 \parallel B_1$, $B \parallel \mathsf{Nil} \equiv B$
- $\vec{B}_1\vec{B}_2[P] \equiv \vec{B}_2\vec{B}_1[P]$
- $\vec{B}^*\hat{\beta}(x : \Gamma)[P] \equiv \vec{B}^*\hat{\beta}(y : \Gamma)[P\{y/x\}]$ provided y fresh in P and $y \notin \mathsf{sub}(\vec{B}^*)$

The laws over β-processes state, respectively, that (i) the structural congruence of processes is reflected at the upper level as congruence of boxes; (ii) the parallel composition of β-processes is a monoidal operation with neutral element Nil; (iii) the actual ordering of elementary beta binders within a composite binder is irrelevant; and (iv) the subject of elementary beta binders is a placeholder that can be changed at any time under the proviso that name clashes are avoided and the well-formedness of β-binder is preserved.

The *reduction relation*, $\longrightarrow$, is the smallest relation over β-processes obtained by applying the following axioms and rules:

$$\text{(intra)}\quad \frac{P \equiv (new\ \tilde{u})\,(x?w.\,P_1 \mid x!z.\,P_2 \mid P_3)}{\vec{B}[P] \longrightarrow \vec{B}[(new\ \tilde{u})\,(P_1\{z/w\} \mid P_2 \mid P_3)]}$$

$$\text{(inter)}\quad \frac{P \equiv (new\ \tilde{u})\,(x?w.\,P_1 \mid P_2) \qquad Q \equiv (new\ \tilde{v})\,(y!z.\,Q_1 \mid Q_2)}{\beta(x:\Gamma)\,\vec{B}_1^*[P] \parallel \beta(y:\Delta)\,\vec{B}_2^*[Q] \longrightarrow \beta(x:\Gamma)\,\vec{B}_1^*[P'] \parallel \beta(y:\Delta)\,\vec{B}_2^*[Q']}$$

where $P' = (new\ \tilde{u})\,(P_1\{z/w\} \mid P_2)$ and $Q' = (new\ \tilde{v})\,(Q_1 \mid Q_2)$

$\Gamma \cap \Delta \neq \emptyset$, $x, z \notin \tilde{u}$, $y, z \notin \tilde{v}$

$$\text{(expose)}\quad \frac{P \equiv (new\ \tilde{u})\,(\mathsf{expose}(x,\,\Gamma)\,.\,P_1 \mid P_2)}{\vec{B}[P] \longrightarrow \vec{B}\,\beta(y:\Gamma)[(new\ \tilde{u})\,(P_1\{y/x\} \mid P_2)]} \qquad y \notin \tilde{u},\ y \notin \mathsf{sub}(\vec{B}),\ y \notin \Gamma$$

$$\text{(hide)} \quad \frac{P \equiv (new\ \tilde{u})\,(\mathsf{hide}(x)\,.\,P_1 \mid P_2)}{\beta(x:\Gamma)\,\vec{B}^*[P] \longrightarrow \beta^h(x:\Gamma)\,\vec{B}^*[(new\ \tilde{u})\,(P_1 \mid P_2)]} \quad x \notin \tilde{u}$$

$$\text{(unhide)} \quad \frac{P \equiv (new\ \tilde{u})\,(\mathsf{unhide}(x)\,.\,P_1 \mid P_2)}{\beta^h(x:\Gamma)\,\vec{B}^*[P] \longrightarrow \beta(x:\Gamma)\,\vec{B}^*[(new\ \tilde{u})\,(P_1 \mid P_2)]} \quad x \notin \tilde{u}$$

(join) $\vec{B}_1[P_1] \parallel \vec{B}_2[P_2] \longrightarrow \vec{B}[P_1\sigma_1 \mid P_2\sigma_2]$

f_{join} defined in $(\vec{B}_1, \vec{B}_2, P_1, P_2)$ and

with $f_{join}(\vec{B}_1, \vec{B}_2, P_1, P_2) = (\vec{B}, \sigma_1, \sigma_2)$

(split) $\vec{B}[P_1 \mid P_2] \longrightarrow \vec{B}_1[P_1\sigma_1] \parallel \vec{B}_2[P_2\sigma_2]$

f_{split} defined in $(\vec{B}, P_1, P_2)$ and

with $f_{split}(\vec{B}, P_1, P_2) = (\vec{B}_1, \vec{B}_2, \sigma_1, \sigma_2)$

$$\text{(redex)} \quad \frac{B \longrightarrow B'}{B \parallel B'' \longrightarrow B' \parallel B''}$$

$$\text{(struct)} \quad \frac{B_1 \equiv B_1' \qquad B_1' \longrightarrow B_2}{B_1 \longrightarrow B_2}.$$

The reduction relation describes the evolution within boxes (intra), as well as the interaction between boxes (inter), the dynamics of box interfaces (expose, hide, unhide) and the structural modification of boxes (join, split).

The rule intra lifts to the level of β-processes any "reduction" of the enclosed process. Note that no reduction relation is defined over processes.

The rule inter models interactions between boxes with complementary internal actions (input/output) over complementary sites (sites with non-disjoint types). Information flows from the box containing the process which exhibits the output prefix to the box enclosing the process which is ready to perform the input.

The rules expose, hide and unhide correspond to an unguarded occurrence of the homonymous prefix in the internal process and allow the dynamic modification of external interfaces.

The rule expose causes the addition of an extra site with the declared type. The name x used in $\mathsf{expose}(x, \Gamma)$ is a placeholder which can be renamed to meet the requirement of well-formedness of the enclosing β-binder.

The rules hide and unhide force the specified site to become hidden and unhidden, respectively. They cannot be applied if the site does not occur unhidden hidden, respectively, in the enclosing β-binder.

The axiom join models the merge of boxes. The rule, being parametric w.r.t. the function f_{join}, is more precisely an axiom schema. The function f_{join} determines the actual interface of the β-process resulting from the aggregation of boxes, as well as possible renamings of the enclosed processes via the substitutions σ_1 and σ_2. It is intended that as many different instances of f_{join} (and hence of join) can be defined as are needed to model the system at hand.

The axiom split formalizes the splitting of a box in two parts, each of them taking away a subcomponent of the content of the original box. Analogously to join, the rule split is an axiom schema that depends on the specific definition of the function f_{split}. This function is meant to refine the conditions under which a β-process can be split in two boxes. Analogously to the case of the join axiom, many instances of split can live together in the same formal system.

The rules redex and struct are typical rules of reduction semantics. They are meant, respectively, to interpret the reduction of a subcomponent as a reduction of the global system, and to infer a reduction after a proper structural shuffling of the process at hand.

C.5 BlenX

We consider here just a subset of BlenX to give the flavor of how the semantics is defined. Let $\mathcal{N}$ be a countably infinite set of *names* (ranged over by x, y, n, x_1, x', $\cdots$) and let $\mathcal{T}$ be finite set of *types* (ranged over by Δ, Γ, Δ', Δ_0, $\cdots$) such that $\mathcal{T} \cap \mathcal{N} = \emptyset$. Moreover, let $\delta : \mathcal{N} \to \mathbb{R}$ be a function that associates stochastic rates to names and $\alpha : \mathcal{T}^2 \to \mathbb{R}^3$ be a symmetric function, called *affinity*, that associates triples of stochastic rates with type pairs. The subset of BlenX that we consider is

$$
\begin{array}{ll}
B ::= \mathsf{Nil} \mid I[P]_n \mid B \parallel B & M ::= \mathsf{nil} \mid \mathsf{rep}\ \pi.\,P \mid \ \pi.\,P \mid M + M \mid \langle C \rangle M \\
I \ ::= S(x,r,\Gamma) \ \mid \ S(x,r,\Gamma)\, I & \pi \ ::= x?y \mid \ x!y \mid \ \mathsf{ch}(x,\Gamma,r) \\
S \ ::= \mathsf{free} \mid \ \mathsf{bound} & C \ ::= (x,\Gamma) \mid \ (x,S) \mid \ op_u\ C \mid \ C\ op_b\ C \\
P ::= M \mid \ P|P. &
\end{array}
$$

A BlenX system, written (B, ξ), is a pair made up of a bio-process B and an environment ξ. We denote with $\mathcal{S}$ the set of all possible systems (ranged over by S, S', S_1, $\cdots$).

Bio-processes are generated by the non-terminal symbol B of the grammar. A bio-process can be either empty (Nil), a *box* ($I[P]_n$) or the parallel composition of bio-processes ($B||B$). In the definition of box $I[P]_n$, I represents its interaction capabilities, P its internal engine and n is used as an *identifier* to address the box at hand.

I is a non-empty string of *interfaces* of the form $S(x, r, \Delta)$, where S denotes the state of the interface, which can be either *free* ($\oplus$) or *bound* ($\otimes$), the name x is the *subject* of the interface and Δ is a type representing the structure of the interface. We sometimes use *active* as synonymous of free, and *complexed* as synonymous of bound. The subject x of the interface $S(x, r, \Delta)$ of a box $S(x, r, \Delta)\, I[P]_n$ is a binding for the free occurrences of x in P. The metavariables I^*, I_1^*, $\cdots$ stay for either an interface or the empty string. Moreover, we use the functions $\mathsf{sub}(I)$ and $\mathsf{sorts}(I)$ to extract from I the set of its subjects and the set of its types, respectively. We say that a box $I[P]_n$ is well-formed if all the subjects and types of the interfaces composing I are distinct. Moreover, we define an equivalence relation $\doteq$ on boxes that allows reasoning up-to renaming of interface subjects permitting a more natural definition of the operational semantics.

We denote with $\doteq$ the smallest equivalence relation on boxes that satisfies the axiom: $I_1^*\, S(x, r, \Delta)\, I_2^*[P]_z \doteq I_1^*\, S(y, r, \Delta)\, I_2^*[P\{y/x\}]_z$ provided $y \notin \mathsf{fn}(P) \cup \mathsf{sub}(I_1^*\, S(x, r, \Delta)\, I_2^*)$. Boxes related by means of $\doteq$ are identified.

The non-terminal symbol P generates *processes* (ranged over by P, P', P_1, $\cdots$). A process can be either the parallel composition of two processes ($P|P$), or the replication of an action-guarded process ($\mathsf{rep}\ \pi.\, P$), or the empty process (nil), or an action-guarded process ($\pi.\, P$), or the non-deterministic choice of action-guarded or replicated action-guarded processes ($M + M$). Replications, deadlock processes, action-guarded processes and non-deterministic choices can be guarded by a condition C ($\langle C \rangle M$). The condition C is a Boolean expression that controls the execution of the process M. Unary and binary operators used in conditions are such that $op_u \in \{\neg\}$ and $op_b \in \{\wedge, \vee\}$, while atoms refer to interfaces via their unique subjects and check whether they have a given type (e.g. Δ) or whether they are in a given state (e.g. $\otimes$). Given an interface I, if C evaluates to *true* on I (denoted with $[\![C]\!]_I = tt$), then M can fire. Considering that the only binder in the definition of processes is the name y in $x?y$, function $\mathsf{fn}(P)$ returns the free names of P and is defined in the usual way.

The *environment* component ξ of the BlenX system (B, ξ) is used to record the bindings of complexes. In detail, ξ is a set of pair sets of the shape $\{\Delta_1 n_1, \Delta_2 n_2\}$ meaning that the two boxes addressed by n_1 and n_2 are bound together through the interfaces with types Δ_1 and Δ_2, respectively.

A BlenX system (B, ξ) is well-formed iff (1) all the boxes composing B are well-formed, (2) all the boxes identifiers are distinct, (3) each element Δx in the environment ξ appears only once in ξ and (4) Δx appears in the environment ξ iff a box in B with identifier x and bound interface with type Δ exists.

Let $\overline{\mathcal{S}} \subset \mathcal{S}$ be the set of well-formed systems and let $\Theta = \{|_0, |_1, +_0, +_1\}^*$ be the set of *addresses* with metavariable ϑ. The reduction relation $\xrightarrow{\omega}_r$ (behavior of systems) is defined using the auxiliary relation $\xrightarrow{\overline{\omega}}_r$ (behavior of boxes) where $r \in \mathbb{R}_{\geq 0} \cup \{\infty\}$ and $\omega \in \Omega$ and $\overline{\omega} \in \overline{\Omega}$ are *labels* needed to identify reductions.

$$
\begin{aligned}
\overline{\Omega} = \; & \{\vartheta x?y \mid \vartheta \in \Theta \wedge x, y \in \mathcal{N}\} \; \cup \{\vartheta x!y \mid \vartheta \in \Theta \wedge x, y \in \mathcal{N}\} \; \cup \\
& \{\vartheta(\Delta, \Gamma) \mid \vartheta \in \Theta \wedge \Delta, \Gamma \in \mathcal{T}\} \; \cup \{\vartheta(\vartheta_1, \vartheta_2) \mid \vartheta, \vartheta_1, \vartheta_2 \in \Theta\} \\
\Omega = \; & \{n\vartheta \mid \vartheta \in \Theta\} \; \cup \{n\vartheta(\vartheta_1, \vartheta_2) \mid \vartheta, \vartheta_1, \vartheta_2 \in \Theta\} \; \cup \{n\Delta \mid \Delta \in \mathcal{T}\} \; \cup \\
& \{n\underline{\Delta} \mid \Delta \in \mathcal{T}\} \; \cup \{n\Delta\vartheta!y \mid \vartheta \in \Theta \wedge y, n \in \mathcal{N} \wedge \Delta \in \mathcal{T}\} \; \cup \\
& \{n\Delta\vartheta?y \mid \vartheta \in \Theta \wedge y, n \in \mathcal{N} \wedge \Delta \in \mathcal{T}\} \; \cup \\
& \{(n_1\Delta_1, n_2\Delta_2) \mid n_1, n_2 \in \mathcal{N} \wedge \Delta_1, \Delta_2 \in \mathcal{T}\} \; \cup \\
& \{(n_1\underline{\Delta_1}, n_2\underline{\Delta_2}) \mid n_1, n_2 \in \mathcal{N} \wedge \Delta_1, \Delta_2 \in \mathcal{T}\} \; \cup \\
& \{(n_1\vartheta_1, n_2\vartheta_2) \mid \vartheta_1, \vartheta_2 \in \Theta \wedge n_1, n_2 \in \mathcal{N}\}
\end{aligned}
$$

Rules (r1) and (r2) describe the execution of output and input prefixes, respectively. Rule (r3) describes how an interface type can be changed through the corresponding action. If Γ does not clash with the types of the other interfaces in I, then the type associated with x is turned into Γ; the label records information about the types involved in the change operation. Rule (r4) describes how guarded replication is unfolded, while (r5) describes *intracommunications*, i.e. communications between processes within the same box. Rule (r5) stores information on the label about the positions of input and output prefixes. Rule (r6) describes the behavior of the conditional operator: if the process guarded by the condition is not in deadlock and the condition is true in I, then the process can proceed and the condition is deleted. Rules (r7–10) collect processes contexts and update labels in the appropriate way.

Rules (r11–14) describe how to lift from $\xrightarrow{\overline{\omega}}_r$ to $\xrightarrow{\omega}_r$ relations, i.e. from box to system level. Outputs are lifted only if the communication object y does not clash with the subjects of the interfaces in I. Since interface subjects are bindings for processes and can be substituted at hand using the $\doteq$ relation, the side condition prevents the extrusion of private information, hence preserving a consistent system configuration. Also, outputs are lifted only if y does not clash with the subjects of the interfaces in I, preventing name captures (i.e. no receiving name has to be captured by interface subjects). Note that since boxes are identified up to $\doteq$ relation, then there always exists a box configuration satisfying the condition. A change action is lifted by applying a substitution on the environment. In particular, since the interface could be involved in a binding, the environment is consistently updated by possibly refreshing the previous value of the interface with Γ. Here notice that the requirement about the freshness of Γ in rule (r3) guarantees the freshness of Γn in the updated environment. Note that for all these rules labels are updated in the proper way, by preserving and adding only relevant information.

(r1) $I[x!y.\,P]_n \xrightarrow{x!y}_0 I[P]_n$ (r2) $I[x?y.\,P]_n \xrightarrow{x?z}_0 I[P\{z/y\}]_n$

$$\text{(r3)}\quad \frac{I = I_1^*\; S(x,r,\Delta)\; I_2^* \wedge \Gamma \notin \mathsf{sorts}(I)}{I[\mathsf{ch}(x,\Gamma,r).\,P]_n \xrightarrow{(\Delta,\Gamma)}_r I_1^*\; S(x,r,\Gamma)\; I_2^*[P]_n}$$

$$\text{(r4)}\quad \frac{I = I_1^*\; S(x,r,\Delta)\; I_2^* \wedge \Gamma \notin \mathsf{sorts}(I)}{I[\mathsf{ch}(x,\Gamma,r).\,P]_n \xrightarrow{(\Delta,\Gamma)}_r I_1^*\; S(x,r,\Gamma)\; I_2^*[P]_n}$$

$$\text{(r5)}\quad \frac{\begin{array}{c} I[P_1]_n \xrightarrow{\vartheta_1 x!z}_{r_1} I[P_1']_n \wedge I[P_2]_n \xrightarrow{\vartheta_2 x?z}_{r_2} I'[P_2']_n \wedge \\ ((S(x,r,\Delta) \in I) \vee (x \notin \mathsf{sub}(I) \wedge \delta(x) = r)) \end{array}}{I[P_1 \mid P_2]_n \xrightarrow{(\vartheta_1,\vartheta_2)}_r I'[P_1' \mid P_2']_n}$$

$$\text{(r6)}\quad \frac{I[M]_n \xrightarrow{\overline{\omega}}_r I'[M']_n \wedge [\![C]\!]_I = tt}{I[\langle C\rangle M]_n \xrightarrow{\overline{\omega}}_r I'[M']_n}$$

$$\text{(r7)}\quad \frac{I[M]_n \xrightarrow{\overline{\omega}}_r I[M']_n}{I[M+N]_n \xrightarrow{+_0\overline{\omega}}_r I[M']_n} \qquad \text{(r8)}\quad \frac{I[N]_n \xrightarrow{\overline{\omega}}_r I[N']_n}{I[M+N]_n \xrightarrow{+_1\overline{\omega}}_r I[N']_n}$$

$$\text{(r9)}\quad \frac{I[P]_n \xrightarrow{\overline{\omega}}_r I[P']_n}{I[P \mid Q]_n \xrightarrow{|_0\overline{\omega}}_r I[P' \mid Q]_n} \qquad \text{(r10)}\quad \frac{I[Q]_n \xrightarrow{\overline{\omega}}_r I[Q']_n}{I[P \mid Q]_n \xrightarrow{|_1\overline{\omega}}_r I[P \mid Q']_n}$$

$$\text{(r11)}\quad \frac{I_1^* \otimes(x, r, \Delta)\, I_2^*[P]_n \xrightarrow{\vartheta x!y}_r I[P']_n \wedge y \notin \mathsf{sub}(I)}{(I_1^* \otimes(x, r, \Delta)\, I_2^*[P]_n, \xi) \xrightarrow{n\Delta\vartheta!y}_r (I[P']_n, \xi)}$$

$$\text{(r12)}\quad \frac{I[P]_n \xrightarrow{\vartheta(\vartheta_1,\vartheta_2)}_r I[P']_n}{(I[P]_n, \xi) \xrightarrow{n\vartheta(\vartheta_1,\vartheta_2)}_r (I[P']_n, \xi)}$$

$$\text{(r13)}\quad \frac{I_1^* \otimes(x, r, \Delta)\, I_2^*[P]_n \xrightarrow{\vartheta x?y}_r I[P']_n \wedge y \notin \mathsf{sub}(I)}{(I_1^* \otimes(x, r, \Delta)\, I_2^*[P]_n, \xi) \xrightarrow{n\Delta\vartheta?y}_r (I[P']_n, \xi)}$$

$$\text{(r14)}\quad \frac{I[P]_n \xrightarrow{\vartheta(\Delta,\Gamma)}_r I'[P']_n}{(I[P]_n, \xi) \xrightarrow{n\vartheta}_r (I'[P']_n, \xi\{\Gamma n/\Delta n\})}$$

Boxes can interact the one with the other in various ways: they can bind together, unbind or communicate when bound. These interactions are based on the existence of the affinity function α which returns a triple of real values representing the binding, unbinding and intercommunication stochastic rates of the two argument values. We use $\alpha_b(\Delta, \Gamma)$, $\alpha_u(\Delta, \Gamma)$ and $\alpha_c(\Delta, \Gamma)$ to mean, respectively, the first, the second and the third projection of $\alpha(\Delta, \Gamma)$. Rules (r15), (r16), (r20) and (r21) describe the dynamics of binding and unbinding, respectively. In both cases the modification of the binding state of the relevant interfaces is reflected in the interface markers, which are changed either from $\oplus$ to $\otimes$ or the other way round. Also, the association $\{\Delta_1 n_1, \Delta_2 n_2\}$ recording the actual binding is either added to the environment or removed from it. The third kind of interaction between boxes, called *intercommunication*, is ruled by (r19). This involves an input and an output action firable in two distinct boxes over interfaces with associated values Δ_1 and Δ_2. Information passes from the box containing the sending action to the box enclosing the receiving process. Notice here that intercommunication depends on the affinity of Δ_1 and Δ_2 rather than on the fact that input and output actions occur over exactly the same name. Intercommunication is enabled only under the proviso that the two interfaces are already bound together. Rules (r17) and (r18) collect bio-process contexts. Labels of binding, unbinding and intercommunication actions record box identifiers and, respectively, the types of the interfaces involved in the bindings or unbindings and the ϑ addresses of the inputs and outputs involved in the communications.

(r15) $(I_1^* \oplus(x,r,\Delta)\, I_2^*[P]_n, \xi) \xrightarrow{n\underline{\Delta}}_0 (I_1^* \otimes(x,r,\Delta)\, I_2^*[P]_n, \xi)$

(r16) $(I_1^* \otimes(x,r,\Delta)\, I_2^*[P]_n, \xi) \xrightarrow{n\underline{\Delta}}_0 (I_1^* \oplus(x,r,\Delta)\, I_2^*[P]_n, \xi)$

(r17) $$\frac{(B_1, \xi) \xrightarrow{\omega}_r (B_1', \xi')}{(B_1 \parallel B_2, \xi) \xrightarrow{\omega}_r (B_1' \parallel B_2, \xi')}$$

(r18) $$\frac{(B_2, \xi) \xrightarrow{\omega}_r (B_2', \xi')}{(B_1 \parallel B_2, \xi) \xrightarrow{\omega}_r (B_1 \parallel B_2', \xi')}$$

(r19) $$\frac{(B_1, \xi) \xrightarrow{n_1\Delta_1\vartheta_1!z}_{r_1} (B_1', \xi) \wedge (B_2, \xi) \xrightarrow{n_2\Delta_2\vartheta_2?z}_{r_2} (B_2', \xi) \wedge \{\Delta_1 n_1, \Delta_2 n_2\} \in \xi}{(B_1 \parallel B_2, \xi) \xrightarrow{(n_1\vartheta_1, n_2\vartheta_2)}_{\alpha_c(\Delta_1,\Delta_2)} (B_1' \parallel B_2', \xi)}$$

(r20) $$\frac{(B_1, \xi) \xrightarrow{n_1\underline{\underline{\Delta_1}}}_{r_1} (B_1', \xi) \wedge (B_2, \xi) \xrightarrow{n_2\underline{\underline{\Delta_2}}}_{r_2} (B_2', \xi)}{(B_1 \parallel B_2, \xi) \xrightarrow{(n_1\underline{\underline{\Delta_1}}, n_2\underline{\underline{\Delta_2}})}_{\alpha_b(\Delta_1,\Delta_2)} (B_1' \parallel B_2', \xi \cup \{\{\Delta_1 n_1, \Delta_2 n_2\}\})}$$

(21) $$\frac{(B_1, \xi) \xrightarrow{n_1\overline{\Delta_1}}_{r_1} (B_1', \xi) \wedge (B_2, \xi) \xrightarrow{n_2\overline{\Delta_2}}_{r_2} (B_2', \xi)}{(B_1 \parallel B_2, \xi \cup \{\{\Delta_1 n_1, \Delta_2 n_2\}\}) \xrightarrow{(n_1\overline{\Delta_1}, n_2\overline{\Delta_2})}_{\alpha_u(\Delta_1,\Delta_2)} (B_1' \parallel B_2', \xi)}$$

Given a well-formed BlenX system S, the following definitions describe how to extract its labeled stochastic transition graph and its continuous time Markov Chain. Let $S \in \overline{\mathcal{S}}$. The labeled stochastic transition graph (LSTG) of S, referred to as Φ^S, is obtained with the following procedure:

$$\begin{aligned}
\Phi^S &= \textstyle\bigcup_n \Phi_n \text{ where} \\
\Phi_0 &= \{S \xrightarrow{\omega}_r S' \mid r > 0 \wedge \nexists \omega' \text{ s.t. } S \xrightarrow{\omega'}_\infty S''\} \cup \{S \xrightarrow{\omega}_\infty S'\} \\
\Phi_{n+1} &= \{S \xrightarrow{\omega}_r S' \mid r > 0 \wedge \nexists \omega' \text{ s.t. } S \xrightarrow{\omega'}_\infty S'' \wedge S \text{ is a state of } \Phi_n\} \cup \\
&\quad\ \{S \xrightarrow{\omega}_\infty S' \mid S \text{ is a state of } \Phi_n\}.
\end{aligned}$$

Let $S \in \overline{\mathcal{S}}$. The continuous time Markov chain (CTMC) $|\Phi^S|$ of S is obtained from Φ^S in the following way:

$$\begin{aligned}
|\Phi^S| = \{\ & S \xrightarrow{r} S' \mid S \neq S' \wedge \exists S \xrightarrow{\omega}_{s \in \mathbb{R}} S' \in \Phi^S \wedge r = \textstyle\sum r_i \text{ s.t. } S \xrightarrow{\omega_i}_{r_i} S' \in \Phi^S\} \cup \\
\{\ & S \xrightarrow{\infty(n/m)} S' \mid n = \#\{\omega \mid S \xrightarrow{\omega}_\infty S' \in \Phi^S\} \wedge m = \#\{\omega \mid S \xrightarrow{\omega}_\infty S'' \in \Phi^S\}\}.
\end{aligned}$$

Note that our CTMCs contain immediate transitions with associated probabilities. Standard algorithms can be used to eliminate immediate transitions from the CTMC, preserving the probabilities of transitions and sojourn times.

C.6 ℓ

This section defines the formal semantics of ℓ. After introducing some basic notions, we define the semantics of commands and expressions that we then use for the definition of the semantics of the rewriting rules.

C.6.1 *Basic notions*

The set of complexes is defined as the set of all multisets over boxes at a location from location, i.e., Cplx $=$ location $\times$ mset Box, and we use $[box_1, .., box_n]@loc$ to denote a generic element of Cplx. We define a *system* $\mu \in$ Sys $=$ mset Cplx as a multiset of complexes. Boxes and complexes are stored at specific *addresses* of the heap. The set of all these addresses is denoted by Address, and we use α to range over it. We define Environment $=$ Id $\rightarrow$ Address, the set of *environments*, i.e., maps assigning addresses to variables, and Store $=$ Address $\rightarrow$ Value, the set of *stores*, i.e., maps assigning values to addresses. The set Value comprises basic type values, locations, box values (represented as maps from fields to values) and complex values. Complex values are represented as multisets of addresses pointing to the contained boxes, the whole multiset being tagged with its location. Moreover, we let Value also include references (Address) and anonymous functions, represented as a list of formal arguments and the function body. By composing Environment and Store we retrieve the values associated with names in the usual way (see Sect. C.2.2).

Given a multiset of complexes $\{C_1, ..C_n\}$, we denote with alloc$(x_1 = C_1, .., x_n = C_n)$ the environment and store obtained by allocating the multiset in store and making variables $x_1, .., x_n$ refer to the complexes. This essentially requires allocating in the store 1) each box, 2) each complex, linking it to its contained boxes and 3) each variable x_i, whose value is the reference to the allocated C_i complex. The environment then maps the variable names to their addresses (that are references pointing to references for complexes). The system configuration

$\mathsf{alloc}(x_1 = C_1, x_2 = C_2, ..) = \langle \rho, \sigma \rangle \in \mathsf{Environment} \times \mathsf{Store}$ is defined as:

$$\rho = [x_1 \mapsto \alpha_{x_1}, x_2 \mapsto \alpha_{x_2}, ..],$$

$$\sigma = \begin{bmatrix} \alpha_{x_1} \mapsto \alpha_{rx_1}, \alpha_{x_2} \mapsto \alpha_{rx_2}, .., & \\ \alpha_{rx_1} \mapsto [\alpha_1^1, .., \alpha_{n_1}^1]@loc_1 & , \\ \alpha_{rx_2} \mapsto [\alpha_1^2, .., \alpha_{n_2}^2]@loc_2 & , \\ .. & , \\ \alpha_1^1 \mapsto box_1^1, .., \alpha_{n_1}^1 \mapsto box_{n_1}^1 & , \\ \alpha_1^2 \mapsto box_1^2, .., \alpha_{n_2}^2 \mapsto box_{n_2}^1 & , \\ .. & \end{bmatrix}$$

where $C_1 = [box_1^1, .., box_{n_1}^1]@loc_1, C_2 = [box_1^2, .., box_{n_2}^2]@loc_2, ..$ and all the α addresses are distinct.

The operation $\mathsf{extractCplx} : \mathsf{Store} \to \mathsf{Sys}$ performs the inverse operation, retrieving the multiset of complexes (i.e., an element of Sys) which are in a store σ (we identify multisets with functions returning natural numbers) and it is defined as:

$$\mathsf{extractCplx}(\sigma) = \lambda C. \left| \left\{ \alpha \,\middle|\, \begin{array}{l} C = [box_1, .., box_n]@loc, \\ \sigma(\alpha) = [\alpha_1, .., \alpha_n]@loc, \\ \forall 1 \leq i \leq n.\sigma(\alpha_i) = box_i \end{array} \right\} \right|$$

that counts the number of instances in σ for a given complex C.

C.6.2 *Semantics of commands and expressions*

We now discuss the *semantics of expressions*

$$\mathcal{E} : \ Exp \times \mathsf{Sys} \times \mathsf{Environment} \times \mathsf{Store} \to \mathsf{Value} \times \mathsf{Store},$$

and the *semantics of commands*

$$\mathcal{C} : \ (Block \cup Cmd) \times \mathsf{Sys} \times \mathsf{Environment} \times \mathsf{Store} \to \mathsf{Store}.$$

Note that $\mathcal{C}$ and $\mathcal{E}$ are partial functions to account for non-terminating programs. For simplicity, we will neglect this issue, assuming termination. Indeed, modeling non-termination in the ℓ semantics would pose no novel challenge, and can be done following standard techniques (e.g., domain lifting) at the expense of verbosity.

Note also that ℓ can be given a static semantics through a standard type system by a routine task, which is easy because each variable is either annotated with its type (e.g., function arguments) or initialized on declaration (var $Ide := Exp$). Hence, no type inference is needed to check whether an ℓ model adheres to the type system.

Since the majority of expressions and commands of ℓ are standard expressions and commands of an imperative programming language, we do not provide a formal definition of the semantics for all of them. Most operations on complexes (e.g., adding or removing boxes from them) pose no technical problems. To exemplify this, we will detail the semantics of the expression count and the command foreach.

C.6.2.1 *Semantics of the* count *expression*

In order to evaluate an expression count($pattern@exp$) in a system $\mu \in$ Sys, environment $\rho \in$ Environment and store $\sigma \in$ Store, we first evaluate all the sub-expressions in *pattern* and *exp*. The evaluation of expressions in *pattern* produces a pattern *pattern′* carrying only literals, while the evaluation of *exp* produces a location *loc*. We then consider the complexes in such a location, and count the ones matching *pattern*. In our semantics, complexes are represented in a partitioned way: the store σ comprises only those complexes being affected by the rule at hand and the newly spawned ones, while the system μ contains all the other complexes. Therefore, we have to count complexes in both σ and μ. The result of the evaluation of count($pattern@exp$) is then the sum of both counts.

We use a set of pairs I which is used as an index set enumerating all the constraints $f_{i,j} = e_{i,j}$ occurring in *pattern*. The set I is assumed to be linearly ordered, and equipped by a successor operation so that $(i, j) + 1$ is the index of the next constraint.

$$\frac{\begin{array}{c} pattern = .., B_i\{..f_{i,j} = e_{i,j}..\}.. \quad \text{with } (i,j) \in I \\ \forall (i,j) \in I.\ \mathcal{E}(e_{i,j}, \mu, \rho, \sigma_{(i,j)}) = \langle v_{i,j}, \sigma_{(i,j)+1} \rangle \\ \sigma_{\min I} = \sigma \quad \sigma' = \sigma_{\max I+1} \\ pattern' = .., B_i\{..f_{i,j} = v_{i,j}..\}.. \quad \text{with } (i,j) \in I \\ \mathcal{E}(exp, \mu, \rho, \sigma') = \langle loc, \sigma'' \rangle \\ m = \sum_{c\ s.t.pattern' \models c} \mu(c@loc) \qquad m' = \mathsf{extractCplx}(\sigma'')(c@loc) \end{array}}{\mathcal{E}(\mathsf{count}(pattern@exp), \mu, \rho, \sigma) = \langle m + m', \sigma'' \rangle}$$

C.6.2.2 *Semantics of the* foreach *command*

In order to execute a command foreach $x : A$ in exp do $Block$ we first evaluate the expression exp which produces the address (α_C) of a complex. We then scan that complex filtering out the boxes of type A. We assign each of them to the variable x and execute the block of commands $Block$.

We use the command $x := \alpha$ to assign address α to x. This is meant to be a run-time syntax, and not to be available to the ℓ user. Finally, note that the order according to which a foreach loop visits the boxes in a complex is left unspecified.

$$\frac{\begin{array}{c}\mathcal{E}(exp, \mu, \rho, \sigma) = \langle \alpha_C, \sigma' \rangle \\ \sigma'(\alpha_C) = [\alpha_1, .., \alpha_n]@loc \\ \{\alpha_{A_1}, .., \alpha_{A_m}\} = \{\alpha_i \mid \sigma'(\alpha_i) = A[..]\} \\ \mathcal{C}(x := \alpha_{A_1}; Block; ..x := \alpha_{A_m}; Block\ , \mu, \rho, \sigma') = \sigma''\end{array}}{\mathcal{C}(\text{foreach } x : A \text{ in } exp \text{ do } Block\ , \mu, \rho, \sigma) = \sigma''}$$

C.6.3 *Syntactic desugaring*

We describe here the simplification of the ℓ rules so that they can all be rewritten by just using the small kernel

dyn $p_1..p_n$ rate custom $rateExp$ react $Block$
dissoc p_1 p_2 rate custom $rateExp$ react $Block$.

First, we handle the non-custom rate expressions corresponding to mass action kinetics (ℓ default). The desugaring is

$$\text{rate } rateExp \Rightarrow \text{rate custom } rateExp \; * MassActionLaw(\text{reagent}_1, .., \text{reagent}_n)$$

where n is the number of reagents of the rule on which the rate expression occurs. $MassActionLaw$ counts the number of submultisets $\{C_1, .., C_n\}$ of the system, where each C_i has the complex species of reagent$_i$. Michaelis–Menten rate expressions rate mm $rateExp1$: $rateExp2$ are handled similarly to mass action law, by inserting a function call to compute the right rate.

The effect of a when $cond$ constraint is obtained by making the rate null when $cond$ is false. We use (as a run-time syntax) a C-style ternary operator $(-\,?-:-)$ acting as an expression-level conditional. We neglect non-custom rates, because they are special cases of custom ones.

$$\text{when } cond \text{ rate custom } rateExp \Rightarrow \text{rate custom } (cond\,?\,rateExp\ :\ 0)$$

The in*loc* constraint restricts the application of the rule to a specific location *loc*. It is imposed by requiring each reagent to be in *loc*. We also record the location in a special variable *currentLoc*, making it visible to the react block.

$$\begin{array}{l} \text{..in } loc \text{ when } cond \text{ rate custom } rateExp \text{ react } Block \\ \Rightarrow \text{..when} \begin{pmatrix} \text{reagent}_1\text{.loc} == loc\ \&\& \\ .. \\ \text{reagent}_n\text{.loc} == loc\ \&\& \\ cond \end{pmatrix} \text{rate custom } rateExp \\ \text{react var } currentLoc := loc; Block \end{array}$$

We now desugar assoc, move and substitute. An assoc rule acts similarly to dyn except that it merges two reagents before executing its *block*, i.e.,

$$\begin{array}{l} \text{assoc } p_1\ p_2\ rateClause \text{ react } Block \text{ end} \Rightarrow \text{dyn } p_1\ p_2\ rateClause \text{ react} \\ \quad \text{reagent}_1\text{.assoc(reagent}_2\text{);} \\ \quad \text{var product} := \text{reagent}_1; \\ \quad Block. \end{array}$$

A move rule changes the location of its reagent. Note that in *loc* is not allowed in move rules, because the from part already defines the source location.

$$\begin{array}{l} \text{move } p_1 \text{ from } loc_1 \text{ to } loc_2 \text{ when } cond \text{ rate custom } rateExp \\ \Rightarrow \text{dyn } p_1 \text{ in } loc_1 \text{ when } cond \text{ rate custom } rateExp \\ \quad \text{react reagent}_1\text{.loc} := loc_2; \text{ end} \end{array}$$

A substitute rule replaces its reagents with its products. We desugar it by removing reagents and spawning the products as new complexes in the same location of reagents. If the reagents are empty, an in *loc* is required in the rate clause to provide a spawning location (via the variable *currentLoc*). The desugaring accounts for both empty/non-empty reagent cases. Note that whenever the reagents are non-empty and there is an in *loc* clause, the variable *currentLoc* is defined twice by our desugaring, but both definitions agree.

$$\begin{array}{l}\mathsf{substitute}\ p_1, .., p_n\ \mathsf{with}\ cplx_1, .., cplx_m\ rateClause\ \mathsf{react}\ Block \\ (n > 0) \Rightarrow \mathsf{dyn}\ p_1, .., p_n\ rateClause\ \mathsf{react} \\ \qquad\qquad \mathsf{var}\ currentLoc := \mathsf{reagent}_1.\mathsf{loc}; \\ \qquad\qquad \mathsf{reagent}_1.\mathsf{kill}(); ..; \mathsf{reagent}_n.\mathsf{kill}(); \\ \qquad\qquad \mathsf{spawn}(cplx_1@currentLoc); ..; \\ \qquad\qquad \mathsf{spawn}(cplx_m@currentLoc); \\ \qquad\qquad Block\end{array}$$

$$\begin{array}{l}\mathsf{substitute\ with}\ cplx_1, .., cplx_m\ rateClause\ \mathsf{react}\ Block \\ \Rightarrow \qquad \mathsf{dyn}\ rateClause\ \mathsf{react} \\ \qquad\qquad \mathsf{spawn}(cplx_1@currentLoc); ..; \\ \qquad\qquad \mathsf{spawn}(cplx_m@currentLoc); \\ \qquad\qquad Block\end{array}$$

The rules having many *RateClauses* are expanded as follows:

$$\begin{array}{l}\mathsf{dyn}\ p_1\ ..p_m\ rateClause_1\ ..rateClause_n\ \mathsf{react}\ Block \\ \quad \Rightarrow \mathsf{dyn}\ p_1\ ..p_m\ rateClause_1\ \mathsf{react}\ Block \\ \qquad .. \\ \qquad \mathsf{dyn}\ p_1\ ..p_m\ rateClause_n\ \mathsf{react}\ Block.\end{array}$$

The same expansion is also applied to assoc, dissoc, move, substitute rules.

C.6.4 *Semantics of rules*

We only consider rules of the form

$$\begin{array}{l}\mathsf{dyn}\ p_1..p_n\ \mathsf{rate\ custom}\ rateExp\ \mathsf{react}\ Block \\ \mathsf{dissoc}\ p_1\ (p_2)\ \mathsf{rate\ custom}\ rateExp\ \mathsf{react}\ Block.\end{array}$$

The semantics of all other rule forms is obtained by gradually desugaring such rules into the small kernel above as shown in Sect. C.6.3.

We extend the matching relation $\vdash$ between patterns and complexes to n-tuples of patterns and n-tuples of complexes:

$$\vec{p} \vdash \vec{C} \iff \forall i.\ p_i \vdash C_i$$

and we use it to define the set of matches of a pattern tuple in a system state μ. Such matches are required to belong to the same location.

The *simple matches* of $\vec{p} = \langle p_1, .., p_n\rangle$ in a system $\mu \in \mathsf{Sys}$ is

$$simple(\vec{p}, \mu) = \left\{\vec{C} \in \mathsf{Cplx}^n \;\middle|\; \{C_1, .., C_n\} \subseteq \mu \ \wedge\ \vec{p} \vdash \vec{C} \ \wedge\ \forall i.\ C_i.\mathsf{loc} = C_j.\mathsf{loc}\right\}.$$

Note that simple matches may contain tuples of complexes, which differ only by their ordering. For instance, take $p_1 = [A, *]$ and $p_2 = [B, *]$, and a system μ with complexes $C_1 = [A, B, X]$ and $C_2 = [A, B, Y]$. Then, $simple(\vec{p}, \mu) = \{\langle C_1, C_2\rangle, \langle C_2, C_1\rangle\}$ since each complex matches with both patterns. This would lead to a semantics in which a rule dyn p_1 p_2 rate 1.0 in a system C_1, C_2 would trigger with rate 2 instead of 1 because of the two different matches. To overcome this issue, when multiple orderings of the same match appear, we will select just one of these (a *canonic* match), and disregard the others.

Let $<$ stand for any total strict ordering relation over Cplx. This induces the lexicographic ordering on n-tuples of Cplx, which we shall also denote with $<$.

The relation $\sim$ holds between two complex tuples differing only by the order of their components, i.e., when the first tuple is a permutation of the second.

$$\vec{C}' \sim \vec{C} \iff \exists \pi : \{1..n\} \to \{1..n\}.\ \pi \text{ bijective} \wedge \forall i.\ C'_i = C_{\pi(i)}$$

The *canonic matches* of $\vec{p} = \langle p_1, .., p_n\rangle$ in a system $\mu \in \mathsf{Sys}$ is given by

$$canonic(\vec{p}, \mu) = \left\{\vec{C} \in\ simple(\vec{p}, \mu)\ \middle|\ \nexists \vec{C}' \in\ simple(\vec{p}, \mu).\ \vec{C}' \sim \vec{C} \wedge \vec{C}' < \vec{C}\right\}$$

We generate a transition between states μ and μ' for each canonic match in μ, which makes the react code block change μ into μ'. The rates of these transitions are computed by evaluating the rate expression, the result of which may depend on the chosen match. More in detail, after choosing a canonic match $\vec{C}$, we remove $\vec{C}$ from the system μ, and name $\hat{\mu}$ the rest of the system. Then, we represent $\vec{C}$ in an environment and store $\hat{\rho}, \hat{\sigma}$, in which we evaluate the rate expression. Finally, we compute the final state μ' by executing the react *block* in the same environment and store, and merging the new complexes spawned in this way with the system $\hat{\mu}$. The rates generated as discussed above are decorated to keep track of their multiplicity. To this aim, we pair rates with the match $\vec{C}$, which generates them.

Given a rule of the form dyn $\vec{p}$ rate $rateExp$ react $Block$ and two systems μ, μ', we generate the (decorated) set of rates of transitions as follows:

$$rates(\mu, \mu', \text{ dyn } \vec{p} \text{ rate } rateExp \text{ react } Block) =$$

$$\left\{ \langle \vec{C}, a\rangle\ \middle|\ \begin{array}{l} \vec{C} = \langle C_1, .., C_n\rangle \in canonic(\vec{p}, \mu)\ \wedge\ \hat{\mu} = \mu \setminus \vec{C}\ \wedge \\ \langle \hat{\rho}, \hat{\sigma}\rangle = \mathsf{alloc}(\mathsf{reagent}_1 = C_1, .., \mathsf{reagent}_n = C_n)\ \wedge \\ a = \mathsf{value}(\mathcal{E}(rateExp)\langle \hat{\mu}, \hat{\rho}, \hat{\sigma}\rangle)\ \wedge \\ \mu' = \hat{\mu} \cup \mathsf{extractCplx}(\mathcal{C}(Block)\langle \hat{\mu}, \hat{\rho}, \hat{\sigma}\rangle) \end{array} \right\}.$$

The rates generated by a rule $\mathsf{dissoc}\ p_1 p_2\ \mathsf{rate}\ \mathit{rateExp}\ \mathsf{react}\ \mathit{Block}$ are defined similarly. First, we consider complexes C_1 matching pattern p_1, as well as their subcomplexes C_2 matching p_2. Then, we define the transition rate a by evaluating the rate expression. For this step we remove C_1 from the starting state, obtaining $\hat{\mu}$, and representing it in the environment and store $\hat{\rho}, \hat{\sigma}$. The rate a is then normalized over the total number of matches of p_2 inside C_1. To compute the ending state μ' we run the react Block. This is done in an environment and store $\hat{\rho}, \hat{\sigma}$ in which the complex C_1 has been dissociated into C_2 and its complement. Every rate generated in this way is decorated to preserve its multiplicity, as in the dyn case. The set of rates of transitions is

$$\mathit{rates}(\mu,\ \mu',\ \mathsf{dissoc}\ p_1 p_2\ \mathsf{rate}\ \mathit{rateExp}\ \mathsf{react}\ \mathit{Block}) =$$

$$\left\{ \langle\langle C_1, \theta\rangle, a\rangle \;\middle|\; \begin{array}{l} \langle C_1\rangle \in \mathit{canonic}(p_1, \mu) \wedge p_2, * \vdash_\theta C_1 \wedge C_2 = \theta(p_2) \wedge \\ \hat{\mu} = \mu \setminus \{C_1\} \ \wedge\ \langle \hat{\rho}, \hat{\sigma}\rangle = \mathsf{alloc}(\mathsf{reagent}_1 = C_1) \wedge \\ a = \mathsf{value}(\mathcal{E}(\mathit{rateExp})\langle \hat{\mu}, \hat{\rho}, \hat{\sigma}\rangle)/|\{\theta \mid p_2, * \vdash_\theta C_1\}| \ \wedge \\ \langle \hat{\rho}', \hat{\sigma}'\rangle = \mathsf{alloc}(\mathsf{product}_1 = C_1 \setminus C_2, \mathsf{product}_n = C_n) \wedge \\ \mu' = \hat{\mu} \cup \mathsf{extractCplx}(\mathcal{C}(\mathit{Block})\langle \hat{\mu}, \hat{\rho}', \hat{\sigma}'\rangle) \end{array} \right\}.$$

Note that the rate expression $\mathit{rateExp}$ can refer to the reagent and the react code Block to the products of the dissociation.

Let R be a set of rules. The corresponding continuous time Markov chain $\mathsf{CTMC}(R) : \mathsf{Sys} \times \mathsf{Sys} \to \mathbb{R}$ is

$$\mathsf{CTMC}(R) = \lambda\mu, \mu'.\ \textstyle\sum_{r \in R} \mathsf{CTMC}(r)(\mu)(\mu')$$

$$\mathsf{CTMC}(r) = \lambda\mu, \mu'.\ \textstyle\sum_{\langle -,\, a\rangle \in \mathit{rates}(\mu, \mu', r)} a.$$

Bibliography

Aimone Marsan, M., Balbo, G., Conte, G., Donatelli, S. and Franceschinis, G. (1995). *Modelling with Generalized Stochastic Petri Nets* (Wiley, New York, NY).

Alberts, B. (2008). *Molecular Biology of the Cell: Reference Edition* (Taylor & Francis, New York, NY).

Alberts, B., Johnson, A., Lewis, J., Raff, M., Roberts, K. P. and Walter, P. (2002). *Molecular Biology of the Cell,* Fourth Edition (Garland Science, New York, NY).

Alon, U. (2007). Network motifs: theory and experimental approaches, *Nature Review Genetics* **8**, pp. 450–461.

Alpern, B. and Schneider, F. B. (1985). Defining liveness, *Information Processing Letters* **21**, pp. 181–185.

Anderson, D. F. (2007a). A modified next reaction method for simulating chemical systems with time dependent propensities and delays, *Journal of Physical Chemistry* **127**.

Anderson, D. F. (2007b). A modified next reaction method for simulating chemical systems with time dependent propensities and delays, *Journal of Physical Chemistry* **127**, 21, pp. 214107–214112.

Auger, A., Chatelain, P. and Koumoutsakos, P. (2006). R-leaping: accelerating the stochastic simulation algorithm by reaction leaps, *Journal of Physical Chemistry* **125**, 8, p. 84103.

Auyang, S. (1999). *Foundations of Complex-system Theories: In Economics, Evolutionary Biology, and Statistical Physics* (Cambridge University Press, Cambridge, UK).

Barbuti, R., Caravagna, G., Maggiolo-Schettini, A. and Milazzo, P. (2011). Delay stochastic simulation of biological systems: A purely delayed approach, *Transactions on Computational Systems Biology XIII* **6575**, pp. 61–84.

Bauer, R. M. and Martinez, H. M. (1974). Automata and biology, *Annual Review of Biophysics* **3**, 1, pp. 255–292.

Blue, J., Beichl, I. and Sullivan, F. (1995). Faster monte carlo simulations, *Physical Review E* **51**, 2, pp. 867–868.

Bock, H., Carraro, T., Jäger, W., Körkel, S., Rannacher, R. and Schlöder, J. (eds.) (2013). *Model Based Parameter Estimation*, Theory and Applications Series: Contributions in Mathematical and Computational Sciences, Vol. 4 (Springer-Verlag, Berlin, Germany).

Bower, J. and Bolouri, H. (2001). *Computational Modeling of Genetic and Biochemical Networks* (MIT Press, Cambridge, MA).

Box, G. E. and Draper, N. R. (1987). *Empirical Model-Building and Response Surfaces*, Wiley Series in Probability and Statistics (Wiley, New York, NY).

Bratsun, D., Volfson, D., Tsimring, L. S. and Hasty, J. (2005). Delay-induced stochastic oscillations in gene regulation, *PNAS* **102**, 41, pp. 14593–14598.

Butcher, J. (2008). *Numerical Methods for Ordinary Differential Equations*, Second Edition (Wiley, New York, NY).

Buxton, J. and Laski, J. (1962). Control and simulation language, *The Computer Journal* **5**, p. 194.

Cai, X. (2007). Exact stochastic simulation of coupled chemical reactions with delays, *Journal of Chemical Physics* **126**, 12, p. 124108.

Cai, X. and Xu, Z. (2007). K-leap method for accelerating stochastic simulation of coupled chemical reactions, *Journal of Physical Chemistry* **126**, p. 74102.

Calder, M., Gilmore, S. and Hillston, J. (2005). Automatically deriving ODEs from process algebra models of signalling pathways, in G. Plotkin (ed.), *Proceedings of Computational Methods in Systems Biology (CMSB 2005)* (Edinburgh, Scotland), pp. 204–215.

Cao, Y., Gillespie, D. and Petzold, L. (2005). Avoiding negative populations in explicit poisson tau-leaping, *Journal of Physical Chemistry* **123**, p. 144917.

Cao, Y., Gillespie, D. and Petzold, L. (2006). Efficient step size selection for the tau-leaping simulation method, *Journal of Physical Chemistry* **124**, p. 844109.

Cao, Y., Gillespie, D. and Petzold, L. (2007). Adaptive explicit-implicit tau-leaping method with automatic tau selection, *Journal of Physical Chemistry* **126**, p. 137101.

Cao, Y., Li, H. and Petzold, L. (2004). Efficient formulation of the stochastic simulation algorithm for chemically reacting systems, *Journal of Physical Chemistry* **121**, pp. 4059–67.

Cardelli, L. (2005). Brane Calculi — Interactions of Biological Membranes, in *Computational Methods in Systems Biology, Int. Conf. CMSB 2004, Paris, France, May 26–28, 2004, Revised Selected Papers, LNCS*, Vol. 3082 (Springer-Verlag, Berlin, Germany), pp. 257–278.

Cardelli, L. (2008a). From processes to odes by chemistry, *IFIP International Federation for Information Processing* **273**, pp. 261–281.

Cardelli, L. (2008b). Molecules as automata, *Computer Science Logic, Lecture Notes in Computer Science* **5213**, p. 32.

Cardelli, L. (2008c). On process rate semantics, *Theory Computer Science* **391**, 3, pp. 190–215.

Cardelli, L. (2009). Artificial biochemistry, in A. Condon, D. Harel, J. N. Kok, A. Salomaa and E. Winfree (eds.), *Algorithmic Bioprocesses* (Springer), pp. 429–462.

Cardelli, L. and Priami, C. (2009). Visualization in process algebra models of biological systems, in J. P. Colins (ed.) *The 4th Paradigm: Data-Intensive Scientific Discovery* (Microsoft Research), pp. 99–105.

Carroll, J. D. and Green, P. E. (1997). *Mathematical Tools for Applied Multivariate Analysis* (Academic Press, New York, NY).

Carsten, M., Rybacki, S. and Uhrmacher, A. (2011). Rule-based multi-level modeling of cell biological systems, *BMC Systems Biology* **166**, 5, p. 166.

Cassman, M., Arkin, A., Doyle, F., Katagiri, F., Lauffenburg, D. and Strokes, C. (2005). International research and development in systems biology, *Technical report*, WTEC Panel on Systems Biology final report.

Ciocchetta, F. and Hillston, J. (2008). Bio-PEPA: an extension of the process algebra PEPA for biochemical networks, *Electronic Notes in Theoretical Computer Science*, **194**, 3, pp. 103–117.

Ciocchetta, F. and Priami, C. (2007). Biological transactions for quantitative models, *Electronic Notes in Theoretical Computer Science* **171**, 2, pp. 55–67.

Ciocchetta, F., Priami, C. and Quaglia, P. (2005). Modelling Kohn Interaction Maps with Beta-binders: an example, *Transactions on Computational Systems Biology* **3**, 33–43.

Ciocchetta, F., Priami, C. and Quaglia, P. (2008). An automatic translation of SBML into beta-binders, *IEEE/ACM Transactions on Computational Biology and Bioinformatics* **5**, 1, pp. 80–90.

Cohen, J. (2008). The crucial role of cs in systems and synthetic biology, *Communications of the ACM* **51**, 5, pp. 15–18.

Curti, M., Degano, P., Priami, C. and Baldari, C. T. (2004). Modelling biochemical pathways through enhanced π-calculus, *Theoretical Computer Science* **325**, 1, pp. 111–140.

D'Ambrosio, D., Lecca, P., Constantin, G., Priami, C. and Laudanna, C. (2004). Concurrency in leukocyte vascular recognition: developing the tools for a predictive computer model, *Trends in Immunology* **25**, 8, pp. 411–416.

Danos, V., Feret, J., Fontana, W., Harmer, R. and Krivine, J. (2007). Rule-based modelling of cellular signalling, in *Proceedings of CONCUR'07*, LNCS 4703.

Danos, V., Feret, J., Fontana, W., Harmer, R. and Krivine, J. (2010). Abstracting the differential semantics of rule-based models: exact and automated model reduction, in *Proceedings of the 25th Annual IEEE Symposium on Logic in Computer Science (LICS)*, pp. 362–381.

Danos, V. and Krivine, J. (2004). Reversible Communicating Systems, in *Proc. CONCUR'04, LNCS* **3170**, pp. 292–307.

Danos, V. and Laneve, C. (2003). Core Formal Molecular Biology, in *European Symposium on Programming (ESOP'03)*.

Danos, V. and Pradalier, S. (2005). Projective Brane Calculus, in *Computational Methods in Systems Biology, International Conference CMSB 2004, Paris, France, May 26–28, 2004, Revised Selected Papers*, LNCS 3082, pp. 134–148.

David, R. and Alla, H. (2005). *Discrete, continuous, and hybrid Petri Nets* (Springer, Berlin, Germany).

Degano, P., Prandi, D., Priami, C. and Quaglia, P. (2006). Beta-binders for biological quantitative experiments, *Electronic Notes in Theoretical Computer Science* **164**, 3, pp. 101–117.

Dematté, L. (2010). *Scaling up Systems Biology: Model Construction, Simulation and Visualization*, Ph.D. thesis, University of Trento, available at www.cosbi.eu/research/publications.

Dematté, L., Larcher, R., Palmisano, A., Priami, C. and Romanel, A. (2010). Programming biology in BlenX, in S. Choi (ed.) *Systems Biology for Signaling Networks*, Vol. 1 (Springer, New York, NY), pp. 777–821.

Dematté, L., Prandi, D., Priami, C. and Romanel, A. (2007a). Effective index: a formal measure of drug effects, in *Proceedings of FOSBE07*, pp. 485–490.

Dematté, L., Priami, C. and Romanel, A. (2008). The BlenX language: a tutorial, *Formal Methods for Computational Systems Biology* **5016** (Springer, New York, NY), pp. 313–365.

Dematté, L., Priami, C. and Romanel, A. (2008a). Modelling and simulation of biological processes in BlenX, *ACM SIGMETRICS Performance Evaluation Review* **35**, 4, pp. 32–39.

Dematté, L., Priami, C. and Romanel, A. (2008b). The Beta Workbench: a computational tool to study the dynamics of biological systems, *Briefings in Bioinformatics* **9**, 5, pp. 437–449.

Dematté, L., Priami, C., Romanel, A. and Soyer, O. (2007b). A formal and integrated framework to simulate evolution of biological pathways, in *Computational Methods in Systems Biology, Lecture Notes in Computer Science*, **4695**, pp. 106–120.

Dematté, L., Priami, C., Romanel, A. and Soyer, O. (2008c). Evolving BlenX programs to simulate the evolution of biological networks, *Theoretical Computer Science* **408**, 1, pp. 83–96.

Eccher, C. and Priami, C. (2006). Design and implementation of a tool for translating SBML into the biochemical stochastic pi-calculus, *Bioinformatics* **22**, 24, pp. 3075–3081.

Edmonds, B. (1996). What is complexity? — The philosophy of complexity per se with application to some examples in evolution, in F. Heylighen and D. Aerts (eds.), *The Evolution of Complexity* (Kluwer, Dordrecht, the Netherlands).

Errampalli, D. D., Priami, C. and Quaglia, P. (2004). A formal language for computational systems biology, *Omics: a journal of integrative biology* **8**, 4, pp. 370–380.

Faeder, R. (2011). Toward a comprehensive language for biological systems, *BMC Biology* **9**, 68.

Faraway, J. (2004). *Linear Models with R*, Chapman & Hall/CRC Texts in Statistical Science (Taylor & Francis, New York, NY).

Fisher, J., Harel, D. and Henzinger, T. A. (2011). Biology as reactivity, *Communications of the ACM* **54**, 10, pp. 72–82.

Fishwick, P. (ed.) (2007a). *Handbook of Dynamic System Modeling* (Chapman & Hall/CRC, Boca Raton, FL).

Fishwick, P. (2007b). The languages of dynamic system modeling, in P. Fishwick (ed.), *Handbook of Dynamic System Modeling*, (Chapman & Hall/CRC, Boca Raton, FL), pp. 1.1–1.10.

Frisco, P., Gheorghe, M. and Pérez-Jiménez, M. (2014). *Applications of Membrane Computing in Systems and Synthetic Biology* (Springer, New York, NY).

Fuchs, A. (2012). *Nonlinear Dynamics in Complex Systems: Theory and Applications for the Life-, Neuro- and Natural Sciences* (Springer, New York, NY).

Fujimoto, R. (2000). *Parallel and Distributed Simulation Systems* (Wiley Interscience, New York, NY).

Fujimoto, R., Perumalla, K. and Riley, G. (2007). *Network Simulation* (Morgan and Claypool Publishers, San Rafael, CA).

Gershenson, C. (2004). Introduction to random boolean networks, in M. Bedau, P. Husbands, T. Hutton, S. Kumar and H. Suzuki (eds.), *Workshop and Tutorial Proceedings, Ninth International Conference on the Simulation and Synthesis of Living Systems (ALife IX)*, pp. 160–173.

Gheorghe, M., Krasnogor, N. and Càmara, M. (2008). P systems applications to systems biology, *Biosystems* **91**, 3, pp. 435–437.

Gibson, M. and Bruck, J. (2000). Exact stochastic simulation of chemical systems with many species and many channels, *Journal of Physical Chemistry* **105**, pp. 1876–89.

Gillespie, D. T. (1977). Exact stochastic simulation of coupled chemical reactions, *Journal of Physical Chemistry* **81**, pp. 2340–2361.

Gillespie, D. T. (1992a). *Markov Processes: An Introduction for Physical Scientists* (San Diego Academics, San Diego, CA).

Gillespie, D. T. (1992b). A rigorous derivation of the chemical master equation, *Physica A* **188**, pp. 404–425.

Gillespie, D. T. (2007). Stochastic simulation of chemical kinetics, *Annual Review of Physical Chemistry*, **58**, pp. 35–55.

Goss, P. J. E. and Peccoud, J. (1998). Quantitative modeling of stochastic systems in molecular biology by using stochastic Petri nets, *Proceedings of the National Academy of Sciences USA* **95**, pp. 6750–6755.

Gostner, R., Baldacci, B., Morine, M. and Priami, C. (2014). Graphical modeling tools for systems biology, *ACM Computing Surveys*.

Gray, M. W., Burger, G. and Lang, F. (2001). The origin and early evolution of mitochondria, *Genome Biology*, **2**, 1018–1018.5.

Guerriero, M., Dudka, A., Underhill-Day, N., Heath, J. and Priami, C. (2009a). Narrative-based computational modelling of the gp130/jak/stat signalling pathway, *BMC Systems Biology* **3**, 40.

Guerriero, M., Heath, J. and Priami, C. (2007a). An automated translation from a narrative language for biological modelling into process algebra, *Proceedings of CMSB07*, Vol. LNBI 4695 (Springer, Berlin, Germany), pp. 136–151.

Guerriero, M., Prandi, D., Priami, C. and Quaglia, P. (2009b). Process calculi abstractions for biology, in A. Condon, D. Harel, J. N. Kok, A. Salomaa and E. Winfree (eds.), *Algorithmic Bioprocesses* (Springer, Berlin, Germany), pp. 463–486.

Guerriero, M. L., Priami, C. and Romanel, A. (2007b). Modeling static biological compartments with beta-binders, in H. Anai, K. Horimoto and T. Kutsia (eds), *Proceedings of Second International Conference, Algebraic Biology* (Springer, Berlin, Germany), pp. 247–261.

Haas, P. (2002). *Stochastic Petri Nets: Modelling, Stability, Simulation* (Springer-Verlag, New York, NY).

Hardy, S. and Iyengar, R. (2011). Analysis of dynamical models of signaling networks with petri nets and dynamic graphs, in I. Koch, W. Reisig and F. Schreiber (eds.), *Modeling in Systems Biology: The Preti Net Approach* (Springer-Verlag, London, UK), pp. 225–250.

Hárs, V. and Tóth, J. (1979). On the inverse problem of reaction kinetics, in *Colloquia Mathematics Society Jànos Bolyai 30: Qualitative Theory of Differential Equations*, pp. 363–379.

Hastie, T., Tibshirani, R. and Friedman, J. (2009). *The Elements of Statistical Learning: Data Mining, Inference, and Prediction, Second Edition*, Springer Series in Statistics (Springer, Berlin, Germany).

Heinemann, V., Xu, Y.-Z., Chubb, S., Sen, A., Hertela, L. W., Grindey, G. B. and Plunkett, W. (1992). Cellular elimination of $2',2'$-difluorodeoxycytidine $5'$-triphosphate: A mechanism of self-potentiation, *Cancer Research* **52**, pp. 533–539.

Hillston, J. (1996). *A Compositional Approach to Performance Modelling* (Cambridge University Press, Cambridge, UK).

Himmelspach, J., Lecca, P., Prandi, D., Priami, C., Quaglia, P. and Uhrmacher, A. (2006). Developing an hierarchical simulator for beta-binders, in *Proceedings of the 20th Workshop on Principles of Advanced and Distributed Simulation* (IEEE Computer Society), pp. 92–102.

Hlavacek, W. S., Faeder, J. R., Blinov, M. L., Posner, R. G., Hucka, M. and Fontana, W. (2006). Rules for modeling signal-transduction systems, *Science STKE* **2006**, 344.

Hood, L. and Galas, D. (2003). The digital code of DNA, *Nature* **421**, pp. 444–448.

Hung, J.-H., Yang, T.-H., Hu, Z., Weng, Z. and DeLisi, C. (2012). Gene set enrichment analysis: performance evaluation and usage guidelines, *Briefings in Bioinformatics* **13**, 3, pp. 281–291.

Iglesias, P. and Ingalls, B. (eds.) (2009). *Control theory and systems biology* (MIT Press, Cambridge, MA).

Indurkhya, S. and Beal, J. (2010). Reaction factoring and bipartite update graphs accelerate the Gillespie algorithm for large-scale biochemical systems, *PLoS One* **5**, 1.

Jensen, K. (1997). A brief introduction to coloured Petri nets, in E. Brinksma (ed.), *Tools and Algorithms for the Construction and Analysis of Systems: Proceedings of the TACAS'97 Workshop, Lecture Notes in Computer Science* **1217** (Springer Verlag, Berlin, Germany), pp. 201–208.

John, M., Lhoussaine, C., Niehren, J. and Uhrmacher, A. (2008). The attributed pi calculus, in M. Heiner and A. Uhrmacher (eds.), *Computational Methods in Systems Biology, Lecture Notes in Computer Science* **5307** (Springer, Berlin, Germany), pp. 83–102.

Jones, D. and Sleeman, B. (2003). *Differential Equations and Mathematical Biology* (Chapman and Hall/CRC, London, UK).

Kahramanoğulları, O., Cardelli, L., Caron, E., Gardner, P. and Phillips, A. (2009). A process model of actin polymerisation, *ENTCS* **229**, 1, pp. 127–144.

Kahramanoğulları, O., Lecca, P., Morpurgo, D., Fantaccini, G. and Priami, C. (2012). Algorithmic modelling quantifies the complementary contribution of metabolic inhibitions to gemcitabine efficacy, *PlosOne* **7**, 12.

Kahramanoğullari, O., Phillips, A. and Vaggi, F. (2013). Process modeling and rendering of biochemical structures: Actin, in *Biomechanics of cells and tissues: experiments, models and simulations, Lecture Notes in Computational Vision and Biomechanics* (Springer Science+Business Media, Dordrecht, the Netherlands) pp. 45–63.

Kauffman, S. (1969). Metabolic stability and epigenesis in randomly constructed genetic nets, *Journal of Theoretical Biology* **22**, pp. 437–467.

Kauffman, S. (1993). *Origins of Order: Self-Organization and Selection in Evolution* (Oxford University Press, Oxford, UK).

Khatri, P., Sirota, M. and Butte, A. J. (2012). Ten years of pathway analysis: Current approaches and outstanding challenges, *PLoS Computational Biology* **8**, 2.

Kitano, H. (2002a). Computational systems biology, *Nature* **420**, 6912, pp. 206–210.

Kitano, H. (2002b). Systems biology: a brief overview, *Science* **295**, 5560, pp. 1662–1664.

Knuth, D, (2012). *The Art of Computer Programming*, Volumes 1–4 (Addison-Wesley, Boston, MA).

Koch, I., Reisig, W. and Schreiber, F. (eds.) (2011). *Modeling in Systems Biology: The Petri Net Approach* (Springer-Verlag, London, UK).

Kremling, A. (2014). *Systems Biology: Mathematical Modeling and Model Analysis* (Chapman & Hall/CRS, London, UK).

Krumsiek, J., Pölsterl, S., Wittmann, D. M. and Theis, F. J. (2010). Odefy — from discrete to continuous models. *BMC Bioinformatics* **11**, 233.

Kuttler, C., Lhoussaine, C. and Niehren, C. (2007). A Stochastic Pi Calculus for Concurrent Objects, in H. Anai, K. Horimoto and T. Kutsia (eds), *Proceedings of Second International Conference, Algebraic Biology* (Springer, Berlin, Germany), pp. 232–246.

Larcher, R. (2001). *Optimizing the Execution of Biological Models*, Ph.D. thesis, University of Trento, available at www.cosbi.eu/research/publications.

Larcher, R., Priami, C. and Romanel, A. (2010). Modelling self-assembly in BlenX, in *Transactions on Computational Systems Biology XII* (Springer, Berlin, Germany), pp. 163–198.

Le Novère, N., Hucka, M., Mi, H., Moodie, S., Schreiber, F., Sorokin, A., Demir, E., Wegner, K., Aladjem, M., Wimalaratne, S., Bergman, F., Gauges, R., Ghazal, P., Kawaji, H., Li, L., Matsuoka, Y., Villéger, A., Boyd, S., Calzone, L., Courtot, M., Dogrusoz, U., Freeman, T., Funahashi, A., Ghosh, S., Jouraku, A., Kim, S., Kolpakov, F., Luna, A., Sahle, S., Schmidt, E., Watterson, S., Wu, G., Goryanin, I., Kell, D., Sander, C.,

Sauro, H., Snoep, J., Kohn, K. and Kitano, H. (2009). The systems biology graphical notation, *Nature Biotechnology* **27**, pp. 735–741.

Lecca, P., Dematté, L., Ihekwaba, A. E. and Priami, C. (2010a). Redi: a simulator of stochastic biochemical reaction-diffusion systems, in *Advances in System Simulation (SIMUL), 2010 Second International Conference on* (IEEE), pp. 82–87.

Lecca, P., Ihekwaba, A. E., Dematté, L. and Priami, C. (2010b). Stochastic simulation of the spatio-temporal dynamics of reaction-diffusion systems: the case for the bicoid gradient, *Journal of Integrative Bioinformatics* **7**, pp. 150–182.

Lecca, P., Laurenzi, I. and Jordan, F. (2013). *Deterministic versus Stochastic Modeling in Biochemistry and Systems Biology* (Woodhead Publishing, Cambridge, UK).

Lecca, P., Morpurgo, D., Fantaccini, G., Casagrande, A. and Priami, C. (2012). Inferring biochemical reaction pathways: the case of the gemcitabine pharmacokinetics, *BMC Systems Biology* **6**, 1, p. 51.

Lecca, P., Nguyen, T.-P., Priami, C. and Quaglia, P. (2011). Network inference from time-dependent omics data, in *Bioinformatics for Omics Data* (Humana Press, New York, NY), pp. 435–455.

Lecca, P., Palmisano, A., Ihekwaba, A. and Priami, C. (2010c). Calibration of dynamic models of biological systems with KInfer, *European Journal of Biophysics* **39**, 6, p. 1019.

Lecca, P. and Priami, C. (2007). Cell cycle control in eukaryotes: a biospi model, *Electronic Notes in Theoretical Computer Science* **180**, 3, pp. 51–63.

Lecca, P. and Priami, C. (2012). Inference and modelling of the gemcitabine pharmacokinetics and resistance, *Journal of the Pancreas* **13**, 5S, p. 567.

Lecca, P., Priami, C., Quaglia, P., Rossi, B., Laudanna, C. and Costantin, G. (2004). A stochastic process algebra approach to simulation of autoreactive lymphocyte recruitment, *SIMULATION: Transactions of The Society for Modeling and Simulation International* **80**, 4.

Lehninger, A., Nelson, D. L. and Cox, M. M. (2008). *Principles of Biochemistry*, Fifth Edition (W. H. Freeman, New York, NY).

Leier, A., Marquez-Lago, T. T. and Burrage, K. (2008). Generalized binomial tau-leap method for biochemical kinetics incorporating both delay and intrinsic noise, *Journal of Chemical Physics* **128**, 20, p. 205107.

Leye, S., Uhrmacher, A. M. and Priami, C. (2008). A bounded-optimistic, parallel beta-binders simulator, in *Proceedings of the 2008 12th IEEE/ACM International Symposium on Distributed Simulation and Real-Time Applications* (IEEE Computer Society), pp. 139–148.

Liu, Z. P., Wang, Y., Zhang, X. S. and Chen, L. (2012). Network-based analysis of complex diseases, **6**, pp. 22–33.

Lodish, H., Berk, A., Matsudaira, P., Kaiser, C., Krieger, M., Scott, M., Zipursky, S. and Darnell, J. (2004). *Molecular Cell Biology*, Fifth Edition (W. H. Freeman, New York, NY).

Lok, I. and Brent, R. (2005). Automatic generation of cellular reaction networks with Moleculizer 1.0, *Nature Biotechnology* **23**, pp. 131–136.

Mangold, M., Angeles-Palacios, O., Ginkel, M., Kremling, A., Waschler, R., Kienle, A. and Gilles, E. D. (2005). Computer-aided modeling of chemical and biological systems: methods, tools, and applications, *Industrial and Engineering Chemistry Research* **44**, 8, pp. 2579–2591.

Mauch, S. and Stalzer, M. (2011). Efficient formulations for exact stochastic simulation of chemical systems, *IEEE/ACM Transactions on Computational Biology and Bioinformatics* **8**, 1, pp. 27–35.

Maus, C., John, M., Röhl, M. and Uhrmacher, A. (2008). Hierarchical modeling for computational biology, in Marco Bernardo, Pierpaolo Degano, Gianluigi Zavattaro (eds.), *Formal Methods for Computational Systems Biology, 8th International School on Formal Methods for the Design of Computer, Communication, and Software Systems, SFM 2008, Bertinoro, Italy, June 2–7, 2008, Advanced Lectures* (Springer, New York, NY), pp. 81–124.

McCollum, J., Peterson, G., Cox, C., Simpson, M. and Samatova, N. (2006). The sorting direct method for stochastic simulation of biochemical systems with varying reaction execution behaviour, *Computational Biology and Chemistry* **30**, pp. 39–49.

Mitra, K., Carvunis, A.-R., Ramesh, S. K. and Ideker, T. (2013). Integrative approaches for finding modular structure in biological networks, *Nature Reviews Genetics* **14**, 10, pp. 719–32.

Morine, M., Tierney, A., van Ommen, B., Daniel, H., Toomey, S., Gjelstad, I. M. F., Gormley, I., Perez-Martinez, P., Drevon, C., Lopez-Miranda, J. and Roche, H. (2011). Transcriptomic coordination in the human metabolic network reveals links between n-3 fat intake, adipose tissue gene expression and metabolic health, *PLoS Computational Biology* **7**, 11.

Morine, M. J., McMonagle, J., Toomey, S., Reynolds, C. M., Moloney, A. P., Gormley, I. C., Gaora, P. Ó. and Roche, H. M. (2010). Bi-directional gene set enrichment and canonical correlation analysis identify key diet-sensitive pathways and biomarkers of metabolic syndrome, *BMC Bioinformatics* **11**, p. 499.

Morine, M. J., Toomey, S., McGillicuddy, F. C., Reynolds, C. M., Power, K. A., Browne, J. A., Loscher, C., Mills, K. H. and Roche, H. M. (2013). Network analysis of adipose tissue gene expression highlights altered metabolic and regulatory transcriptomic activity in high-fat-diet-fed il-1ri knockout mice, *The Journal of Nutritional Biochemistry* **24**, 5, pp. 788–795.

Mura, I. (2011). Stochastic modeling, in I. Koch, W. Reisig and F. Schreiber (eds.), *Modeling in Systems Biology. The Preti Net Approach* (Springer, New York, NY), pp. 121–150.

Mura, I., Prandi, D., Priami, C. and Romanel, A. (2009). Exploiting non-Markovian bio-processes, *Electronic Notes in Theoretical Computer Science* **253**, 3, pp. 83–98.

Nguyen, T.-P., Scotti, M., Morine, M. J. and Priami, C. (2011). Model-based clustering reveals vitamin d dependent multi-centrality hubs in a network of vitamin-related proteins, *BMC Systems Biology* **5**, 195.

Nikolić, D., Priami, C. and Zunino, R. (2012). A rule-based and imperative language for biochemical modeling and simulation, in *Software Engineering and Formal Methods* (Springer, Berlin, Germany), pp. 16–32.

Noble, D. (1960). Cardiac action and pacemaker potentials based on the hodgkin-huxley equations, *Nature* **188**, pp. 495–497.

Novak, B., and Tyson, J. (2003). Cell cycle control, in *Computational Cell Biology* (Springer, New York, NY), pp. 261–284.

Orton, R. J., Sturm, O. E., Vyshemirsky, V., Calder, M., Gilbert, D. R. and Kolch, W. (2005). Computational modelling of the receptor-tyrosine-kinase-activated mapk pathway, *Biochemical Journal* **392**, 2, p. 249.

Pahle, J. (2009). Biochemical simulations: stochastic, approximate stochastic and hybrid approaches, *Briefings in Bioinformatics* **10**, 1, pp. 53–64.

Palmisano, A. (2010). *Modeling and Inference strategies for Biological Systems*, Ph.D. thesis, University of Trento, available at www.cosbi.eu/research/publications.

Palmisano, A., Mura, I. and Priami, C. (2009). From odes to language-based, executable models of biological systems, *Pacific Symposium on Biocomputing* **14**, 12, pp. 239–250.

Papadimitriou, C. H. (1994). *Computational Complexity* (Addison-Wesley, Boston, MA).

Păun, G. (2002). *Membrane Computing: An Introduction* (Springer-Verlag, New York, NY).

Pettigrew, M. and Resat, H. (2007). Multinomial tau-leaping method for stochastic kinetic simulations, *Journal of Physical Chemistry* **126**, p. 84101.

Phillips, A. and Cardelli, L. (2004). A correct abstract machine for the stochastic pi-calculus, in *BioConcur '04, Workshop on Concurrent Models in Molecular Biology*.

Phillips, A., Cardelli, L. and Castagna, G. (2006). A graphical representation for biological processes in the stochastic pi-calculus, *Transactions on Computational Systems Biology* **4230**, pp. 123–152.

Plotkin, G. (1981). A structural approach to operational semantics, Technical report, DAIMI FN-19, Aarhus University, Denmark.

Plotkin, G. (2013). A calculus of chemical systems, in *In Search of Elegance in the Theory and Practice of Computation, Lecture Notes in Computer Science*, Vol. 8000 (Springer, Berlin, Germany), pp. 445–465.

Prandi, D., Priami, C. and Quaglia, P. (2005). Process calculi in biological context, *Bulletin of the EATCS* **85**, pp. 53–69.

Prandi, D., Priami, C. and Quaglia, P. (2006). Shape spaces in formal interactions, *ComPlexUs* **2**, 3–4, pp. 128–139.

Prandi, D., Priami, C. and Quaglia, P. (2008). Communicating by compatibility, *JLAP* **75**, p. 167.

Priami, C. (1995). Stochastic π-calculus, *The Computer Journal* **38**, 6, pp. 578–589.

Priami, C. (1996). Stochastic π-calculus with general distributions, in *4th Workshop on Process Algebras and Performance Modelling (PAPM '96)* (CLUT Torino), pp. 41–57.

Priami, C. (2002). Language-based performance prediction of distributed and mobile systems, *Information and Computation* **175**, pp. 119–145.

Priami, C. (2006). Process calculi and life science, *Electronic Notes in Theoretical Computer Science* **162**, pp. 301–304.

Priami, C. (2009a). Algorithmic systems biology, *Communications of the ACM* **52**, 5, pp. 80–88.

Priami, C. (2009b). Algorithmic systems biology, *CACM* **52**, 5, pp. 80–88.

Priami, C. (2012). Algorithmic systems biology: an opportunity for computer science, in G. Rozenberg, T. Bäck, J. N. Kok (eds.), *Handbook of Natural Computing* (Springer, New York, NY).

Priami, C., Ballarini, P. and Quaglia, P. (2009a). BlenX4bio: BlenX for biologists, in *Proceedings of CMSB09, LNBI*, Vol. 5688 (Springer, New York, NY).

Priami, C. and Quaglia, P. (2004). Modeling the dynamics of bio-systems, *Briefings in Bionformatics* **5**, 3, pp. 259–269.

Priami, C. and Quaglia, P. (2005a). Beta Binders for Biological Interactions, in V. Danos and V. Schächter (eds.), *Proceedings of 2nd International Workshop on Computational Methods in Systems Biology, CMSB '04, LNBI*, Vol. 3082 (Springer, New York, NY), pp. 21–34.

Priami, C. and Quaglia, P. (2005b). Operational patterns in beta-binders, in *Transactions on Computational Systems Biology, LNCS*, Vol. 3380 (Springer, Berlin, Germany) pp. 50–65.

Priami, C., Quaglia, P. and Romanel, A. (2009b). BlenX – static and dynamic semantics, in *Proceedings of CONCUR09, LNCS 5710* (Springer-Verlag), pp. 37–52.

Priami, C., Quaglia, P. and Zunino, R. (2012). An imperative language of self-modifying graphs for biological systems, in *Proceedings of the 27th Annual ACM Symposium on Applied Computing* (ACM), pp. 1903–1909.

Priami, C., Regev, A., Shapiro, E. and Silverman, W. (2001). Application of a stochastic name-passing calculus to representation and simulation of molecular processes, *Information Processing Letters* **80**, pp. 25–31.

Raczynski, S. (2006). *Modeling and Simulation. The Computer Science of Illusion* (Wiley, Chichester, UK).

Ramaswamy, R., González-Segredo, N. and Sbalzarini, I. (1995). A new class of highly efficient exact stochastic simulation algorithms for chemical reaction networks, *Journal of Physical Chemistry* **130**, 24, p. 244104.

Raval, A. and Ray, A. (2013). *Introduction to Biological Networks*, Chapman & Hall/CRC Mathematical and Computational Biology Series (CRC Press, Boca Raton, FL).

Regev, A., Panina, E. M., Silverman, W., Cardelli, L. and Shapiro, E. Y. (2004). BioAmbients: an Abstraction for Biological Compartments, *Theoretical Computer Science* **325**, 1, pp. 141–167.

Reisig, W. (1985). *Petri Nets: An Introduction*, EATCS Monographs on Theoretical Computer Science (Springer, Berlin, Germany).

Reisig, W. (2011). Petri nets, in I. Koch, W. Reisig and F. Schreiber (eds.), *Modeling in Systems Biology. The Preti Net Approach* (Springer, New York, NY), pp. 37–56.

Reisig, W. and Rozenberg, G. (eds.) (1998). *Lectures on Petri nets I: Basic models, LNCS*, Vol. 1491 (Springer, New York, NY).

Rodrìguez, J. V. (2009). *Stochasticity in Signal Transduction Pathways*, Ph.D. thesis, Universiteit van Amsterdam.

Romanel, A. (2010). *Dynamic Biological Modelling: a language-based approach*, Ph.D. thesis, University of Trento, available at www.cosbi.eu/research/publications.

Romanel, A. and Priami, C. (2008). On the decidability and complexity of the structural congruence for beta-binders, *Theoretical Computer Science* **404**, 1–2, pp. 156–169.

Romanel, A. and Priami, C. (2010). On the computational power of BlenX, *Theoretical Computer Science* **411**, pp. 542–565.

Roos, D. (2001). Bioinformatics — trying to swim in a sea of data, *Science* **291**, pp. 1260–1261.

Rosen, R. (1991). *Life Itself* (Columbia University Press, New York, NY).

Saito, A., Nagasaki, M., Matsumo, H. and Miyano, S. (2011). Hybrid functional petri net with extension for dynamic pathway modeling, in I. Koch, W. Reisig and F. Schreiber (eds.), *Modeling in Systems Biology. The Preti Net Approach* (Springer, New York, NY), pp. 101–120.

Saraiya, P., North, C. and Duca, K. (2005). Visualizing biological pathways: requirements analysis, systems evaluation and research agenda, *Information Visualization* **4**, 3, pp. 191–205.

Schiff, J. (2008). *Cellular Automata: A Discrete View of the World* (Wiley, Chichester, UK).

Schulze, T. (2008). Efficient kinetic Monte Carlo simulation, *Journal of Computational Physics* **227**, 4, pp. 2455–2462.

Schwartz, R. (2008). *Biological Modeling and Simulation. A Survey of Practical Models, Algorithms, and Numerical Methods* (MIT Press, Cambridge, MA).

Shoval, O. and Alon, U. (2010). Snapshot: Network motifs, *Cell* **143**, 2, pp. 326–e1.

Sibani, P. and Jensen, H. (2013). *Stochastic Dynamics of Complex Systems: From Glasses to Evolution* (Imperial College Press, London, UK).

Sible, J. and Tyson, J. (2007). Mathematical modeling as a tool for investigating cell cycle control networks, *Methods* **41**, pp. 238–247.

Sigismund, S., Algisi, V., Nappo, G., Conte, A., Pascolutti, R., Cuomo, A., Bonaldi, T., Argenzio, E., Verhoef, L. G. G. C., Maspero, E., Bianchi, F., Capuani, F., Ciliberto, A., Polo, S. and Di Fiore, P. P. (2013). Threshold-controlled ubiquitination of the EGFR directs receptor fate, *The EMBO Journal* **32**, 15, pp. 2140–2157.

Sipser, M. (2005). *Introduction to the theory of computation*, Second Edition (Thompson Course Technology, Boston, MA).

Slepoy, A., Thompson, A. and Plimpton, S. (2008). A constant-time kinetic Monte Carlo algorithm for simulation of large biochemical reaction networks, *Journal of Chemical Physics* **128**, 20.

Sokolowski, J. and Banks, C. (eds.) (2009). *Principles of Modeling and Simulation — A multidisciplinary approach* (Wiley, Chichester, UK).

Spengler, S. (2000). Bioinformatics in the information age, *Science* **287**, pp. 1221–1223.

Spicher, A., Michel, O., Cieslak, M., Giavitto, J.-L. and Prusinkiewicz, P. (2008). Stochastic p systems and the simulation of biochemical processes with dynamic compartments. *Biosystems* **91**, 3, pp. 458–472.

Spitalnic, S. (2004). Test properties 1: Sensitivity, specificity and predictive values, *Hospital Physician* **27**, pp. 27–31.

Szallasi, Z., Stelling, J. and Periwal, V. (eds.) (2006). *System Modeling in Cellular Biology. From Conceots to Nuts and Bolts* (The MIT Press, Boston, MA).

Thanh, V. and Zunino, R. (2012). Tree-based search for stochastic simulation algorithm, in *Proceedings of SAC'12* (ACM).

Thanh, V. and Zunino, R. (2014). Adaptive tree-based search for stochastic simulation algorithm, *International Journal of Computational Biology and Drug Design* (accepted for publication).

Tian, T. and Burrage, K. (2004). Binomial leap methods for simulating stochastic chemical kinetics, *Journal of Physical Chemistry* **121**, 21, p. 1035664.

Trivedi, K. and Kulkarni, V. (1993). FSPNS: Fluid stochastic petri nets, in M. Ajmone Marsan (ed.), *Application and Theory of Petri Nets 1993, Lecture Notes in Computer Science*, Vol. 691 (Springer, Berlin, Germany), pp. 24–31.

Uhrmacher, A. M. and Priami, C. (2005). Discrete event systems specification in systems biology-a discussion of stochastic pi calculus and devs, in *Proceedings of the 37th conference on Winter simulation* (Winter Simulation Conference), pp. 317–326.

Veltkamp, S. A., Pluim, D., van Eijndhoven, M. A., Bolijn, M. J., Ong, F. H., Govindarajan, R., Unadkat, J. D., Beijnen, J. H. and Schellens, J. H. M. (2008). New insights into the pharmacology and cytotoxicity of gemcitabine and 2′,2′-difluorodeoxyuridine, *Molecular Cancer Therapeutics* **7**, 8, pp. 2415–2425.

Versari, C. and Busi, N. (2007). Efficient stochastic simulation of biological systems with multiple variable volumes, in *Proceedings FBTC 07*.

Vo, T., Priami, C. and Zunino, R. (2014). Efficient rejection-based simulation of biochemical reactions with stochastic noise and delays, *Journal of Chemical Physics* **141**, 13, 134116.

Voit, E. (2013). Biochemical systems theory: a review, *Biomathematics* **2013**, p. Article ID 897658.

Volterra, V. (1926). Fluctuations in the abundance of species considered mathematically, *Nature* **118**, pp. 558–560.

Wallin, I. E. (1927). *Symbionticism and the Origin of the Species* (William & Wilkins Company, Baltimore, MD).

Weaver, W. (1948). Science and complexity, *American Scientist* **36**, 4, p. 536.

Weimar, J. (1998). *Simulation with Cellular Automata* (Logos-Verlag, Berlin, Germany).

Weiss, J. (1997). The Hill equation revisited: uses and misuses, *FASEB* **11**, pp. 835–841.

Wing, J. (2006). Computational thinking, *Communications of the ACM* **49**, 3, pp. 33–35.

Wu, J., Vidakovic, B. and Voit, E. O. (2011). Constructing stochastic models from deterministic process equations by propensity adjustment, *BMC Systems Biology* **5**, 1.

Yi, N., Zhuang, G., Da, L. and Wang, Y. (2012). Improved delay-leaping simulation algorithm for biochemical reaction systems with delays, *Journal of Physical Chemistry* **136**, 14, p. 144108.

Zámborszky, J. (2010). *Compositional Modeling of biological Systems*, Ph.D. thesis, University of Trento, available at www.cosbi.eu/research/publications.

Zámborszky, J. and Priami, C. (2010). BlenX-based compositional modeling of complex reaction mechanisms, *EPTCS* **19**, pp. 85–102.

Zeigler, B., Praehofer, H. and Kim, T. (2000). *Theory of Modeling and Simulation*, Second Edition (Academic Press, San Diego, CA).

Index

Printed in the United States
By Bookmasters